MEASURED TONES

The interplay of physics and music

MEASURED TONES

The interplay of physics and music

Second edition

Ian Johnston

Institute of Physics Publishing
Bristol and Philadelphia

British Library Cataloguing-in-Publication Data

A catalogue record for this book is available from the British Library.

ISBN 0 7503 0762 5

Library of Congress Cataloging-in-Publication Data are available

First published 1989
Reprinted with corrections 1993
Reprinted 1994, 1997, 2001
Second edition 2002

Commissioning Editor: James Revill
Production Editor: Simon Laurenson
Production Control: Sarah Plenty
Cover Design: Frédérique Swist
Marketing Executive: Laura Serratrice

Published by Institute of Physics Publishing, wholly owned by The Institute of Physics, London

Institute of Physics Publishing, Dirac House, Temple Back, Bristol BS1 6BE, UK

US Office: Institute of Physics Publishing, The Public Ledger Building, Suite 1035, 150 South Independence Mall West, Philadelphia, PA 19106, USA

Printed in the UK by Antony Rowe Ltd, Chippenham, Wiltshire

Acknowledgments

The production of a book of this character would be impossible without a large number of illustrations, particularly of musical instruments. We have tried to obtain permission for all illustrations which are still in copyright, but there are one or two where we have been unsuccessful. My thanks are due to the following organizations and individuals who gave their permission to use material:

H. Bagot, p.276*l*; Boston Symphony Orchestra, p.178; Cordon Art, Baarn, Holland (for M.C. Escher heirs) p.111*u*, p.112*u*; Diagram Visual Information Ltd, p.224*u*, p.270*u*, p.303*u*; Dover Publications Inc, p.57*u*, p.59*ul*, p.60*ul*, p.118*ul*, p.137, p.167; and (from *Music, A Pictorial Archive of Woodcuts and Engravings* by Jim Harter 1980) p.41, p.43*ul*, p.54*u*, p.118*m*, p.120*u*, p.123*l*, p.125*u*, p.211*l*, p.213, p.224*l*, p.226*l*, p.227*ul*, p.230*ul*, p.231, p.232, p.269*ul*, p.273, p.274; Harper and Row Inc (from *Harper's Dictionary of Music* by Christine Ammer, drawings by C.M. Ciampa and K.L. Donlon, 1972) p.87*l*, p.89*u*, p.218*l*, p.270*l*, p.271*m*, p.272*u*, p.275*u*; S. Levarie, p.2, p.275*l*; Lord Rennell of Rodd, p.338; Macmillan Publishers Ltd (from *The New Grove Dictionary Of Music and Musicians* ed. Stanley Sadie, 1980) p.303*l*; McGraw-Hill Publishing Co. (from R.D. Chellis, *Pile Foundations*, 1961), p.63; Oxford University Press (from *The New Oxford Companion to Music*, ed Denis Arnold, 1983) p.128, p.307; The Ann Ronan Picture Library, p.21, p.22, p.24, p.27, p.72, p.93, p.154, p.185, p.236, p.237; W.J. Strong, p.276*u*, p.277*l*; Unwin Hyman Ltd (from C.A. Ronan *The Astronomers*, 1964) p.11, p.23; John Wiley and Sons (from J.W. Kane and M.M. Sternheim, *Life Science Physics* 1978) p.241; Michael Nicholson/CORBIS, p.107; Bettmann/CORBIS, p.107.

Contents

List of Tables

Prologue

The purpose of this book is easy to explain. I am a physicist and not a musician. Yet I am interested in music and am aware of just how much the two fields have to offer to one another. I would like to make this more widely known. I believe that your enjoyment of music can be increased thereby and so can your understanding of physics. Whatever your interest, a little bit of extra knowledge can only help. It is as simple as that.

There are of course many books on this subject. They are usually written by physicists (I have never seen one written by a musician telling scientists what they should know about music). Most of them have what seems to me a grave fault: they presuppose a certain frame of mind in their readers. Their early chapters are devoted to simple physics—vibrations, waves, Fourier analysis, etc. The second half deals with musical instruments, architectural acoustics, scales, electrical recording and reproduction. It is a classical scientific approach—theory first and applications later.

The trouble is that this approach really only suits scientifically minded readers. Most musical people that I know, find the physical concepts quite obscure, and have great difficulty recalling them for later applications. So I want to try a different approach, to develop the physical concepts and the musical applications together. This approach will make for rather an unusual ordering. For example, it is possible to understand a lot about brass instruments with relatively little physical knowledge, so I will talk about them quite early. But woodwinds are much more complicated theoretically, and must wait till towards the end. The very first topic will be musical scales and consonance, which is an extremely subtle *musical* question, but can be discussed with only a few very simple *scientific* observations.

So my plan has been firstly to explain the physics that is involved in music, trying all the time to simplify the details as much as possible. To this end I have scrupulously avoided algebraic mathematical equations, in the belief that they tend to frighten non-scientists. It seems to me that if they are written out in words, they look more user-friendly. Secondly, I have tried at each stage to make it clear where the physics is important in a musical context. However I am not a expert on music or musicology—although I think that everything I have to say is perfectly well known and accepted, and can be checked in one of the standard musical encyclopedias. So please keep in mind that, when I talk about musical matters, I am trying to say that the physical concepts have application here, without being definitive.

One word of warning. I have arranged the material historically, because that seems a natural and logical ordering for the approach I want to take. Physical ideas have to be understood in sequence, the easy ones first and later ones building on the earlier. And by and large, that is how they developed. So this book will, in parts, read like a history of the subject. But it is important, I think, to make it clear that this does not claim to be a serious historical work.

Scientists, when trying to explain science, have always used an 'historical' approach—thumbnail sketches of great figures, the dates and circumstances of various discoveries, and so on. A lot of what they say is probably apocryphal. Did Archimedes really jump out his bath shouting "Eureka!"? Did Röntgen really discover X-rays by leaving a lump of pitchblende on some unexposed film? In a way, literal historical accuracy is not important. The message usually is: "This is a simple and easy to observe fact, but it is not trivial. There was certainly a time when people did not know it or did not realize its significance".

It may be that my approach in this book should be put into this 'apocryphal history of science' category. To give you one example, many historians and philosophers of science (though not working scientists) seem to dislike the idea that science 'progresses'. The story, as I tell it, very definitely progresses from the time of Pythagoras to the present day. I have chosen people and ideas to talk about in order to build up a complete picture, each concept resting on the foundation of the previous ones. I deliberately chose not to talk about the dead ends, the ideas that didn't lead anywhere, regardless of whether or not this presents an accurate picture of how science is actually done. Nevertheless I have tried to be as accurate as I can in my use of historical information, and I have been careful to include a reference bibliography from which you can check on what I have said or can follow it further if you wish.

But most importantly, I have *never* attempted to explain a physical idea in any but modern terms. To present a correct explanation of some effect by talking about an incorrect explanation first, usually just increases confusion in the reader's mind, no matter how valid that approach might be in historical terms. In the end, the history is there as a context for the physics and the music. It is also there because I find it interesting. I hope you do too.

Lastly, my treatment of musical instruments, for reasons that I have hinted at earlier, will not fit into this simple historical ordering. So I have called the chapters dealing with them **interludes**, and you can skip over them if you are looking for a connected story. Within these interludes I have talked about the historical development of each class of instrument because it is usually easier to appreciate the physics in the simpler instruments of the past and to see how the solutions of problems which arose led to the very elaborate models that are in use today. My approach has been unashamedly 'evolutionary'.

Now I am aware that many musicologists tend not to approve of this. They argue, rightly, that instruments do not beget other instruments, so they cannot be said to 'evolve'. But to my mind the evolutionary model has a lot going for it. The mere fact that certain instruments exist means that people will want to play them, and some of those players will teach other players, who then need more instruments. In that sense they *do* beget. Similarly the way new instruments are introduced is very similar to the biological idea of filling an ecological niche—only those survive which are fittest for the musical tastes of the day. So I remain an evolutionist.

I think this attitude is particularly important today, at the start of the 21st century. Technology is changing at a rate seldom, if ever, matched in the past. And this includes musical technology. When I wrote the first edition of this book, in 1988, computers had been around for a generation. Digital sound recording, electric guitars and synthesizers were where we were at. Very few people, myself included, were aware of the enormous effect that the personal computer would have on instrumentation, and what the world wide web would mean to communications. So in this new edition I have to talk about the handling of sound files, digital interfaces, fractals. Whether they will still be of great importance in the next decade is anybody's guess. If I get to do a third edition, it will all be different. That much is sure. But the physics and the musical concepts will be the same. And they are what is important.

When I was two-thirds of the way through the first edition of this book, I had the opportunity to turn some of the material into a series of radio programmes, to be broadcast nationally by the Australian Broadcasting Corporation. In the process of preparing those programmes, I talked to many musicians, historians and scientists: and if they ever get to read this book, they will not be surprised to find some of their ideas incorporated herein. So I would like to record my thanks to John Aranetta, Graham Pont, David Russell, Nicholas Routley, Winsome Evans, Nicholas Lomb, Bill Elliott, Alan Saunders, Donald Hazlewood, Peter Platt, Jamie Kassler, John Hopkins, Fergus Fricke, Lloyd Martin, Carole Adams, Harvey Dillon, Alan Marett, Ian Fredricks, Martin Wesley-Smith and Eric Gross.

Finally I would reserve special thanks for Robyn Williams who gave me so much encouragement in his role as radio producer; and to my family for their patience and forebearance during the long gestation period of this work, especially my daughter, Helen, who read each and every word as it emerged painfully from my labour of love.

Ian Johnston

Sydney, December 2001.

Chapter 1

Why these and not others?

The story is often told of the tourist from the city asking the country villager how to get to the local church. "If that's where you want to go," comes the reply, "then you shouldn't start from here."

It seems to me that that answer might be wiser than it looks. In this book, the point we are aiming for is where appreciation and enjoyment of music is increased by understanding the fundamental role that science plays in it. But how you get to that point depends very much on where you are now.

Obviously we must begin with the easy bits first. Typically, performing musicians start off playing scales, studies, simple pieces, eventually building up an extensive repertoire. It is only at the end (if at all) that they might ask where physics impinges on what they already know. Science students on the other hand, start by studying vibrations and waves, measuring and experimenting until a complete theoretical framework is established. Only then will they try to apply what they know to music. But I want to try a third path.

Few would disagree that as our science (particularly physics) progressed from earliest recorded times to what it is today, it had a great influence on the way music changed. But what isn't so well known is that the influence was two-way. Music, in its turn has influenced the way physics developed. So I want to develop musical and scientific ideas together, by tracing the strands of this parallel development. I will follow the path of those who asked the critical questions and discuss the answers they arrived at. It will mean that I will sometimes be led away from the main track, into astronomy and quantum theory. But for me that is part of the charm of the subject.

It also gives me a sensible point to start from. We know that music is constructed from notes. Not just any notes, but a special set of notes called a **scale**. But why just these particular notes? Why these and not others? That is the earliest musical/scientific question that we have record of, and its answer begins in an experimental observation that first seems to have come to light in ancient Greece in about 500 B.C.

Division of a stretched string

You can do this experiment on a simple device called a **monochord**. It consists of a string stretched between two supports on the top of a hollow sounding box. The string is tightened or loosened with an adjusting nut; and a movable bridge **stops** the string at different points so that when you pluck it, only part of the string vibrates.

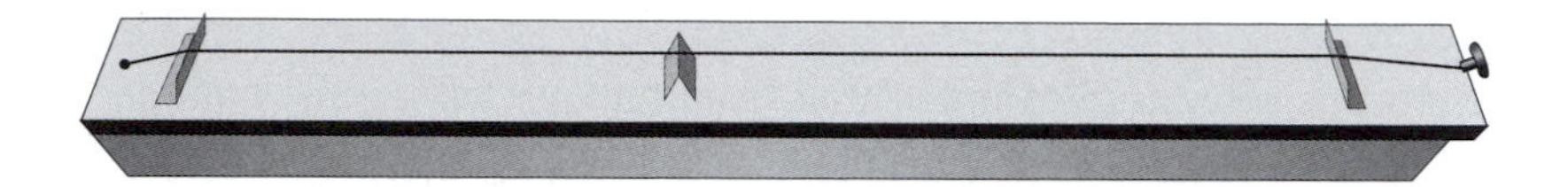

Playing around with this device will tell you the basics known to every player of a stringed instrument. Pluck the open (that is, unstopped) string and it sounds a note. Move the bridge to shorten the string and the pitch of the note rises—the shorter the string the higher the pitch. Tightening the nut also raises the pitch.

The experiment you want to do consists of listening to the two notes sounded when the bridge is placed at two different positions. The degree to which one note is higher than the other—the **interval** between them—involves your perception of what they sound like, together or one after the other. It is difficult to describe this perception in words, but your brain clearly measures it in some way, because most people can easily sing the second note after being reminded of the first. What is interesting is that this perceptual interval is related to the *ratio* of the two lengths of string you plucked. You can test this by tightening the adjusting nut or by altering the points at which the string was stopped. So long as the ratio of lengths is the same, even though the pitch of the notes changes, the interval between them stays the same.

Now compare *two* notes from the string, firstly unstopped and secondly stopped exactly in the middle, so that the length ratio is 2/1. The interval you hear is that which is called the **octave** (the first two notes of "*Somewhere*, Over the Rainbow", which is how I remember things). Musically speaking this is the most important interval there is. When men and women sing together in 'unison' they are actually singing an octave apart, even though their perception is that they are singing the same notes. There is a real sense of sameness about two notes separated by this interval. To make the example concrete, if the unstopped note was a D, say, then the note from half the string would be the next D up.

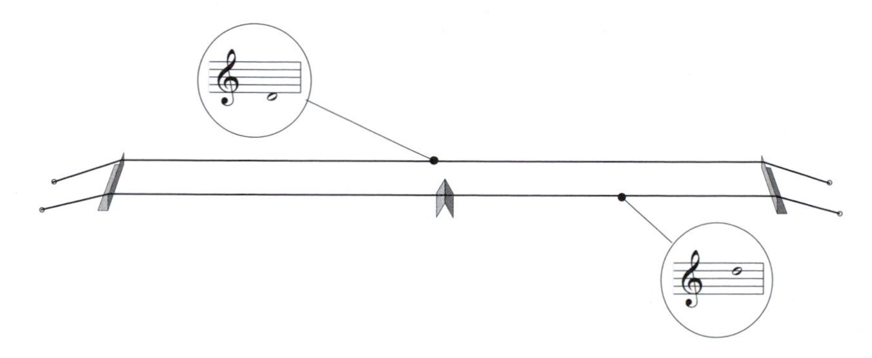

(Incidentally, if you don't know the ordinary notation for writing musical notes, I have given a brief overview in Appendix 1).

The name 'octave' is misleading. We call it that because the Western tradition is to further divide this interval by eight different notes. But the number 8 is a cultural accident and has no validity in nature, whereas the interval itself has. All musical cultures base their scales on this interval, but the Chinese divide it into 5 parts, the Arabs into 17 and the Indians into 22. In ancient Greece they called this interval the **diapason**, which literally meant 'through all'. I hope you will agree that it is a more suitable name and I want to use it, at least for the rest of this chapter.

The crucial experimental fact then, is that this all-important musical interval corresponds to the numerical ratio 2/1. And not just approximately so, but as accurately as you can determine. Try another simple experiment. Ask someone to place the bridge in the exact middle of the string without measuring, judging distances by eye only. The chances are that they will be indifferently successful. But ask them to do it by ear, by comparing the sounds of the divided and the undivided string. Anyone at all musical will place it exactly in the centre. This proves that the ear is a more sensitive measuring instrument than the eye. But it also demonstrates the precise relationship between the diapason and the whole number 2.

The next most important musical interval is that which we call the **perfect fifth** and which the ancient Greeks called the **diapente** ("*Baa*, Baa, *Black* Sheep"). You will hear this interval if you stop the string a third of the way from one end. The notes from the unstopped string, and from the longer divided part will be exactly a fifth apart. So, if the unstopped note was a D, the part of the string that is 2/3 of the length will sound an A. The length ratio this interval corresponds to is 3/2 (note that the convention is always to put the shorter length last). The other part of the string (which is 1/3 of the length) will sound another A, a diapason higher.

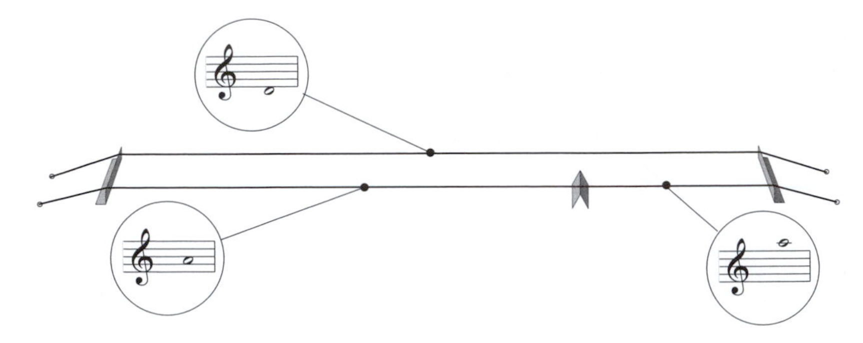

Again this interval occurs naturally. It is typically the separation between a tenor and a bass voice, or between a soprano and an alto. Untrained singers will often find themselves singing a fifth apart without being aware of it. Again there is a perception of sameness about two notes separated by this interval—not as much as with the diapason, rather it is a feeling of blending together. In fact this gave rise to one of the earliest forms of harmony in Western music when, in medieval monasteries, the monks often sang their chants with the bass and tenor voices a constant fifth apart. This was to be a precursor to a whole range of singing techniques called **organum**.

The same division of the string contains another important interval, between the note from 2/3 of its length and from 1/2 (the A and the high D). We call this interval the **perfect fourth** ("*Hark the* Herald Angels Sing") and its old name was the **diatesseron**. It corresponds exactly to the ratio 4/3 and you won't be surprised to learn that conventional music theory says there is a very close affinity between the two intervals, the fifth and the fourth.

These then are the earliest recorded musical/scientific facts—that the most fundamental intervals in musical harmony correspond exactly to ratios of the whole numbers 1, 2, 3 and 4. The philosophical implications of this are clearly immense, and to think about them we have to go back to the school of Pythagoras, in the city of Crotona in southern Italy in the 6th century B.C.

Pythagoras

Every school student will recognize this name as the originator of that theorem which offers, in the words of W.S. Gilbert, "many cheerful facts about the square on the hypotenuse". Yet we know very little about the man himself. Tradition has it that he was born on the island of Samos in about 580 B.C. and travelled and studied widely throughout all the centres of Greek civilization around the Mediterranean. At the age of fifty he established a school at Crotona where he taught arithmetic, geometry, astronomy and music. By all accounts this school was more like a monastery. Students had to observe long periods of silence, to refrain from eating meat, eggs and beans, to dress modestly and to avoid swearing oaths. But it also took in women students as well as men, which was remarkably enlightened for that time.

The school survived for some thirty years. Though Pythagoras himself left no written works, his teachings were passed on by his students, of whom the best known was Philolaus, later to become an important philosopher in his own right. As time went on, the band of scholars became more like a secret society. They believed that the most suitable people to be in charge of things were philosophers, so naturally they became embroiled in politics. Eventually in some local difference of opinion the school was burned down and the society dispersed, though it survived in scattered groups throughout Greece for hundreds of years. It is said that Pythagoras himself was captured and killed while trying to escape, because he couldn't bring himself to run across a field of beans.

Pythagoras is often called the father of science and philosophy in Europe, but many of the ideas of the Pythagoreans seem very strange to scientists today. They didn't believe that the order of the universe was to be understood through observation and experiment: they held that the laws of nature could be deduced by pure thought. It was to be a defect in most of Greek science, the idea that the use of technical equipment was the job of slaves and not a suitable pastime for gentlemen. As a result, few of those who had the leisure to do experimental science had the inclination.

But back to music. It is a much told story that one day the young Pythagoras was passing a blacksmith's shop and his attention was caught by the ringing sounds from the anvil which he judged to be separated by regular musical intervals. When he found that the hammers were of different weights, it occurred to him that the intervals might be related to those weights. Contrary to his usual practice, when he got home he performed experiments on the monochord, which confirmed his supposition. Other observations he made led to the same conclusion.

Pythagoras's influence lasted for over 2 000 years. Right up to Renaissance times his ideas were still being taught. The illustration on the next page is from a manuscript on the theory of music dated 1492.

It attributes the perception of the consonances of weighted hammers to Jubal, the biblical father of music; and it depicts Pythagoras and Philolaus demonstrating the tones of bells of different sizes, glasses filled with water to different levels, strings stretched with different weights and pipes of different lengths all in the so-called harmonic ratios—4, 6, 8, 9, 12, 16. The implication is that all of these demonstrations will sound intervals of diapasons, fifths and fourths.

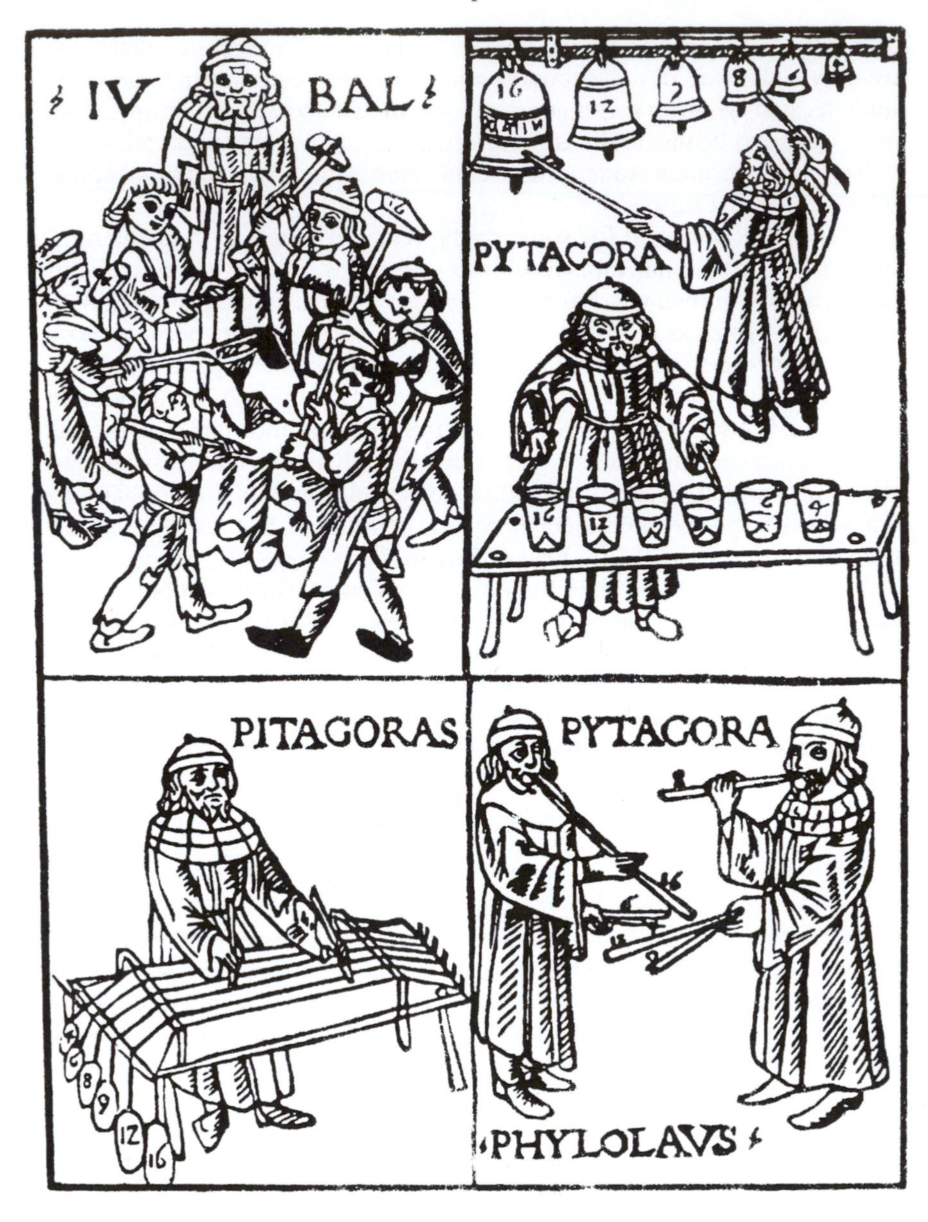

What it does show is that the bad habit of not checking things experimentally also lasted 2 000 years. For in fact several of those demonstrations will *not* show what they are supposed to. Hammers or bells whose weights are in such ratios won't in general sound those intervals; nor will strings stretched by those weights, as you can verify for yourself after reading Chapter 3 of this book. But the pipes will show the effect reasonably accurately, and the water glasses won't be too bad (though it isn't the volume of water which is relevant, but that of the air above it); and of course, strings of those lengths will do so exactly.

In a way then, Western science and music is lucky that Pythagoras' attitude to experimentation was as it was. His insight was quite correct, but if he had known that he didn't have as much supporting evidence as he thought he had, perhaps he wouldn't have pushed ahead to philosophize about it quite so extensively. For philosophize he did, into the realms of music and science, and neither would be the same again.

The Pythagorean scale

Since Pythagoras left no writings it is difficult to reconstruct exactly how he went about applying mathematical reasoning to the question of musical scales. What follows is my own version of how the argument might have gone, but the general idea is right. We know about four musical ratios already—1/1, 4/3, 3/2 and 2/1. Let me rewrite these in decimal form—1.000, 1.333, 1.500 and 2.000. An obvious question we want to answer is: are there any other decimal numbers between 1 and 2 which correspond to singable intervals? Well, using the correspondence between numerical ratios and harmony, the Pythagoreans evolved a simple recipe for creating new notes from old.

> **Note generating procedure:** *Take an existing ratio and multiply or divide it by 3/2. If the number you get is greater than 2 then halve it; if it is less than 1 then double it.*

In case this seems a bit abstract, I will illustrate on a musical stave what is happening as you apply this procedure. Multiplying or dividing by 3/2 is the same as raising or lowering the note by a fifth. Multiplying or dividing by 2 just keeps the notes within the required range, which is a legitimate thing to do because any two notes separated by a diapason are considered the 'same'. But please remember that it's only a picture. The Pythagoreans were operating mathematically and not musically; and the notes you get on a modern piano will not be exactly the same as they got from their numerical string lengths.

The notes we already have can be imagined to be generated by this procedure from a single starting note:

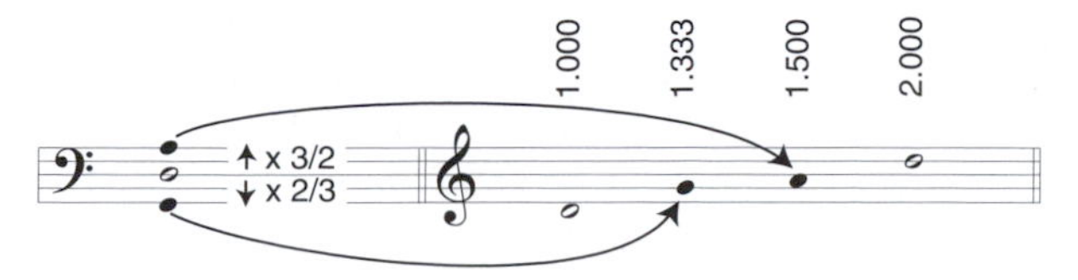

A pleasant set of four notes which Orpheus was supposed to have used to tune his lyre; but not yet a scale.

Now apply the procedure again and construct two new notes:

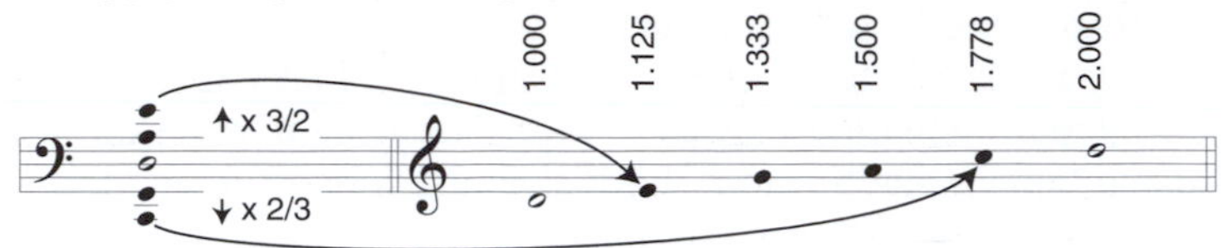

Now this is a scale. It has five different notes. It is called a **pentatonic scale**. You might notice that if you play these notes on a piano, but go down four keys from what is written, you get the five black notes on the keyboard. So any melody that can be played on the black keys is written in this scale.

Now the extraordinary thing about this scale is that it has existed in many musical cultures from the very earliest recorded times; and is still widely used, particularly in Eastern music. Many old folk songs which originally came from other continents are like this. Think for example of "Nobody Knows the Trouble I've Seen" (Black African), "There is a Happy Land" (North American Indian), or "Auld Lang Syne" (Celtic). Think also of the kind of music that Western composers write when they paint tone pictures of oriental lands—that splendid pseudo-Japanese march from *The Mikado*, "Miya sama, miya sama", or the pseudo-Chinese choruses from *Turandot*, "Dal deserto" and "Ai tuoi peidi". All of these melodies can be played on the black notes of the piano.

There is a lot more that I could say about pentatonic scales and the way they are constructed, but rather than interrupt the flow of my story here, I've put all that in Appendix 6. It's enough for now just to note how truly ubiquitous pentatonic scales are in musical cultures the world over.

Let us press on and apply the procedure again:

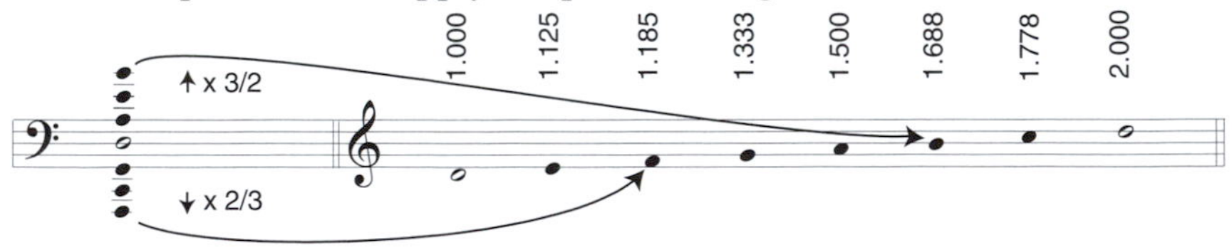

Now we have created a scale of seven different notes—a **septatonic** scale—and it will not have escaped your attention that they are the same as the notes of our own scale, except that they start on 'ray' rather than 'doh'. To give it its technical name it is the **Dorian mode**.

You will also notice, if you care to check it with a pocket calculator, that all of the eight notes are separated by only two different intervals.

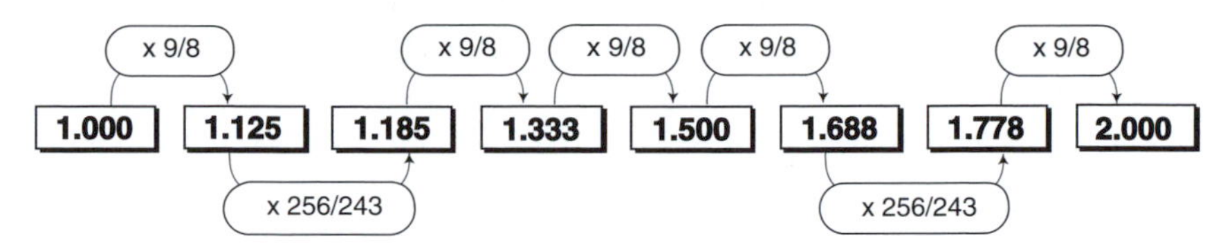

The larger of these ratios, 9/8 or 1.125, is an interval which is called a **tone**. The smaller ratio, 256/243 or 1.0535, is called a **semitone**, since your ear judges it to be about half of the other. The arrangement of intervals in the Dorian mode is thus:

T s T T T s T.

You can see why the Pythagoreans felt they had constructed an elegant scale which divided the diapason naturally and symmetrically into seven parts.

Let us pause here to think about the implications of all this. Remember that I stressed that this was an abstract mathematical exercise. There was never any suggestion that real musical scales actually originated in this way. Presumably in very early human civilizations, musicians found from experience that some intervals were pleasing and others not; that only some steps in pitch were good to sing. And when they found that their audience also liked them, then these steps were remembered. Over time they became the common language of musical communication, passed from one generation to the next. Surely thus must musical scales have originated.

What the Pythagoreans found so wonderful was that their elegant, abstract train of thought should have produced something that people everywhere already knew to be aesthetically pleasing. They had discovered a connection between arithmetic and aesthetics, between the natural world and the human soul. Perhaps the same unifying principle could be applied elsewhere; and where better to try than with the puzzle of the heavens themselves.

The music of the spheres

Astronomy is probably as old as music. Our earliest myths and legends tell of primitive peoples' attempts to account for the splendour of the night sky. It was observed very early that the whole vault of the heavens apparently revolved around the earth once a day. The ancient Egyptians were aware of an additional yearly revolution, and they could predict important annual events like the flooding of the Nile by observing when different constellations were in the sky. In Babylon it was known that the sun and the moon moved slowly through the fixed stars along a well-defined path, which they called the **Zodiac**. In the absence of other information they attributed mystical significance to the constellations which lined the zodiac, a belief system which somehow survives today in the practice of astrology. But it was the Greeks who gave most attention to a group of five stars which also ambled along the zodiac, the so-called **planets** or 'wanderers'.

It was Pythagoras who first realized that the morning star and the evening star were the same object (and one of the planets), and during the next five or six hundred years an enormous store of knowledge was amassed. They came to understand where the moon's light came from, what caused eclipses, that the earth was a sphere and that it rotated once each day. They even had a good idea of the distance to the moon and sun. But above all they were interested in how, and why, the whole thing worked. They imagined there was some sort of giant machine causing the earth to rotate inside the sphere of the fixed stars. The sun, moon and planets therefore had to be on other, independently moving spheres inside the first. These were made of transparent crystal, and those closest to the earth revolved the fastest.

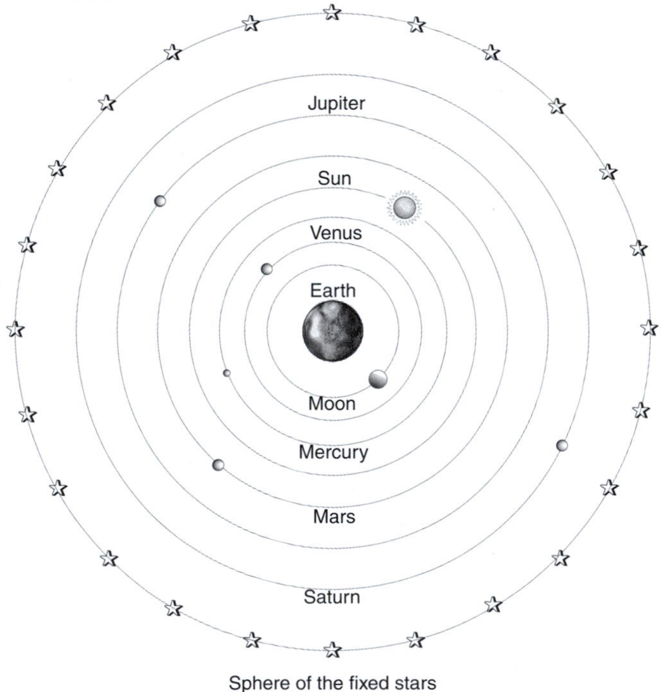

Sphere of the fixed stars

But what they really wanted to know was *why* the universe was like this. They felt they would understand if only they could find answers to a few simple questions. Why were there only seven crystal spheres, and not (say) six or eight? Why did each take its own particular time to complete one revolution? Why these and not others? It cannot escape your attention, and it did not escape theirs, that this same question had been asked before by Pythagoras about music—and answered.

To whomever it was who actually first thought of it, the idea must have seemed a blinding revelation. The order which the astronomers had observed was a manifestation of some underlying principle of cosmic harmony. The arrangement of the planets was nothing more or less than a *musical scale*. Even the separation of earth and sun corresponded to a perfect fifth. So was born the **music of the spheres**.

Initially the concept was, to our minds, rather unsophisticated. It was argued that the crystal spheres must make a sound as they revolved, and since their speeds, as judged by their distances, were in the ratios of musical consonances, those sounds must be harmonious. They were audible but unnoticed by mortals who hear them from birth, in much the same way presumably as adults today lose their ability to hear the background squeal of a television set. Somewhat later, Plato even suggested that there was a siren sitting on each of the planets and singing. Eventually however it came to be seen as a metaphor for the laws that governed the way the cosmos was ordered. It was studied as a branch of mathematics describing the relationships between the laws of motion and the structure of the heavens. It is probably fair to say it was a precursor of what physicists know as Rational Mechanics.

There is not much point in following the Music of the Spheres through its later elaborations by Roman and Christian philosophers. There was never very much of what we would call science in it. But the beautiful conception of a world bound together in a harmony has entranced poets and philosophers in every age. The Bible tells us (Job, 38.7) that there was a time,

> "When the stars sang together,
> And the sons of God shouted for joy."

Even in our own century the Music of the Spheres continues to strike a chord within the human spirit, and it is particularly evident in the music of Paul Hindemith. You can see it also in J.R.R. Tolkien's attempt to invent a completely new mythology for a whole world, *The Silmarillion*. This is how he describes the very act of creation:

> "A sound arose of endless interchanging melodies woven in harmony that passed beyond hearing into the depths and into the heights; and the music and the echo of the music went out into the Void, and it was not void."

The Ptolemaic system

The pinnacle of achievement in the astronomy of the ancient world took place in about 140 A.D. when a book which collected together all current knowledge in the field was published by an astronomer in Alexandria called Claudius Ptolemy. One of the shadowiest figures in the history of science and music, nothing at all is known about him as a person. He was not related to the royal family of Ptolemies who ruled Egypt after Cleopatra; he probably got his name from the city of Ptolemais on the Nile where he was thought to have been born in about 75 A.D. No one knows when he died. It is not even sure what nationality he was, though he wrote in Greek.

The model of the solar system which he described was an elaboration of the simple picture I gave earlier. By this time observations had shown that the motions of the planets were not in simple circular paths, so Ptolemy imagined them to be travelling in circles whose centres moved around the crystal spheres—called **epicycles**, wheels within wheels. His theory might seem cumbersome and difficult to apply, but it worked. It was reliable enough for hard-headed sailors and explorers to navigate by. Greek science, just before its end, had learned the lesson that it wasn't enough for theories to be aesthetically pleasing, they also had to match observational facts.

As the Roman empire started to crumble and the centres of Greek culture withdrew more and more to Constantinople, Ptolemy's work (like so much of Greek science) survived among the Arabs. They called his book *The Almagest*, which means, simply and appropriately, *The Greatest*. It was not to be rediscovered by the west until the time of the crusades, when it was at long last translated into Latin.

One of the happiest legacies of the Greek tradition, which unfortunately we seem to have discarded in the last couple of hundred years, was the expectation that eminent natural philosophers (i.e. physicists) would also be interested in musical theory. Another of Ptolemy's major books was called *Concerning Harmonics.* , It contains, among much else, some ideas which offer a new light on the understanding of musical scales.

You will recall that the Pythagorean 'explanation' of the musical scale depends on the fact that dividing a string into 2, 3 or 4 gives pleasing musical intervals. What is new in Ptolemy's work is the suggestion that there is no need to stop at 4, but that 5 should be included. Go back to the monochord and put the bridge one fifth of the way from one end. The interval between the note from the unstopped string and that from the longer portion—which corresponds to a ratio of 5/4—is what we call a **major third** ("*While Shep*-herds Watched their Flocks by Night"). It is *not* one of the intervals in the Dorian mode and it was not considered harmonious by the Greeks, though it is today. Nevertheless a scale can be constructed using it.

Again you begin with one note. Then raise it by a third and also by a fifth. This gives a set of three notes which we call a **major triad**.

This threesome will play a crucial part in later developments in harmony, but for now just think of it as a mathematical entity constructed from the numbers 1–5.

The next step of course is to generate new notes by the recipe we used earlier. Firstly multiply by 3/2 (raise the notes by a fifth):

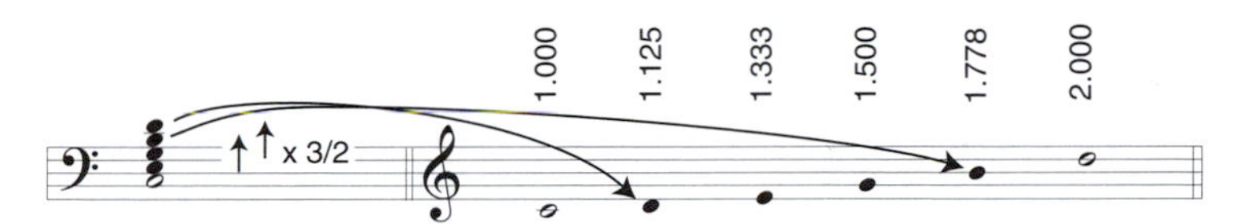

Then divide by 3/2 (or lower the notes by a fifth):

So the process leads again to a seven note scale, which apparently also consists of the same notes we use today. But they aren't exactly the same—remember I warned that playing these notes on a modern piano will only give an approximation to what you would hear from a string divided into these exact ratios.

The important point about this scale is that, although its notes are very close to those of the Pythagorean scale, they are not quite the same. To see this, look at the intervals between successive notes. Get out your pocket calculator again and check.

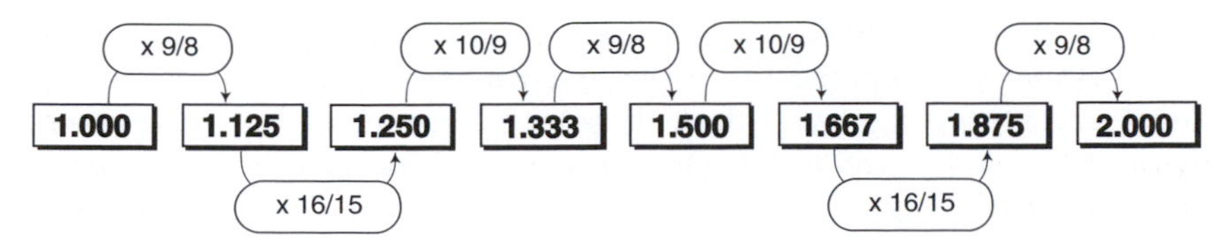

There are two intervals of similar sizes, one equal to 9/8 (or 1.125) which you will remember was the **Pythagorean tone**, and the other equal to 10/9 (or 1.111). A reasonably acute ear can tell these two are different intervals, but an untrained one may not notice. So they are both considered whole tones. The former is called a **major tone** and the latter a **minor tone**.

There is also an interval of 16/15 (or 1.0686). Again this obviously should be called a **semitone** because two of them together make roughly a tone; but it is not quite equal to the Pythagorean semitone (1.0535). The complete scale consists of these intervals:

M m s M m M s.

where M represents a major, and m a minor tone. It is very close indeed to our modern scale, and it even starts on 'doh'

Again we should pause to consider just what are the implications of all this. The tremendous insight of the Pythagoreans was that there was a connection between the aesthetic quality of music and the simple logic of arithmetic. What Ptolemy's work showed was that the connection was not unique; very similar logic would lead to another, different scale. Possibly there were still others. The connection between music and mathematics was clearly nothing like as simple as was once thought.

The Greek legacy to Western music

Whenever I listen to early medieval music, particularly church music, I cannot help being impressed by how long-lasting was the influence of the Greek theorists. More than a millennium and a half after Pythagoras, everyday music still showed evidence of his ideas.

During the intervening centuries, which we loosely call the Dark Ages, the ancient traditions had been kept alive mainly through the writings of a sixth century philosopher and mathematician, Anicius Manlius Severinus Boethius. He was born in Rome of a noble family in about 480. In adulthood he was befriended by Theoderic the Great, ruler of the Ostrogoths, who had deposed the last of the Roman emperors in 476, and he rose to a public position of some importance. But he was never tactful in his criticisms of his patron's barbarian friends and they retaliated by accusing him of being a spy for the Eastern emperor in Constantinople. They won, and in 524 he was imprisoned, tortured and executed without trial.

Boethius was reputed to be the last Roman writer who understood Greek and his writings were the only source from which Europeans could learn of Greek science and musical theory until the time of the crusades. However he did more than simply preserve the ideas of the past. He tried to reinterpret them in terms of current thinking. He divided all music into three types: *musica instrumentalis* or the actual playing and singing of music, *musica humana* or the harmony of the human soul, and *musica mundana* or the music of the spheres. The first category was the lowliest of the three, being considered an imperfect reflection of the other two. The true musician knew music as part of the celestial order; the mere performer was only a servant—not a popular view among western musicians today, I think.

But most of the musical theory he passed on was pure Pythagoreanism and this forced music to develop in a certain way. If you recall my discussion of the Pythagorean scale, you might have been struck by the fact that, though it seemed a logical enough process, it was really quite arbitrary. I created the notes of the scale by alternately multiplying and dividing by 3/2 and ended up with the Dorian mode. But there was no compelling reason why I had to proceed in exactly that way. I could for example have multiplied each time and never divided at all.

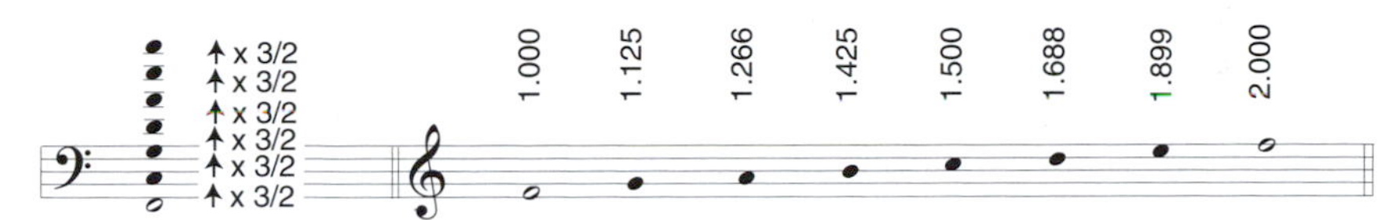

The resulting scale is called the **Lydian mode**, though you will notice that, to end up with the same notes, I had to start on an F. Similarly, had I started on a B and *divided* six times, I would have got a scale with the same notes in it starting on a B, which is called the **Locrian mode**.

What I am trying to show is that, for these seven notes tuned in exactly this way, there is no mathematical reason to single out any one of them as the most important, the natural starting point. The same is true aurally. They are all related to one another by intervals of fifths, so no one note strikes our ears as being the most fundamental. These scales do not really have a *tonic* in the sense that we normally use that word. That means that there are really seven different scales, or modes, in the Pythagorean scheme, one beginning on each of the seven different notes.

It is worth noting in passing that this is not true of the Ptolemaic scale. If you recall how that was constructed, you will realize that it consists of three triads:

The lowest notes of each of these triads are clearly of more importance than the other notes, both mathematically and, as we will see later, aurally. These three notes were later given the names **tonic**, **subdominant**

and **dominant** respectively. However this is a bit of a sidetrack for now, because it was the Pythagorean tradition that Boethius passed on—the tradition that stressed the uniqueness of diapasons, fifths and fourths as the only acceptable consonances.

I think it is clear that many of the features of early medieval music come directly from the underlying musical theory, and indirectly from the scientific ideas on which that theory was based. Much of that music was vocal and meant for liturgical use. The principles laid down by Boethius passed into canon law and the seven modes, with some modifications, became known as the **Ecclesiastical modes**. Each mode developed its own emotional connotations so that, even though polyphony had yet to be developed and harmony was frowned upon, this vocal music was rich and rewarding, as anyone who has sung or listened to Gregorian chant can testify.

Let me quote you an ancient plainchant melody which is still in use today. It is the "Dies Irae", a part of the latin mass for the dead sung in Roman Catholic churches. Its sombre tone and apocalyptic connotations had a particular fascination for 19th century composers—Berlioz, Liszt, Saint-Saëns, Tchaikovsky and particularly Rachmaninov have all written works which incorporate the plainchant "Dies Irae". The currently used words, in rhyming triplets, are due to Thomas of Celano (13th century), but the melody is almost certainly much older.

You will note that it uses the white notes of the modern piano keyboard, and each bar comes to rest on the note D. It is therefore in (one of) the Dorian modes.

From my point of view the most important part of the Greek legacy to survive was the nexus between music and science. When the great European universities were established in the 12th and 13th centuries, in Paris, Bologna and Oxford, they offered students of natural philosophy a curriculum of subjects called the **quadrivium** ('the four ways')—the same four subjects that Pythagoras had taught at Crotona: Arithmetic, Geometry, Astronomy and Music.

Chapter 2

Music and scientific method

In terms of what I originally said this book was going to be about, all I have done so far is to establish that there is some sort of fundamental connection between science and music; but exactly what is the nature of this connection, or why it exists, I have not touched on at all. Because I discussed things in the context of ancient history, it isn't even clear what matters we should talk about next to help us answer these very difficult questions. So what we have to do is to begin looking at things from a modern point of view, which means, in historical terms, we've got to move at least beyond the Renaissance.

Strictly speaking, there was an unbroken progression in all areas of the arts and sciences during the time in which ancient Christendom was transforming itself into modern Europe. But there was a period, which we call the Renaissance, when new ideas emerged so rapidly that the intellectual climates before and after might have belonged to different planets. The actual event which you might consider to have triggered it off is largely symbolic. Whether you take it to be the capture of Constantinople by the Ottoman Turks, or the invention of the movable type printing press by Johannes Gutenberg, or whatever, the date you will come up with is somewhere around 1450.

All human intellectual activities are interconnected, so it isn't surprising that both physics and music should have undergone great changes during this period. Music became increasingly secularized; and its centre of interest shifted from melody to harmony. The study of physics also blossomed and left the control of the church; and its centre of interest changed from philosophy to experimentation. So for the purposes of my story it makes sense to symbolize this intellectual watershed in two figures: Josquin des Prez and Nicholas Copernicus.

Ars nova

The ancient tenets of musical theory lasted right through the Middle Ages, largely owing to the conservative influence of the Church. As late as 1324, Pope John XXII from his palace at Avignon, promulgated the edict which permitted only "... concords such as the octave, fifth and fourth, that enrich the melody and may be sung above the simple ecclesiastical chant".

In many ways the church fathers were really only worried about preserving the integrity of the liturgical chants which were inextricably part of the older musical tradition—the so-called *ars antiqua*. But things were changing despite them. Secular music was gradually gaining importance. From the time of the crusades, minstrels and troubadours roamed the countrysides with their songs of chivalrous romance and political satire. There arose schools and guilds (like the famous Mastersingers), where musicians could learn their art and acquire some measure of job security. Even within sacred music, organum had become common during the 9th and 10th centuries, and church singing gradually became more and more elaborate; so that, by the 14th century, the great age of polyphony had begun (which was exactly what Pope John was getting annoyed about).

Musical theorists gave their rather belated approval to the changes that had occurred when in 1316 one Phillipe de Vitry published a treatise entitled *Ars Nova* ("The New Art"); and that name came to be applied to the music of the whole period. It is interesting that de Vitry's work, which was mainly about rhythmic innovations, still called on Pythagorean mathematical doctrines as the ultimate justification for what he was doing.

But more important were the changes taking place in attitudes to harmony. There was a growing acceptance of thirds and sixths as consonant intervals. (You will recall that I discussed the third when I was talking about the Ptolemaic scale. The **major sixth** also occurs in that scale. It corresponds to a ratio of 5/3 and you can hear it in "*My Bon*-nie lies over the Ocean".)

English theorists were among the first to sanction these harmonies; and the famous song "Sumer is Icumen in", which dates from about 1240, is a splendid example of the use of major thirds as consonances.

This is a **round**. It is meant to be sung by several singers, each one starting a bar later than the person before them. But for this to work, at least the first notes of all the bars must sound good when sung together. In this melody you will notice that the first notes of the six bars form a major triad: C E G C.

Gradually the Ecclesiastical modes themselves changed. At first it was done in an *ad hoc* fashion. In the Dorian mode, for example, there was a tendency to sharpen the third note (the F)— which produced a major third. This was usually done to avoid having three whole tones in a row—always considered a particularly difficult interval to sing. The sharp sign was never actually written down—that would have been against canon law: but singers understood that it was meant to be there. This subterfuge went by the name of **musica ficta**.

The need for such devices showed that all the modes were *not* musically equivalent. Two of them in particular were more useful than the rest because they contained many intervals which were very close to the newly acceptable consonances. In time the other modes fell into disuse and these two changed into our present day major and minor scales. Furthermore it was the tuning of the Ptolemaic intervals which matched the new harmonies rather than the Pythagorean ones, so when the major scale finally settled down it was the same scale that Ptolemy had described a thousand years earlier—now rechristened the **just scale**.

Clearly the tradition was arising that composers shouldn't be strait-jacketed by abstract philosophical rules, but should be guided by their own ears and musical judgement. But it doesn't answer the question of *why* they were so clear in their preference for just tuning. Why for example was the interval of a major tone plus a minor tone, so much more like the major third they wanted to hear, than two Pythagorean tones? I can think of two naturally occurring effects which might throw some light on that.

The first concerns the natural tones which are produced by certain wind instruments: natural trumpets, clarions, post horns and bugles—instruments that don't have valves, slides or finger-holes and which were common in Medieval times. Such instruments can only play a set of notes like these:

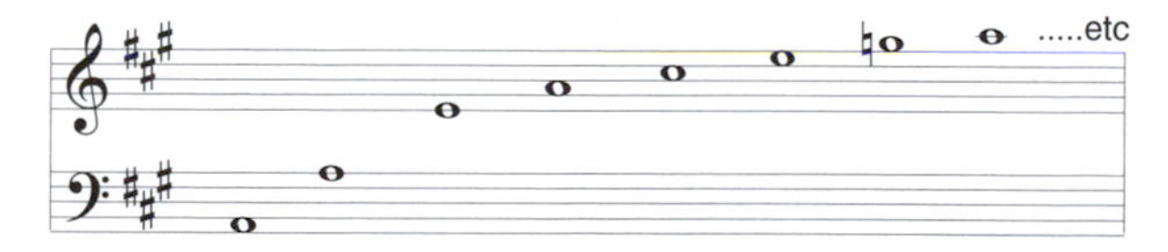

This set is called a **harmonic series**. (There are higher notes, which I haven't shown and which are more difficult to play). If you were to check them very carefully against a monochord, you would find that the first five intervals are exactly the octave, perfect fifth, perfect fourth, major third and minor third (this last corresponds to the ratio 6/5—"*Go-in'* home"). Why these instruments should give just these intervals was not understood, but nevertheless it meant that there was some sense in calling them 'natural'

intervals. Any scale which contained them, like the just scale, had a claim to having more 'natural' validity than one which did not.

Secondly, there is a musical idiom well known to choral singers, that it is somehow more satisfying to end a piece of music on a chord containing a major third in preference to a minor one. By way of example, here are the last six bars of the 16th century motet "O Magnum Mysterium" by Victoria, which you will still hear sung by choirs at Christmas time.

You will notice that the soprano line (top stave, sticks pointing up) and the bass line (bottom stave, sticks pointing down) come to rest on an A. Except for right near the end, all parts use only the white notes, with the occasional G♯. If you know about these things you will recognize that the key is A minor, of which G♯ is the 'leading note'. You might expect then that the last chord would be the minor triad A-C-E-A. But in fact it is A-C♯-E-A—the major triad.

You might try the experiment of playing it with each C♯ replaced by a C♮. It just doesn't sound as good, especially if you're singing in a building like a church where the sound reverberates for a long time. It is almost as though the building itself sings the major third with you. Again there is the suggestion that it is a 'natural' consonance, and therefore a scale that contains it is 'natural' too.

This idiom is known as the **tierce de Picardie** or **Picardy third**. The reason for the name is unclear, but it was probably because it first came into fashion among the renowned Flemish school of composers in the 15th century, of whom the very greatest was Josquin des Prez.

Josquin was born around 1440 in the district near the present Franco-Belgian border. His musical career took him to the ducal courts of Milan, Modena and Ferrara, the papal chapel in Rome and the court of Louis XII in Paris. Towards the end of his life he returned to Flanders where he died in 1521. His work was immensely popular during his lifetime but went out of fashion after his death. It wasn't rediscovered until nearly 300 years later, yet today most music historians count him among the greatest composers of all time.

When you look back at the music of that time, Josquin stands out as a milestone from which you measure what happened next. He was the first composer whose music is listenable to in the same terms as we listen to the music of the classics. The language he invented is one we can understand. His mournful music sounds mournful, and his joyful music sounds joyful. He didn't do it single-handedly of course, and a lot of his techniques had been used by his immediate predecessors; but he's the one we notice the changes in, and he's the one his successors copied.

What we respond to in his music is its harmoniousness. It stems from the way he conceived each piece of music as a whole—melody and harmony together. In his work there is the beginnings of tonal organization: all the harmonies in the whole piece are in one way or another related to one single, ultimate **tonic**, or **key**. Even when it goes off into another key, the aware listener can tell it is only a temporary deviation and the original tonic will return. This principle, which was to become such a large part of Western music, is known as **tonality**. And it's in Josquin's music that you can most clearly see it starting.

Just how the listener detects this musical structure, and why it should be aesthetically pleasing is one of those very difficult questions that theorists debated about for centuries. But what I find most interesting is the fact that (much later) the word 'gravity' came to be used to describe the relation of a piece of music to its tonal centre. It demonstrates what seems to me an undeniable parallel between what Josquin was doing in music and what others at the same time were doing in science.

Scientia nova

At the time when Josquin was about 30 years old, in 1473, Nicholas Copernicus was born of a wealthy family in the north of Poland. An extraordinarily gifted young man, he gained an early reputation in mathematics, medicine, law and astronomy. He took minor orders in the Church but seemed to spend most of his working life as a diplomat, a doctor, an economist or simply running his family estates until he died in 1543.

His interest in astronomy began during his student days at the universities of Cracow and Bologna, and stayed with him all his life. These were the times when the classical works of astronomy had come back to the West from the Arabs, and for over 200 years there had been talk of the need to reform the calendar. Copernicus set himself the task of updating Ptolemy's astronomical calculations. He realized early that the elaborate geometrical picture of the universe that Ptolemy had devised was too cumbersome. It would make calculations easier, and the explanations of certain observations simpler, if he imagined the sun to be at the centre of the universe rather than the earth. So for the rest of his life his spare time was spent checking and rechecking figures derived from the assumption that the old model of the universe, which I drew on page 10, should be replaced by this:

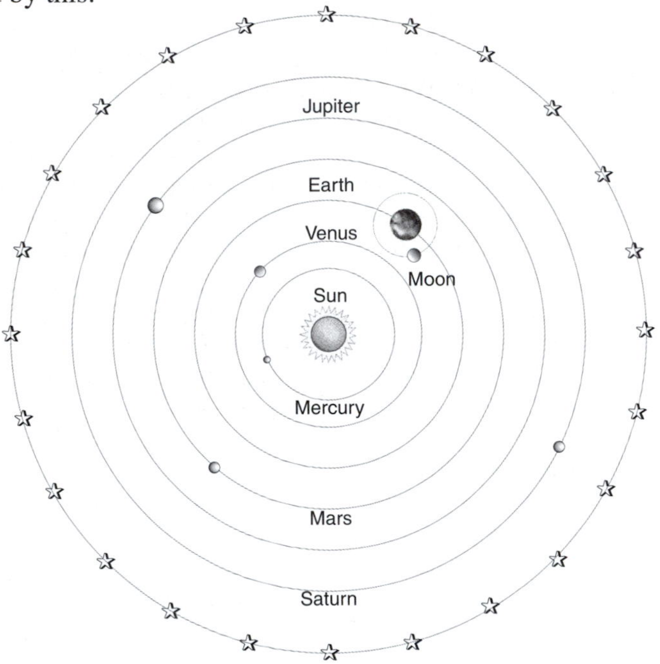

Sphere of the fixed stars

It is not clear just when he came to consider the idea of the sun-centred universe as more than a mere calculational device. He was clear-headed enough to realize that, if true, it would have enormous ramifications, because if the earth were not at the centre of creation, then neither were human beings. He was also clear-headed enough to know that he was on dangerous religious ground. The protestant reformation, just beginning, was placing more emphasis on the Bible as the prime source of faith; and if the Bible said the sun stood still so that Joshua could win the battle of Gibeon, it would be foolhardy to suggest otherwise. Given later developments, how right he was!

Therefore, although his friends urged him to publish what he was doing, it wasn't until a few weeks before his death in 1543 that he saw the first copy of his monumental book *De Revolutionibus Orbium Coelestium*—"On the Revolutions of the Heavenly Bodies".

So began the great scientific upheaval, rightly called the Copernican Revolution, quietly at first, so that very few knew about it; and it would be nearly 100 years before it came into open conflict with the Church. But there is no doubt that the medieval world view changed with the publication of *De Revolutionibus*, and the way was paved for the next important figure in this story.

Johannes Kepler was born in Lutheran Germany in 1572, and must surely be one of the most paradoxical figures in the history of science. Sickly from birth and nearly blinded by smallpox in childhood, he trained for the priesthood but gave it up, formed an unhappy marriage with a rich widow at the age of 25, was widowed at 40, spent some years in demand among prosperous matrons with marriageable daughters, married again and fathered a total of fifteen children before dying at the age of 59. He was a mathematician of genius but never seemed to question the logical basis of astrology which he practised professionally (at one time his mother was imprisoned for witchcraft). He was a poor public speaker and an appalling teacher, but he ended up in court circles under the patronage of the Emperor Rudolf II. He was a mystic and a romantic—he wrote what is probably the first ever science fiction story; yet he helped to introduce into science the self-correcting philosophy that if theory does not agree with observation, no matter how small the discrepancy, then the theory must be discarded.

From very early on, this strange man developed a definite goal for his astronomical endeavours. He wanted to reinterpret the ancient doctrine of the music of the spheres in the light of Copernicus' sun-centred universe. At first his attempts were purely mystical. He represented each of the planets by a note on a musical stave, with a pitch corresponding to its speed. As they moved in their orbits these pitches changed and so, in principle, each gave rise to an endlessly repeating line of melody. The Earth for example sang a sad minor second, which Kepler compared with the first such interval on the standard 'doh ray' scale—'fa/mi'—and he was apparently impressed that this was also the Latin word for famine. But this approach to the problem didn't really lead anywhere, so it is lucky (for us) that eventually he gained a position where real astronomy was done—in Prague, with the world famous Danish astronomer Tycho Brahe.

Tycho died not long after and Kepler found himself heir to a collection of planetary observations of unparalleled accuracy. So he began an analysis of this data which was to last for many years. After a decade he published some results in a treatise called *The New Astronomy*. For the first time in history, the real shapes of the orbits of the planets were known. They were not circles, and were not on crystal spheres, but were ellipses with the sun slightly off centre. As the planets move around these ellipses, their speed changes in a beautifully predictable manner. These results we still call Kepler's first and second law.

However his main aim was not achieved. It was to be another ten years before he came to his third law, the harmonic law—a precise connection between the size of a planet's orbit and the period for it to go once around the sun. He described this in what he considered his most important book, *The Harmonies of the World*; and in this book he wrote:

> "The heavenly motions are nothing but a continuous song for several voices, to be perceived by the intellect, not by the ear; a music which, through discordant tensions, through syncopations and cadenzas as it were, progresses towards certain predesigned six-voiced cadences, and thereby set landmarks in the immeasurable flow of time."

But in the end, elegant though his harmonic law might be, it clearly had nothing to do with musical scales as we know them. Posterity has judged that there is no such thing as the music of the spheres. Instead, what Kepler had found was knowledge of an entirely different character—an exact description of how the heavens worked.

Nevertheless the emotional appeal of his ideas remains strong. You can feel it in Paul Hindemith's 1957 opera about the life of Kepler (the only person in the history of science whose life has been so treated), called—what else?—*The Harmonies of the World.*

The new method

It was during the 16th century that the whole of physics, and not just astronomy, felt the full effect of the Copernican revolution. More than anything else it was a change of attitude. No longer would it be felt that we could never hope to attain the splendour of the ancients, but rather we should be considered more fortunate than they because we are heir to all they had discovered.

No one stated this view more successfully than the English writer Francis Bacon with the publication in 1624 of what was essentially a work of science fiction, *The New Atlantis*. In it he argued that science should be systematically studied and applied to the welfare of society. In order to achieve this he proposed a 'new method', based on the idea that the store of knowledge can be increased, not by consulting the ancient philosophers, but only by performing new experiments and deducing the laws of nature from them. Bacon's **scientific method** was to prove a powerful concept, even in fields other than science.

During this time, the centre of the musical world had shifted to Italy, away from the Low Countries. But the Flemish tradition was carried on by Adriaan Willaert, a pupil of a pupil of Josquin's, who occupied the influential position of Director of Music at St Mark's in Venice; and by his pupil and successor in the post, Geoseffo Zarlino, who was recognized as Italy's leading theorist.

Now Zarlino belonged to the old school. For him the harmony of music was still a reflection of the beauty of arithmetic, and in particular the major triad shared semi-mystical attributes with the set of numbers 1–6, which he called the **senarius**. But change was on the way. A small group of Florentine musicians, scientists and noblemen, who called themselves the **Camerata**, began advocating that music had reached a state of impasse.

They argued that the practice of polyphony, once so fresh and rewarding, had become impossibly complex and academic. They called for a return to the simplicity of the old ideals, in particular to the musical declamation of poetry as in the ancient Greek tragedies. The elaborate polyphonic music of the Church was obviously unsuited to this, so there arose the idea of **recitative**—a solo melodic line which could imitate the accents of speech, with the simplest possible accompaniment, consisting of little more than a few chords. The natural venue for such music was opera, and the historical landmark was the first performance in 1607 of Monteverdi's *Orfeo*.

But it is what happened in musical theory that concerns me. One of the members of the Camerata was a minor composer and performer, Vincenzo Galilei, who had studied under Zarlino and had himself gained a reputation as a theorist. At his instigation—and here's where the new method comes in—the Camerata conducted a comprehensive series of experiments to try out for themselves the musical relationships based on the senarius. Their results were disappointing, and Vincenzo concluded that the ancient theories didn't square with musical practice. What is more, he wrote to Zarlino and told him so. There followed an acrimonious and polemical quarrel between the two, which lasted till they were both old men. But it changed musical theory. From then on, abstract mathematics must always defer to what musicians actually hear.

Galileo

The two threads of this story, the idea that a study of the nature of sound is vital for an understanding of musical harmony, expounded by Vincenzo, and the new experimental approach to scientific method, characterized by Kepler, came together triumphantly in Vincenzo's son—Galileo Galilei.

Born in Pisa in 1564, Galileo (he is universally referred to by his first name only) was the epitome of the Renaissance man: a scholar of genius, an accomplished singer and lutenist, a passable artist and a writer of charm and skill. In fact, it was because he was such a good writer that his writings describing his experiments and theories became fashionable; and in the end this was an important factor in what happened to him.

He made his first contribution to science while still a teenager. The legend is that one day, while sitting through some tedious service in the Cathedral at Pisa, he found himself watching the gently swinging sanctuary lamp. It seemed to him that the time it took to swing backwards and forwards was the same whether it was swinging through a large arc or a small one. He verified it, as best he could, against his own pulse beat. From this small observation, he was led to the study of pendulums and vibrating systems in general, and into the whole question of the measurement of time. It led eventually to the invention of the pendulum clock and later in his life to the theory of harmony.

At the age of 25, he took a teaching position at the University of Pisa and started on his important studies of motion and of falling bodies, which we will come back to in the next chapter. Here also he might, or might not, have performed that legendary experiment of dropping different sized cannon-balls from the leaning tower to prove that they hit the ground together. He was a short tempered, acid-tongued individual who couldn't stand fools. So, because of his habit of publishing his opinions in well written articles which the public enjoyed, when he moved from Pisa in 1592, he left enemies behind.

At the University of Padua he became interested in astronomy and started corresponding with Kepler. He didn't actually invent the telescope as many believe, but on hearing that one had been invented in Holland, he realized how it must work and made his own. He also managed to sell the idea to the merchants and military authorities in Venice, never being averse to making money out of pure science.

His greatest achievements came when he used his telescope for astronomy. He had been converted to Copernicanism by Kepler, and now he had proof that the old view of the heavens was wrong. He observed that there were sun-spots on the face of the sun, so it wasn't the unsullied celestial body that the ancients believed. He observed that Jupiter had moons of its own going round it, so it couldn't be fixed to a crystal sphere. He observed that Venus showed phases just like our moon, from crescent to full, so it must travel round the sun. It is a sad irony that these observations damaged his eyes, and in old age he went blind.

In 1623, at the age of 59, he published his masterpiece, *The Dialogue of the Great World Systems*. It was written— in Italian of course, not Latin—in the form of a conversation in which two characters, one presenting the views of Ptolemy and the other speaking for Copernicus, expound their arguments before an intelligent layman. Needless to say, the brilliant Copernican won the battle; but Galileo's enemies—and there were many by this time—won the war. They persuaded the pope, Urban VIII, that the character speaking for Ptolemy was a deliberate and insulting caricature of himself.

The rest of the story is well known. After several warnings, Galileo was arraigned before the Inquisition on a charge of heresy, for teaching that the earth moves round a stationary sun whereas the Bible clearly teaches the opposite. After being shown their instruments of torture, he chose to recant—after all, he was 70 years old by this time. However it is a fondly held belief that as he rose from his knees, he muttered, "*Eppur si muove!*" ("But it still moves!").

For the rest of his life he was confined under house arrest in his villa outside Florence and forbidden to write further on astronomical questions. He wasn't ill treated, and was allowed to work with students. His fields of enquiry being constrained, he returned to the enthusiasms of his youth—in particular his researches into pendulums, vibrating bodies and musical notes. The results of this study, which I will describe a little later, were published in his last important work, *Discourses on Two New Sciences*.

He died in 1642. His enemies, vindictive to the last, refused to bury his body in consecrated ground. The *Dialogue* was not removed from the Index of Prohibited Books until 1835; and it was as recently as 1980 that the Catholic Church formally admitted that he had been wrongly convicted. Luckily the rest of the world was not so slow to recognize the contributions to science made by this towering intellect.

On the nature of sound

At this point I want to make a complete change of course. Up till now I have been talking about the development of ideas in a more or less strict chronological sequence. But now I want to talk about the ideas themselves; and it is very difficult to do this in a strictly historical setting because, although the concepts themselves haven't changed since then, the way we understand them has.

Clearly, what we need to look into now, following Vincenzo's suggestion and his son's explanations, are those physical properties of a sound which the ear perceives as being musical. But if I were to try and describe these properties in Galileo's mental images it would be very confusing, simply because today we can see some things more clearly than he could, and it's very difficult to explain something the wrong way if you know what the right way is. As Bacon had pointed out, scientific knowledge is cumulative.

Therefore it will be necessary, if a little tedious, to spend some time laying out clearly just what areas I am going to talk about—making sure of the appropriate terminology. And there is no point in doing this in any but modern terms. So I will make a complete break at this point, and return to the historical sequence some time later.

From a purely mechanistic point of view, a study of music starts with a study of the notes which make it up. There are four properties which most obviously distinguish one note from another. Three of these don't need much comment—**pitch**, **loudness**, and **duration**. The meanings of those words are clear enough. The fourth property is that which distinguishes notes of the same pitch, loudness and duration, but which are played on different instruments. This is not easy to describe, and various names are given to it—for example 'quality' or 'tone colour'. The most widely accepted term, which I will use, seems to be **timbre**.

A musical note, or indeed any sound, comes from a **source**, travels through a **medium** (most often the air) and is picked up by a **receiver** (usually your ear, but could well be a microphone). Each of these three can affect the sound, and it makes sense to consider their effects separately, which we will do eventually. For now, the source is the most obviously important and the key observation to make is that whenever a source gives out a sound, it **vibrates**. Sometimes this vibration is not easy to detect, especially when the sound is short and sharp like a handclap, but with a sustained musical note it is. So it would be useful to keep in your mind a picture of a sounding tuning fork or violin string.

There are at least two quantities I can use to characterize a vibration. Firstly there is the rapidity of the motion, the number of swings that one of the tines of the fork (or the string) makes in a given time. Secondly there is the physical distance that it moves through as it swings back and forth. Since these quantities will come up many times, it is worth defining very carefully the terms which describe them:

- The **period** of a vibration is the time it takes to go through one complete swing.
- The **frequency** is the number of complete swings it makes in one second. Frequency is the inverse of the period.
- The **amplitude** is the distance it moves to either side from its rest position.

There are two important interrelationships which are not difficult to establish. Firstly, the amplitude of the vibration is directly connected with the loudness of the note it gives out. Strike the fork hard, or bow the string firmly, and you get a large amplitude and a loud sound. That much is simple. Secondly, pitch and frequency are related, though it is less easy to observe this because the vibrations are usually too fast for your eye to follow. However the strings of a double bass, for example, move slowly enough that your eye can almost make out the movement; and if you feel with your fingertips, you can certainly tell that they vibrate much more slowly than those of a violin. Clearly a small frequency corresponds to a low pitch and a large frequency to a high pitch.

Those then are the most direct correspondences between what you hear when listening to a musical note and what the source of the note is doing. To get any further understanding we obviously have to talk about simpler and slower vibrators.

Pendulums and skipping ropes

A pendulum in its simplest form is a heavy mass, or bob, swinging on the end of a light string. So if you want to check any of the statements I am about to make, it is very easy to do so.

The way the bob moves as it swings through an arc is beautifully symmetric. It moves fastest at the bottom and slows down continually as it nears the highest point. It stops there instantaneously, and then does exactly the reverse motion as it falls back down. It passes the bottom with exactly the same speed as it had before and then it executes identical behaviour on the other side. This regular and symmetric pattern of movement, for future reference, is called **simple harmonic motion**.

Subsequent swings follow the same pattern, but as time goes on the amplitude gets smaller and the vibration tends to die away. Nevertheless the period stays the same (and therefore the frequency also). This was the observation that Galileo made in the Cathedral of Pisa. Put into the formal language of a physical law, it says:

- The frequency of vibration of a pendulum is independent of its amplitude.

Just for accuracy, it is worth mentioning that this 'law'—which is often referred to as the Law of Isochronicity—is not absolutely true. If the amplitude gets very large, the period tends to get a bit longer, but in most ways a pendulum is used, you won't notice the difference.

This property of a pendulum is not at all obvious, which is why it wasn't widely appreciated before Galileo's time. There are plenty of vibrating systems that don't do this. Most things that rock, and most that bounce, tend to speed up as their vibrations die out. (In fact, I can't think of any simple vibration that slows down as it dies out).

But there are others that behave like pendulums, and when you come to think of it, musical instruments must be in this class. One characteristic you would look for in a well designed instrument would be that you should be able to alter the loudness of the note you are playing, without its pitch changing. (One reason, for example, why cheap recorders are difficult to play really well is that, whenever you blow too hard, the note tends to go out of tune).

There are other interesting features about a pendulum. So far as the period is concerned, the weight of the bob is not really important, provided only that it is much heavier than the string which supports it. But the length of the string is. If you shorten the string, the period decreases. However the dependence is not direct (in the mathematical sense). In fact the period depends on the *square root* of the length. That means that for a string 1 metre long you will find the period is very close to 2 seconds. If you make it 2 m the period will be 2.8 seconds; and 4 m, 4 seconds.

It should be clear now why pendulums made such good regulating devices for clocks, once their properties had been established. They keep very accurate time under a wide variety of conditions, and if they were running fast or slow they could be adjusted simply by lengthening or shortening the support. Shortly after Galileo's first experiments, they were used in laboratories all over Europe to keep track of small units of time; but they weren't incorporated into the escapement mechanisms of clocks until later. Galileo was supposed to have designed such a device but it was never built.

It was the Dutch scientist Christiaan Huygens, of whom I'll have more to say later, who constructed the first fully functional pendulum clock in 1656, and for nearly 400 years they were the world's most accurate time-keepers. Since the invention of crystal clocks in this century they have been superseded and you don't see them around much any more; though the **metronome** is a close relative and is still in common use.

There is another vibrating system of obvious relevance to us, which shares some of the properties of a simple pendulum, and that is a **stretched string**. It needs to be long and heavy before its motion is slow enough to follow, so if you want to set one up for yourself, it is best to use a thick cord.

It can be made to vibrate either from side to side like a violin string, or round and round like a skipping rope. The frequency with which it makes these vibrations depends on how taut it is, (you might remember from school-yard games that to make a skipping rope go faster you had to pull on it harder). But if you set it up carefully with a constant force pulling it tight and get it going gently by hand, it will behave in some sense like a pendulum. There is one frequency at which it will vibrate normally, whether you pluck it and let go or whether you try to keep the vibration going.

This too has the property that its frequency is independent of amplitude (strings are, after all, used a lot in musical instruments). Its frequency also depends on the length of the rope. If you shorten the rope you decrease the period, and increase the frequency. But no square root is involved. If you halve the length you double the frequency, and this result is worth writing down as another formal experimental law:

- The normal frequency of vibration of a stretched string is inversely proportional to its length (other things being equal).

You recall the experiments with the monochord of course. There we observed that halving the length of the string raised the pitch an octave: here we observe that the same operation doubles the frequency of vibration. Similarly, the operation which raised the pitch by a fifth (selecting 2/3 of the string), multiplies the frequency by 3/2. Previously we intuited that frequency and pitch were related. Now it becomes clear that frequency can be used to *measure* pitch; and musical intervals can be identified on the scale of numbers that this measurement provides.

The puzzle of consonance

Before I talk about what the actual frequencies of musical notes are, I would like to go right back to the question that was behind the quarrel between Vincenzo and Zarlino, the so-called 'puzzle of consonance'. Why should two notes separated by (say) a fifth sound well together? From the time of the Pythagoreans musical intervals had always been associated with ratios of string lengths, but by the end of Galileo's life the connection with frequency had been established and he was able to see the puzzle like this: why should two audible vibrations whose frequencies are in the ratio of 3:2 combine pleasingly? In this form he came up with an answer.

His argument went like this. Imagine that you have two pendulums, whose frequencies are in the ratio of 2:3. For example, you could choose one of length 1 metre and the other 44.44 cm (i.e. 4/9 m). If you start them swinging together, they will immediately get out of step. However in the time it takes the slower one to make 2 complete swings (4 seconds) the other makes exactly 3 swings; so they will come back into step at that instant. And they will continue to come into step every 4 seconds thereafter. Needless to say it only works if their frequencies are exactly in that ratio.

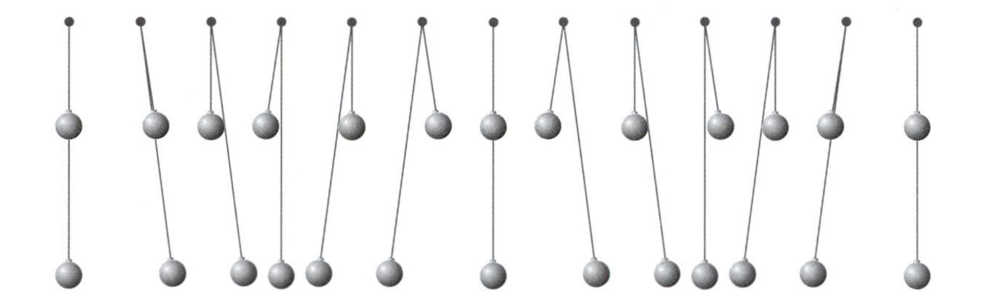

(In case you can't interpret what the diagram is showing, observe that the long pendulum has swung only half of its cycle in the first four 'snapshots'—i.e. out to one side and back to vertical. But in that same time the shorter pendulum, being one and a half times as fast, has swung to the right, back to vertical and out to the left.)

Now make it more complex. Add a third pendulum whose length is 64 cm. The frequencies are now in the ratios 4:5:6—the ratios of the major triad, you might remember. The motion of these three will look most intriguing. Different pairs will continually move in and out of step with one another, and after every 8 seconds all three will momentarily be perfectly in step. If the ratios were not exact of course, the motion would be, and would remain, a jumble.

Could it be that, when judging some combination of sounds consonant, your ear is detecting some such pattern of regularity?

This was Galileo's suggestion. I'm sure you will agree that it was ingenious and very suggestive. Whether there really is any connection between the intellectual satisfaction of seeing patterns in moving objects, and hearing a pleasant combination of sounds, is not at all obvious. So there is no way in which it can be said that Galileo 'explained' the puzzle of consonance. Yet his analogy seems so apt that it is impossible to believe that it is completely meaningless. It is just what philosophers had been saying for so long, that harmony lies in the perception of order.

What frequency is sound?

This is the big question that I have left until last. Any audible vibration is too fast for your eye to follow, but that doesn't say much. After all, your eye can't discern even a flicker in a motion picture image which changes 24 times a second. Audible frequencies were first measured around Galileo's time, which is understandable because that was when the great improvements in mechanical clocks occurred. Until it was possible to mark off one second accurately, there was no way that the number of vibrations per second could be measured.

There is one thing it would be wise to get correct at the start. The unit in which frequency should be measured, as laid down by the *Système Internationale* in 1960, is the **hertz**, named after the 19th century German physicist who discovered how to generate radio waves. Now even though you might feel that it is more straightforward to talk in terms of 'vibrations per second', as do most music books, standardization of technical terms is always sensible. Therefore I will always use the approved S.I. unit. Its formal definition is:

- Something which makes 1 complete vibration every second has a frequency of 1 hertz (or 1 Hz).

So, for example, a pendulum of length 1 m has a frequency of 0.5 Hz; and a motion picture image a frequency of 24 Hz.

The earliest recorded attempts to measure a musical frequency were made in 1648 by Marin Mersenne, of whom I will have more to say later. He took a string long enough that its vibrations could be counted, and then shortened it till it was in unison with the note he wanted to measure. After that it was just a matter of multiplying the counted frequency by the ratio of original to final lengths. A more direct 'measurement' was done in 1672 by the English physicist and amateur musicologist, Robert Hooke. He constructed a cog-wheel with a fixed number of teeth around its circumference, and set it rotating at a rate he could control. Then by holding a card against the edge, he could associate the pitch he heard with the known vibration frequency. This remains a popular demonstration in physics classes to this day.

Nowadays this measurement is done electronically. A microphone changes the sound vibrations of the air into fluctuations of an electric current. These are simply counted in one of the myriad ways that electronic circuits can be arranged to do this kind of thing. Recent advances have miniaturized such devices very successfully and it is now possible for ordinary musicians to carry round their own battery operated frequency meters, every bit as accurate as a digital watch and not much larger.

Anyhow, the results of such measurements are what is important. This diagram should help you to connect a few standard pitches with measured frequencies:

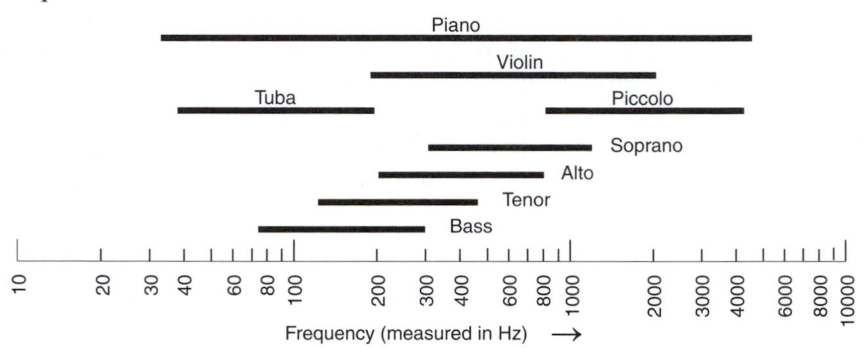

By and large, frequencies that can be heard are in the range 50 Hz to 20 000 Hz. Typically a bass voice can't get below about 80 Hz and a soprano can't get much above 1100 Hz; and since musical instruments have always been designed around the human voice, the range they cover is a bit larger than this, but only by a couple of octaves on either side.

You will notice that on the numerical scale along the bottom of this diagram, the numbers are not evenly spaced. This is called a **logarithmic scale**. It is chosen that way so that intervals which sound the same to your ear, will look the same to your eye. Check this by observing the distance that corresponds to an octave (which you get by multiplying the frequency by 2). This kind of numerical scale will occur again — it seems to be the natural one to use whenever you measure anything to do with human perception. (If you want to know more about the arithmetic involved with logarithms, refer to Appendix 2).

The standard of pitch

Measurement of the physical quantity, frequency, is relatively straightforward; but assigning values to the musical quantity, pitch, needs special conventions for identifying notes. To discuss this I will restrict attention to one particular note, in about the middle of the soprano's range or at the top of the tenor's. It is the fifth note of the Dorian mode and the sixth of the major septatonic scale, and very early was symbolized by the letter A. Today this particular note is often written as A_4 because it is in the fourth octave on the modern piano keyboard, and is six white notes up from **middle C**, which is written C_4. (There is more information about these conventions in Appendix 1 if you are interested).

As you can well imagine, in the Middle Ages when Europe was little more than a collection of independent kingdoms and communication was extremely slow, there was no unanimity about the exact pitch of this note. That there should have been as much agreement as there was, probably reflects the fact that human voices are much the same everywhere and so was the ecclesiastical music they sang. The only instruments of the time whose pitch was absolutely determined were organs, and a study of early organs shows that the lowest tuning generally was in France, where A_4 measured around 380 Hz, while it was highest in Germany at around 500 Hz. As you can see the ratio of these two is 1.32, so they were nearly a fourth different. English church pitch was not as extreme as German, but still high—around 480 Hz. You can notice the effects of this in the church music of the 16th century composer Orlando Gibbons which usually has to be transposed up for modern choirs (back towards what it was originally), otherwise the lowest notes are too difficult to sing.

As instrumental music became more important, the requirements of pitch range became more varied and the old church pitches were often found to be too high or too low. As new organs were built their tuning tended to be somewhere between the two extremes. Somehow or other there arose a **mean pitch** which varied only within very narrow limits and prevailed throughout Europe from about 1670 to about 1820. A fair representative value is that of Handel's A_4 tuning fork, which still exists and is tuned to 422.5 Hz.

It is important to realize that most of the composers of the Baroque and Classical periods composed their music and arranged their voice parts to be played and sung in this pitch, which is almost a semitone lower than today's standard. So, the next time you go to *The Magic Flute*, pity the poor soprano straining for those top F's in the Queen of the Night's aria. In Mozart's time she would only have had to get to an E.

During the 19th century pitches rose again, mainly because symphony orchestras were gaining prominence. There is a good reason why singers would prefer a lower pitch than most instrumentalists. By and large, when orchestral instruments heat up during performance they tend to go sharp. But this does not happen to a singer's voice, which has a tendency to flatten as it gets tired. So they would prefer not to have to chase after excessively high notes.

Also in the 19th century it was fashionable to use military bands to accompany operas; and the makers and players of brass instruments tend to like the brilliance that comes from a higher pitch. The result of these influences was that by about 1850 there was more variation between the pitches of the opera houses and concert halls of Europe than at almost any time for the preceding 200 years. So in 1858 the French government adopted a standard pitch of 435 Hz, which was known as the **diapason normal**. In 1896 the Royal Philharmonic Society, believing that the French pitch had been defined at too low a temperature, adjusted this upwards to the **Philharmonic pitch** of 439 Hz. And in 1939 the International Standards Association (the same body that is responsible for S.I. units) agreed on a world wide standard of 440 Hz for the note we now call **concert A**.

It is interesting that at the same conference there was a move, by scientists, to change to the so-called **philosophical pitch** which puts Middle C at 256 Hz and this makes A_4 426.7 Hz, almost back to where it was in the 18th century. The reason for the number 256 is that it is got by multiplying 2 by itself 8 times, so in this tuning Middle C would be exactly 8 octaves above the absolute standard of 1 Hz. Rather sensibly (in my opinion) the musicians decided they had had enough of being pushed around by mathematicians over the last 2500 years, and stuck to their choice.

Well, once the frequency of A_4 is settled, that determines the frequencies of all the other notes we use, provided only that we can agree on what scale we want. I haven't come to the end of that story yet, but for now I can at least write down the frequencies of the notes in one octave of the **just scale**, from C_4 to C_5 :

JUST FREQUENCIES	
C_4	264 Hz
D_4	297 Hz
E_4	330 Hz
F_4	352 Hz
G_4	396 Hz
A_4	440 Hz
B_4	495 Hz
C_5	528 Hz

Interlude 1

Brass instruments

Despite what the ancient philosophers thought, music is not purely abstract. It cannot be divorced from the sounds which make it up nor from the instruments which make those sounds. We've already seen how the physical properties of some instruments, in particular the human voice, have influenced the way music has developed. So no scientific study of music can be complete without some understanding of musical instruments.

Most musical instruments are exceedingly sophisticated devices. They have been around for a long time and many creative men and women have contributed to their development. So to understand them fully might be a tall order. But it can be done on two levels. Firstly, there are the general principles of how any instrument makes its notes, and these are often straightforward, because there aren't really many different ways of making pleasing tones. Secondly, there are the later refinements whose purpose is to make them easier to play and more pleasant to listen to, but which often don't involve any real advance in understanding. It is towards the first of these levels that I will be aiming.

Therefore I intend, at points throughout this book when the necessary physical concepts have been assembled, to break into the general story and talk about specific instruments; and I have called these chapters **interludes**. But please remember that I will not be giving an exhaustive description of the instruments in these interludes, only as much as can be understood with the physics we have at our disposal right now.

In this light then, the group of instruments I want to discuss first are the brasses, because you can understand a great deal about them with no more background than a knowledge that some long vibrating systems—other than stretched strings—will vibrate with a frequency which is inversely proportional to their length.

What is a brass instrument?

The easiest way to understand why musical instruments are classified as they are is to consider, in broad simplified terms, the functions which all instruments share. There are three. Firstly there has to be some way for the player to get a vibration going. Secondly there has to be something which ensures that the vibration occurs at exactly the right frequency. And thirdly the vibration must be communicated to the outside air so that its note can be heard. In other words, in any instrument you ought to be able to identify a **generator**, a **resonator** and a **radiator**. It is in terms of these functions that instruments are usually classified.

That group of instruments which we call the 'brasses' developed from simpler forms. Possibly the earliest were certain large shells, conches or triton shells, used since the Stone Age by native peoples living near the sea.

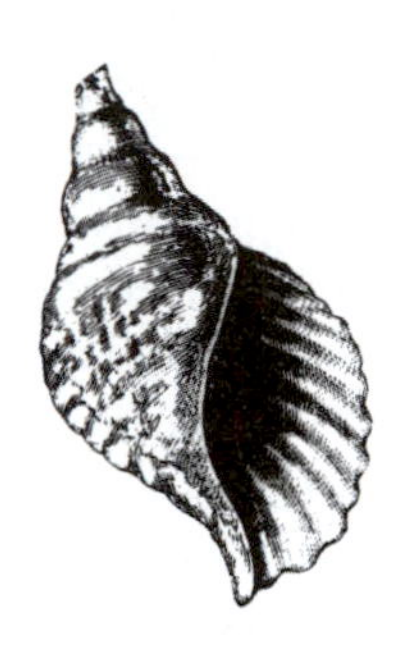

A blowing hole was made by knocking off the point of the shell, and the coiled cavity inside made an expanding wind-tube of flattened oval cross-section. The player held the hole against lightly closed lips and simply blew. With a bit of practice, a satisfying, if not particularly musical, blast could be obtained. Presumably they owed their importance to their being an easily portable sound-maker that could deliver a really loud noise which could be heard over distances of some kilometres. There is a beautiful passage in William Golding's *Lord of the Flies* describing how the young hero first blows on a conch he has found. It is only fiction of course, but it nevertheless gives a feeling of the wonder of producing such a majestic sound and the awe it inspired in the other boys.

Other very early instruments were made from animal horns. A cow's horn, for example, could be hollowed out by dissolving the core in hot water or by simply leaving it to scavenging insects. Then the narrow end was cut off to form the blowing hole.

Different horns gave different quality notes, and in particular their length determined the pitch. African tribes, for example, produced very low notes by using long elephant tusks. But the most important feature was always the loudness and carrying quality of the note. Throughout the millennia such horns have always been used in sounding alarms, calling armies to war, announcing solemn festivities, keeping in touch while hunting and controlling the flocks of countless Little Boys Blue.

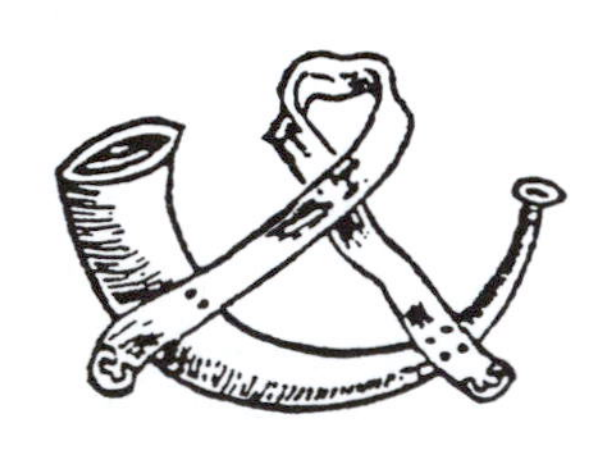

Long wooden trumpets also have a venerable history. The Australian **didjeridu** is a surviving example. It is made from a smooth tree branch up to a metre and a half long, hollowed out by termites or the use of a fire stick. They are more or less cylindrical or conical, depending on what sort of branches are available, and if the blowing end is too wide to be covered by the player's lips, it is often made narrower with bee's wax or resin. They give a low booming note which, when sustained, is used as a drone accompaniment to tribal ceremonies.

In Northern Europe, shepherds and herdsmen made somewhat similar instruments which we now classify by the generic name of **alphorns**. They were made by sawing lengthwise the trunk of a young tree, hollowing it out and binding the two halves together again. Often a mouthpiece, rather like a cotton reel, was fitted separately. Usually over 2 m long, and sometimes as much as 5 m, they give out a very low note indeed; and because the tube expands to a very wide bottom opening, it is also extremely loud. They used to be employed, particularly in Switzerland, for calling cattle home from the summer pastures in the evenings. Hence their popular image of being played exclusively from the tops of mountains.

With the development of metal working, similar instruments were made of bronze, either from rolled metal sheets or by casting. Short metal trumpets appear on Egyptian reliefs dating from the 15th century B.C., obviously about half a metre long with a broad funnel-shaped bell. They were apparently used for military and ceremonial purposes and their use spread to surrounding nations. We know that the Israelites, for example, adopted the military trumpet on their return from exile, for in the Bible (Numbers 10) we read of the Lord saying to Moses:

> "Make thee two trumpets of beaten silver, wherewith thou mayst call together the multitude when the camp is to be removed."

And everybody knows that they were used with notorious success outside the walls of Jericho.

In Roman times there was the long trumpet, or **tuba**, a handsome instrument with a long (about 1 m) narrow tube ending in a wide curved bell. It was used for imperial funerals and other major state events. It is the *tuba mirum* which tradition says will be played by the avenging angel on the day of judgment. Another instrument, the **buccina**, of similar length, was bent nearly into a circle for ease of carrying on the shoulder. It's the one you see in movies being played as the Roman legions march off to conquest.

By the Middle Ages these had developed into the medieval straight trumpet, or **buisine**, an imposing instrument over 2 m long made of brass in jointed sections with a flared bell. There was also a shorter version of the same thing, less than 1 m in length, called the **clarion**. The long trumpets were ideal for ceremonial occasions. A dozen or more buisine players, with standards flying from their instruments, made a splendid impression on the eye as well as the ear, as Kings and popes were quick to appreciate. The buisine came to be more or less reserved for use on occasions of special pomp and circumstance—the most elite of musical instruments.

These then were the early forms from which developed the group of instruments we call the 'brasses'—bugles, cornets, horns, trumpets, trombones, tubas and a few less well known like serpents, euphoniums and sousaphones. But it is in the precursor forms that you can see most clearly what characterizes the group. In each, the player's lips serve as the generator of the note: and the mouthpiece or cup exists primarily to make it easy for the lips to vibrate in a controlled way. The resonator function is performed by the unbroken column of air inside the tube. This is largely what controls the pitch of the note being played, the longer the column the lower the frequency. Then it is the bottom opening, or the **bell**, which acts as the radiator. By ensuring that the air inside the tube which is vibrating, and the air outside the tube which is not, are in contact over a large area, it efficiently radiates the sound and gives these instruments the loudness which has always been such an important feature of their use.

Of these three functions, the one we can say something useful about with the background we have developed so far, is the second.

Air columns as vibrators

Most people know, from everyday experience, that long regular columns of air can be made to vibrate, either from hearing the wind moaning in chimneys or open drainage pipes, or from blowing across the tops of open bottles. Now it is possible to observe many features of such vibrations with a minimum of equipment.

If you can't get hold of a long thin piece of plastic tubing, you could just roll a sheet of cardboard into a long thin open cylinder. Now blow across the top of the tube and listen to the note that it produces. You'll have to rely on your ear to judge the pitch; but if you do it carefully you should be able to verify for yourself the following observations.

- The frequency of vibration depends on the length of the tube in a similar way as for a stretched string. If the tube is 75 cm long, you will get a note very close to Concert A (440 Hz). If you make it 150 cm, the note will drop an octave to A_3 (220 Hz): while if you make it 37.5 cm, the note will rise an octave to A_5 (880 Hz).

- Provided the tube is much thinner than it is long, the note it produces is pretty well independent of its diameter. Only when the tube is wide is relation to its length will the note become noticeably lower. Similarly the pitch is largely independent of the shape of the cross-section of the tube, whether it is circular or oval or square.

- The pitch will also hardly change if you roll the cardboard into a cone rather than a cylinder (provided again that the diameter of the wider end is still small). It doesn't matter if the diameter of the narrow end is as small as you can make it, or the same size as the other end; if the length is 75 cm, the note will still be close to 440 Hz.

Clearly, in some way the air column is behaving like a stretched string or a skipping rope. But with those, it was obvious what actually happened during the vibration. The string itself moved sideways. You will recall that I represented that fact by drawing this diagram.

There is, I hope, no difficulty about interpreting this drawing. The curved lines represent the most extreme positions that the string takes up. At any particular instant during the vibration, the actual shape of the string is somewhere between these two extremes. Note too, that the end held by the hand, which is actually causing the vibration, hardly moves at all compared with the middle of the string.

Inside your cardboard cylinder however, the air cannot be moving from side to side like this. It can only move *along* the tube, and in and out the ends. As it does this, there will be too much or too little air at various points inside the tube and the *pressure* will go up or down. But at the ends of the tube the air pressure cannot change because, being unconstrained, it must be at ordinary atmospheric pressure. And this is determined by the local weather conditions: nothing you can do will change that.

Therefore, what must be happening inside the tube is that the pressure oscillates above and below the atmospheric value, most violently in the middle and less so towards the ends. In that sense it is like the string, and the same diagram will serve to describe both vibrations.

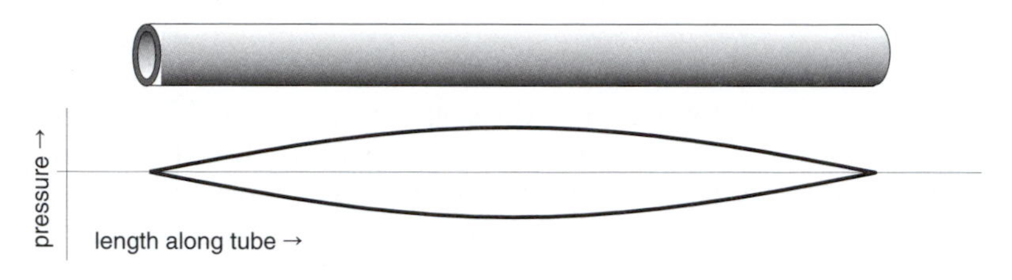

But now this diagram must be interpreted as a *graph* in which vertical distances represent pressure changes and horizontal distances represent lengths along the tube. Again the curved lines signify the extreme values that the pressure changes can attain. At any instant the actual pressure pattern is somewhere between these two.

It is not clear just *why* the pressure should vary like this, though I trust it is plausible from the experimental evidence that it does so. As long as the tube is perfectly cylindrical, I suppose it is reasonable to say that the processes taking place in the air are to some extent similar to what is happening in the string. But when the tube is conical, things are a lot less obvious. It is not easy to guess how the pressure varies inside a conical tube. I will have to come back to this point later, in Interlude 6.

Now the air column in any real brass instrument is more complicated still. One end of the tube is sealed by the player's lips, so there is no way it can behave as though it were open to the atmosphere. The pressure cannot remain constant at this end. So what effect does this have?

Well, back to the experiment. If you cover one end of the cardboard cylinder tightly with your hand, you will find that the note drops by a whole octave. This is an unexpected result, but what must be happening is this. The air now vibrates as though its most violent pressure variations occur, not in the middle of the tube, but right near the player's lips. Or, looking at it another way, it behaves as though it were one half of an air column twice as long as the tube. And because we know that an air column of twice the length would have half the frequency of vibration, this 'explains' why the pitch of the note fell by an octave when you blocked one end.

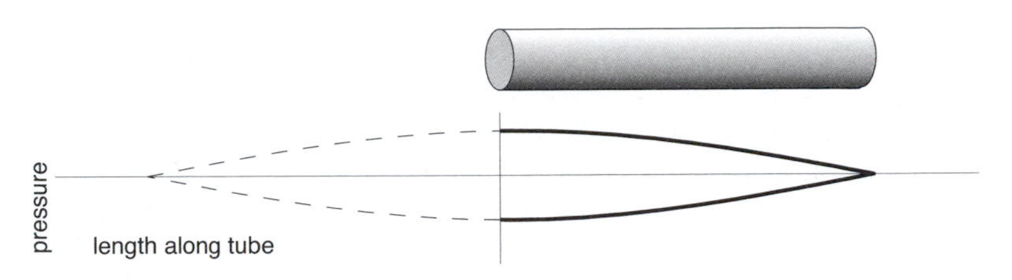

This idea of what must be going on also explains the action of the player's lips. Behind the lips the pressure is slightly higher than atmospheric, because the lungs are trying to force the air out. Just in front of them the pressure is oscillating—sometimes higher, sometimes lower. When it falls, air will be pushed out and the lips will be forced open (provided they are held sufficiently relaxed). When the pressure rises, air cannot come out and the lips close. The lips are behaving just like an automatically controlled valve, opening and closing in time with the vibrating air column. And this matches exactly what all brass players have to learn to do. They start off by forcing their lips to vibrate at what they think is the note, but once they have found it they let the instrument take over.

But, returning to the experiment, what about a conical tube? If the narrow end is very narrow, it is hard to believe that putting your hand over it is going to have much effect. This is getting more difficult to test with your rolled sheet of cardboard, but experimentally it is found that, for a cone which comes right to a point, the pitch doesn't change at all. But as the narrow end is made wider (still keeping your hand over it), the pitch drops more and more till you get to the stage where the cone has become a cylinder and the note has dropped the whole octave.

So let me summarize. For a tube of air closed at one end and open at the other (as in any brass instrument), the natural frequency of vibration depends inversely on the length. For perfect cylinders and cones, the tube lengths needed to produce some typical frequencies, accurate to the nearest centimetre, are given in this table; for tubes of more general shape the appropriate length will lie somewhere between the pair of values.

FREQUENCY	LENGTH	
	cylinder	cone
55 Hz	151 cm	302 cm
110 Hz	75 cm	151 cm
220 Hz	38 cm	75 cm
440 Hz	19 cm	38 cm
880 Hz	9 cm	19 cm

Overtones

There is one all-important feature of brass instruments which I have deliberately avoided mentioning until now, and that is that a column of air can be made to give out more than one note. I have already mentioned that brass players play their notes by first vibrating their lips at about the right frequency. But all players know that if they start with their lips vibrating at a higher frequency, at least an octave up, they can find other notes that their instruments will play. In fact, there are a whole series of such notes—a kind of natural scale—and special tunes have been composed to be played on them. But before I talk about the music, let me discuss why this occurs.

Go back to thinking about a long stretched rope. To get that vibrating you have to jiggle one end at about the right rate until you lock on to the natural vibration. But jiggle your hand a lot faster— at least twice as fast—and you will find that there is a frequency at which the rope will vibrate like this:

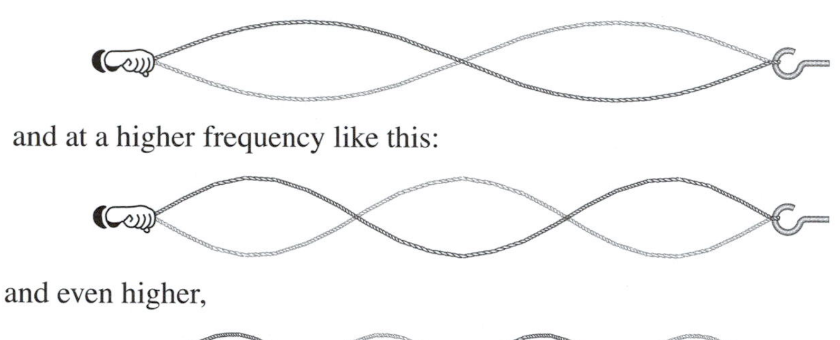

and at a higher frequency like this:

and even higher,

and so on.

Since we are obviously going to be talking about this phenomenon a lot, let me introduce you to some technical words straight away.

- The different ways in which an extended body (like the rope) can vibrate are called the **modes of vibration**. (Obviously the word 'mode' has a different meaning here from where I used it when talking about scales.)

- When vibrating in one of its modes, there are some points on the body which essentially do not move at all. These points are called **nodes**. (The word comes from the phrase 'no displacement'.)

  The positions at which the vibratory movement is greatest are called **antinodes**.

- The mode of vibration which has the lowest frequency is called the **fundamental**, and the modes with higher frequencies are called **overtones**, obviously because they were first discussed in the context of musical notes.

  The three diagrams on the page opposite were of the first, second and third overtones of a stretched rope.

I have already argued that an open cylindrical tube of air can be expected to vibrate in an analogous way to a rope, so this must be what gives rise to the overtones in a brass instrument. I can therefore draw a pressure diagram to show, as an example, what must be happening inside an open cylinder sounding its **first overtone**.

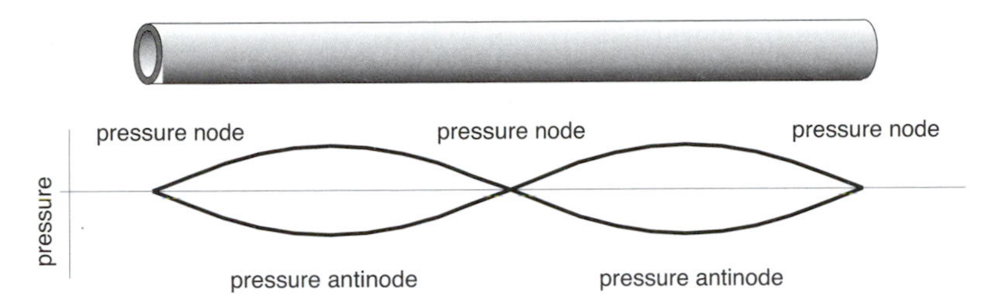

Now the interesting question is: what frequencies do these overtones vibrate with? And we know enough about air columns to answer that. You will notice from the diagram that the pressure in the middle of the tube is behaving just like the pressure at the ends. So you could imagine cutting the tube through at that point and pulling the two halves apart. (This is a bit of mental gymnastics much beloved of physicists, called a **thought experiment**. It is an 'experiment' which would be essentially impossible to perform—how, for example, would you keep the vibration going while you were cutting?—but you know what would happen if you could do it.)

The result should be that this mode of vibration would have the same frequency as these vibrations:

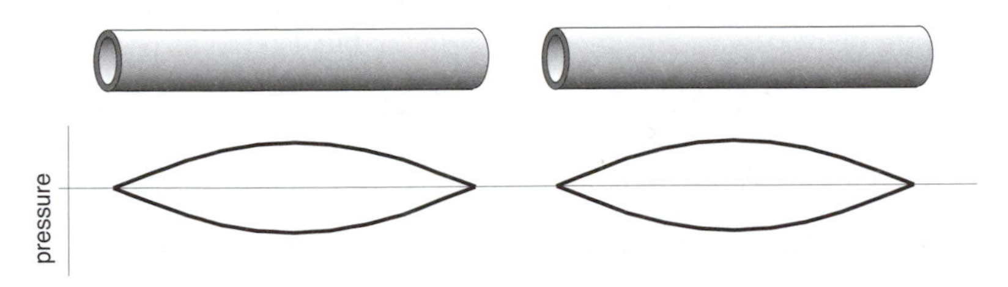

But you know what frequency each of these tubes will vibrate with—twice the frequency of the full pipe, because they are each half its length. So the frequency of this first overtone is twice that of the fundamental—i.e. an octave higher. In the technical jargon of musicians you would say that an open cylinder 'overblows into the octave'.

Now the argument I just went through will not only apply to the first overtone. In thinking about the second overtone you would imagine cutting the pipe into three parts, the third overtone four parts, and so on. And thus you would find the frequencies of these overtones to be respectively three times, four times, etc, that of the fundamental. So just as a concrete example, if the cylinder were 151 cm long, then the notes it could play would be these:

	FREQUENCY	NOTE
fundamental	110 Hz	A_2
1st overtone	220 Hz	A_3
2nd overtone	330 Hz	E_4
3rd overtone	440 Hz	A_4
4th overtone	550 Hz	$C\sharp_5$
5th overtone	660 Hz	E_5
...etc		

Or, on a musical stave:

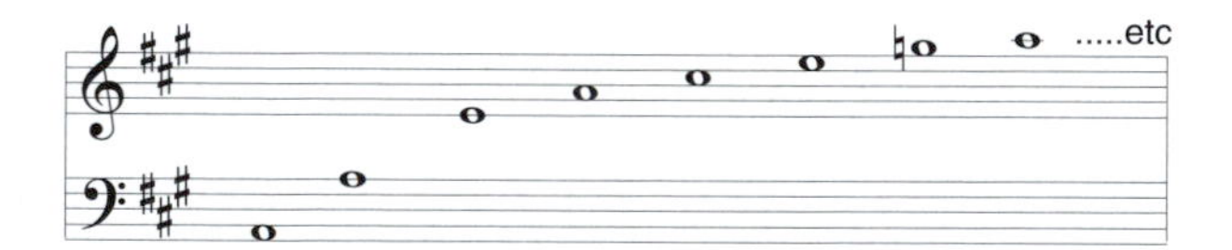

This is exactly what earlier I called a **harmonic series**. I only deduced this series of overtones for an open cylinder but I have already pointed out that a cone vibrates like an open cylinder, therefore the overtones of a conical pipe should also be a harmonic series.

But again, brass instruments have the complication that to play them you have to cover one end with your lips. An instrument in the shape of a perfect cone, coming down to as small an opening as you could still blow into—a kind of idealized horn—would have a harmonic series as its overtones; but trumpets and trombones tend to be cylindrical over the greater part of their length. So how does a cylinder behave? Luckily we can answer that question since we have already intuited the key observation that the player's lips correspond to a *pressure antinode*. Hence the diagram for the first overtone of a cylinder closed at one end must look like this:

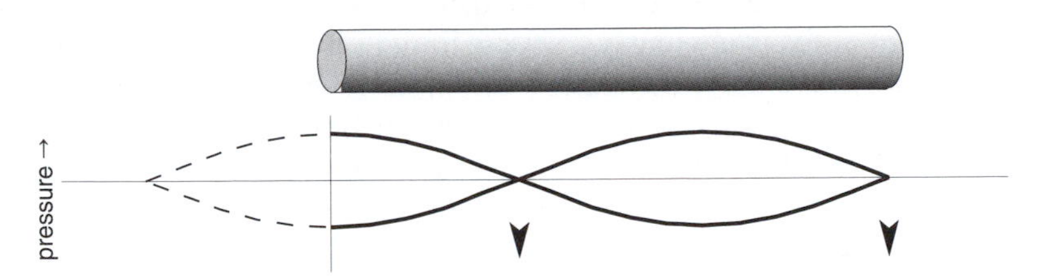

To deduce the frequency of this mode, I would use the same trick as before, and conclude that the vibration is the same as occurs in this open cylinder:

Its length is 2/3 of the length of the closed pipe, or 1/3 of the length of the 'equivalent' open pipe; hence its frequency is 3 times that of the fundamental. This overtone will therefore produce a note which is an octave plus a fifth higher than the fundamental frequency (it is multiplied by 2 and by 3/2). Again in technical musical jargon, you say that an open-closed cylinder overblows into the twelfth.

You can probably see further, without my spelling it out, that the second overtone will have 5 times the frequency of the fundamental, the third 7 times, and so on. But I don't want to worry about those exact values now.

Any real instrument will be neither a perfect cone nor a perfect cylinder, but something which is, in a sense, 'in between'. So the frequencies of both fundamental and overtones will also be 'in between', but this may not be of much use. As we have seen, Western music uses only notes which have definite relationships to one another; so the art of the brass instrument maker is to alter judiciously the bore of the tube so that a musically useful set of overtones will result. Ever since Roman times they have striven to achieve exactly the overtones of a perfect cone—the harmonic series.

Over centuries of trial and error, it was found that small changes, particularly to the shape of the mouthpiece, would enable the higher overtones to be tuned individually; while the lower ones could be tuned by altering the shape of the bell. But it is difficult to do much to the fundamental without altering the whole pipe, so you will often find that it is out of tune and never used, or just not playable at all. Nevertheless, the instruments we use today are all successful realizations of the brass workers' striving. They are easy to play, they give a good loud sound, and their overtones form (at least part of) a harmonic series.

Music for natural horns and trumpets

The fact that all the instruments I have discussed so far can only play a restricted set of notes is obviously a severe limitation to their use for making music. The most extreme example still in use today is perhaps the **didjeridu**, which can usually only play two notes. Since they are often about a metre and a half long and usually closer to cylindrical than conical, their fundamental is somewhere around 110 Hz (A_2); and their first overtone is roughly a tenth higher (i.e. an octave plus a third), though the more cylindrical they are, the closer to a twelfth it gets. Musically this seems very limiting, but a good player can get a very impressive array of sounds by speaking or humming into the pipe while blowing it. So whereas Western music is largely constructed from notes of different pitch, Aboriginal music is built from notes of different timbres.

A more traditional instrument to Western ears, the **alphorn**, can usually play at least notes 2–8 of the harmonic series; and traditional melodies written for these instruments use these notes. The following is a so-called *ranz des vaches*, a "cow call" played in rural Switzerland to summon the cattle home at evening. (Apparently cows really can learn to recognize their own tune and to come when it is played.) You might recognize the melody because Beethoven used it as a theme in the last movement of his *Pastoral Symphony*.

You will note that it uses only three notes: notes 3, 4 and 5 in the harmonic series based on an F. To play the melody as written on an alphorn, the fundamental has to be F_2, which has a frequency of 88 Hz. Since all horns are closer in shape to cones than to cylinders, its length would probably be just under 1.9 m.

Shorter instruments are more restricted, because they must use only the lower notes of the series if their melodies are to stay within the 'normal' singing range. This is not to say that they cannot make interesting music. In the *Gloria, ad Modum Tubae*, written in 1420 by Guillaume Dufay, two medieval trumpets perform a fanfare using only the notes G_3, C_4, E_4 and G_4: and that is splendid music indeed.

But the florid trumpet works of the 17th and 18th centuries demand much more than three or four notes. Pieces like the Toccata from Monteverdi's *Orfeo* (1607) or Bach's *Brandenburg Concerto, No 2* (1720), were written to be played on the very high overtones, numbers 8 to 15. This poses something of a problem, because some of those overtones do not coincide exactly with notes of the ordinary scale. Notes 7, 11, 13 and 14 of the harmonic series are related to the other notes by intervals like 7/6, 11/10, 13/12, 14/13 and so on. None of these intervals are considered consonant in our musical culture.

As an example, here is Handel's idea of what the avenging angel will play on the day of judgment. It is the great trumpet solo (transposed to the key of C) from Part 3 of *The Messiah*—"The Trumpet Shall Sound".

I have indicated here which are the 'non-consonant' notes of the harmonic series by filled ovals. In particular, the 11th note lies somewhere between F_5 and $F\sharp_5$. If the player sounds this note naturally it will seem dissonant to the audience. So in practice he or she must sometimes force the note up to $F\sharp_5$, and sometimes down to $F\natural_5$—as indicated by the score. As you might guess, this kind of playing demands exceptional skill in performance, and it is interesting that trumpeters who specialized in this range were always paid more than the others.

However, the most obvious example of this limitation of this scale is to be found in the repertoire of tunes especially written for the **bugle**, the only natural trumpet or horn still in wide use today. It is compact (about 40 cm) for ease of carrying, but the tube itself measures about 120 cm, and its fundamental (which cannot actually be played) is C_3.

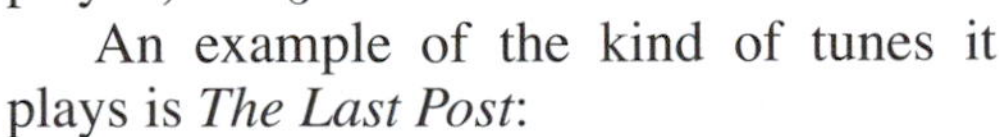

An example of the kind of tunes it plays is *The Last Post*:

Most of the well known bugle calls emerged over the last two centuries and were used by the military to mark the time of day, to call armies to battle and retreat, and so on. Musical values had nothing to do with their use. They were in essence messages in code, and the bugle itself was really only a kind of pre-electronic 'walkie-talkie'. That is why, in musical terms, it is an evolutionary dead end.

Filling the gaps

With the great advances in metal working of the Renaissance came a determination on the part of instrument makers to overcome the limitations of natural horns and trumpets, by increasing the number of notes that any one instrument could play—by filling in the wide gaps between the lower notes of the harmonic series. In broad outline of course the answer was simple. Since it was the length of the pipe which determined which notes it would play, more could be added by finding a way to change the length at will. Over the succeeding centuries, four different solutions were found to this problem.

Solution 1: Finger-holes

If a hole of suitable diameter is drilled into the side of the pipe, it exposes the air inside the tube to the atmosphere at that point, which makes it very difficult for the internal pressure to fluctuate there. So it is almost the same as if the pipe had been cut, and only that part of the air column between the hole and the player's lips is free to vibrate. This column has a shorter length and therefore will play a higher set of notes. So with suitable holes drilled along the pipe, and agile fingers to cover or uncover these holes, the player can get a much greater selection of notes.

Such instruments were very popular during the Renaissance, the most widespread being the **cornett** (note the spelling!). It was usually made of wood, in more or less the shape of an animal horn, and covered with leather. It came with an ordinary brass player's mouthpiece, but it didn't have a bell.

However they didn't survive to modern times as a popular instrument, and I think the reason is clear. I said that opening a hole had a similar effect to cutting off the tube, but there is one important point of dissimilarity. The vibrating air column is no longer in contact with the outside air over the wide area of the bell, but only over the much smaller area of the hole. Hence it can't radiate sound as efficiently, and the note is much quieter.

So this solution to the problem of filling the gaps sacrificed the very feature which was absolutely characteristic of the brasses—their loudness and carrying power. I think that is why the cornett and its kind were another evolutionary dead end.

Solution 2: Crooks

It was realized very early that the simplest way to increase the length of the pipe was to take it apart at one point and insert an extra length of tubing. Therefore players began to carry round one or two such detachable pieces, neatly coiled into manageable sizes, with which they could lengthen their instrument and lower the notes it could play. These were known as **crooks** and are still occasionally used today; and, for example, a horn in B♭, which typically has a tube of 273 cm, can be transposed to a horn in F by inserting a crook about 100 cm long.

However the crook was a very limited solution to the problem. Only one series of overtones could be played at a time, though the player could change crooks in the middle of a piece; and it became common for composers to indicate on the score which crook was the most suitable at any time. But this, in turn, had an unfortunate consequence. Composers got into the habit of writing all this music in the same key, merely specifying which crook was to be used instead of writing out the actual notes they wanted. Horns and trumpets therefore became what are called **transposing instruments**, and on an orchestral score they appear to be playing in a different key from everyone else. I will never understand why this ridiculous system should have lasted so long.

Solution 3: Slides

Another way to alter the natural frequencies of a tube is to change its length by a sliding mechanism. Towards the end of the Middle Ages there appeared the medieval **slide trumpet**, an instrument with a long cylindrical pipe folded into a flattened S-shape and a telescopic mouthpiece. The player would steady the mouthpiece against the lips and slide the whole thing in and out. It was a simple and elegant solution to the problem, but it was rather difficult to play. The movable part was unwieldy, and the amount of movement necessary comparatively great: I can imagine its being rather hard on a careless player's front teeth. Any fast passages must have been very tricky indeed.

In the 15th century it developed into an instrument called the **sackbut**.

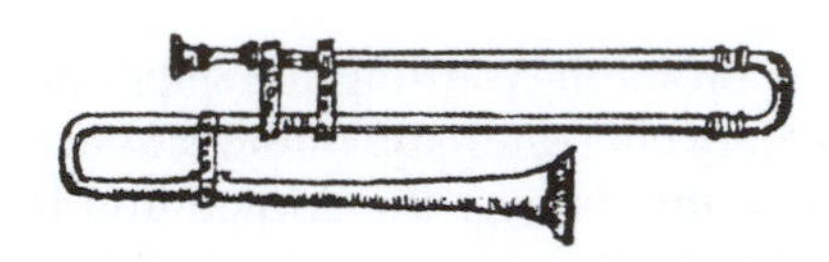

In this, the sliding section was a U-shaped tube—a kind of adjustable crook. Its advantage over its predecessor was that a double tube needs only half the movement of a single tube to get the same change of pitch, and so is much less cumbersome to slide.

These instruments haven't changed much since the 16th century. The modern **trombone** has a greatly increased flare to its bell, but apart from that is essentially the same. They come in at least three sizes. The most common is the tenor trombone, which in its unextended form has a total length of tubing is about 275 cm. The fundamental (usually referred to as the **pedal note** when talking about brass instruments—no one seems to know why) is $B\flat_1$. But, as I said before, you often cannot play the fundamental at all.

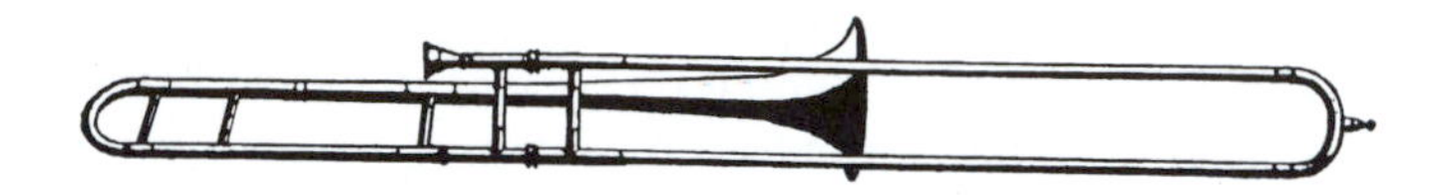

Apart from members of the violin family, the trombone is the only modern instrument which can produce a continuous glide in frequency, or **glissando**, between notes. So the player must have a very accurate sense of pitch and develop the ability to move the slide instantly to the correct position. There are seven standard positions which have to be learned, each a semitone apart, to fill in the gap of three and a half tones between the lowest two (usable) notes of the instrument. But there is one slight complication. Since lowering a note by a semitone involves *multiplying* its length by about 1.06, the extra length of tubing that has to be *added* increases as the note drops, and the standard positions look like this:

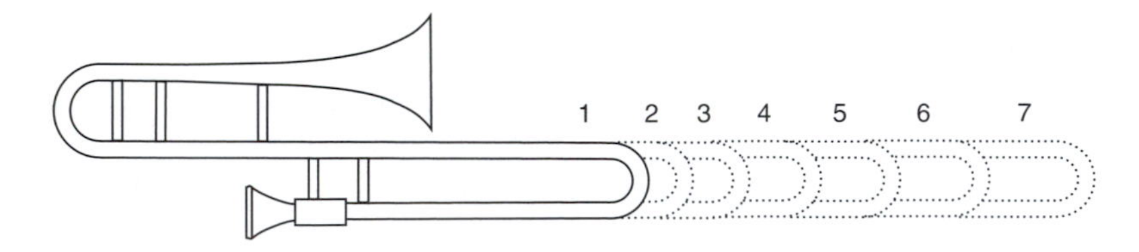

Perhaps these difficulties are responsible for the fact that there are not more trombone-like instruments around today. In the classical repertoire, the trombone is not much used for solo work (the "Tuba mirum" from Mozart's *Requiem* is a notable exception); and it did not even become a regular member of the orchestra much before the 19th century. It had previously been considered a 'special effects' instrument—as exemplified by its use in the Underworld scene in Monteverdi's *Orfeo*, or in the final scene of Mozart's *Don Giovanni*. But in the 20th century it gained a new prominence in jazz groups and swing bands, mainly for the thing it can do that others cannot, play spectacular glissandos.

Solution 4: Valves

In 1815 a new invention was announced, which was without doubt the most successful solution to the old problem—a special piston-operated valve which enabled a crook to be connected to the main tube, and removed again, all at the press of a key. Although over the years many different versions of this device have arisen, some of amazing complexity, the basic idea is very simple. Pressing the key closes off the tube at one point and causes the air stream to detour through the extra tubing, as illustrated by this schematic diagram of a modern rotary valve:

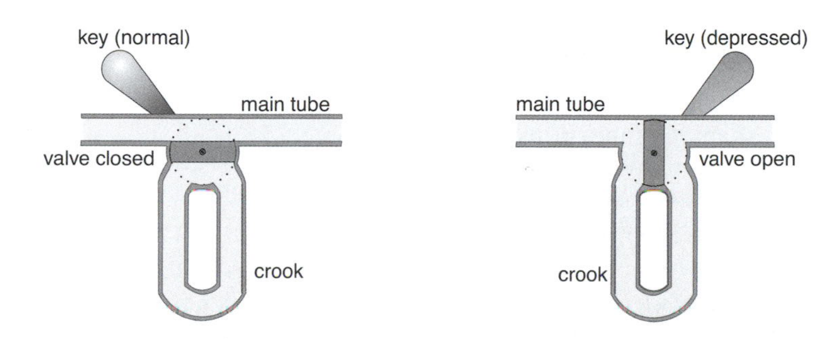

To appreciate the musical effect of this revolutionary new invention, it is interesting to consider a rather eccentric instrument especially designed for the first performance in 1871 of Verdi's *Aïda*, commissioned, not to mark the opening of the Suez Canal, as is sometimes said, but of the Cairo Opera House a year or so later.

For the famous "Triumphal March", Verdi—ever the showman—wanted the visual spectacle of six buisines announcing the return of the victorious Egyptian army. But a natural trumpet can only play a very limited range of tunes, especially in the hands of inexperienced players. (I don't know what kind of players the Cairo Opera Company had, but they probably weren't the flower of the Baroque tradition). So he had constructed a long straight trumpet with just one valve, small enough to be hidden in the player's hand, but which would lower the pitch of the instrument by a whole tone. Then he had available *two* harmonic series, separated by a tone, on which to construct what must surely be one of the best known trumpet melodies in Western music:

But that's not the end of the story. Verdi must have been the despair of his producer, because in the middle of the piece he *changes key*! The melody is played a minor third higher. The trumpets can't play it any more. So there had to be another sets of trumpets, to play the melody in the new key. Luckily the scales which the two sets of trumpets play have a few notes in common; and at the end they play in duet, with the first trumpets sounding a repeated E♭5.

The single valve trumpet was much too idiosyncratic an instrument to catch on, but the idea of *three* valves did. With one valve to lower the pitch a semitone, the second a whole tone and the third a tone and a half, the player could achieve any note on the diatonic scale; since combinations of these three can lower the pitch by any number of semitones up to six. All modern instruments use this principle. Even modern versions of the Aïda trumpet are made with three valves, so they can be used to play something else other than one tune, however well known.

Although the three-valve solution to the old problem is undoubtedly very neat, there is still one difficulty with it. For the same reason that I mentioned when talking about the trombone, the adding of three crooks of fixed lengths introduces a slight mis-tuning.

To see why this is, imagine you are trying to lower the fundamental pitch of a tube by a semitone. To do this you have to increase the length of the tube by exactly 1/15 of its length (you remember, of course, that the ratio corresponding to an interval of a (just scale) semitone is 16/15). But the length of the tube you 'start from' is longer if the other valves are open than if they are closed. On the other hand, the length of tubing that is added when you open the semitone valve cannot depend on the state of the other valves. Therefore opening the semitone valve when the other valves are closed will lower the note more than when they are open.

So in real instruments some compromises are necessary. The first two valves are just a little longer than they need be. When either is used alone the note is a little flat; but when they are used together it is a little sharp. Players have to be aware of these difficulties and try to compensate, forcing the frequency of the vibration a fraction by using their lips.

There is not much more I can say here about the modern instruments which have developed since the invention of the valve. Regular members of our symphony orchestras are the **trumpet**, with a typical overall tube length of 140 cm and lowest note (which is theoretically the first overtone) $B\flat_3$:

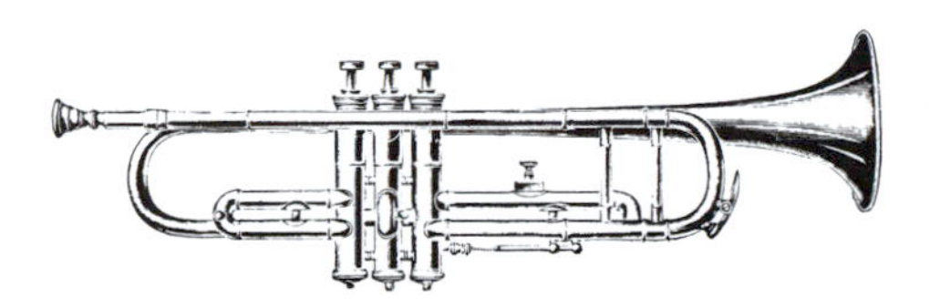

the **French horn** (375 cm, F_2):

and the **tuba** (536 cm, $B\flat_1$):

This by no means exhausts the members of the family, nor the interesting physics involved in their use (some of which I will return to later, particularly in Appendix 5). If you want more information, I would refer you to one of the detailed books in my bibliography. But at this stage we have got just about all we can from the simple observation that a column of air will vibrate with a frequency which is inversely proportional to its length; so I will return to the main theme of my book.

Chapter 3

Harmonies of a mechanical universe

It is no coincidence that astronomers were extending our understanding of the heavens at exactly the same time as the great seafaring explorers were extending our knowledge of the Earth. Navigators need reliable star charts, and by the 1600s Kepler's new way of calculating planetary orbits made the drawing up of these charts easier and more accurate. But the next question was: why do the planets move in these orbits? I do not believe that any scientist with normal curiosity could ignore that question, any more than Columbus could have stopped himself wondering what was on the other side of the New World.

So with music. We know that harmony is related to simple numerical rules connecting the frequencies of musical vibrations. It is not clear that investigating further will help the ordinary player make better music, but we can't leave it there. We must ask: why do things vibrate as they do? And the answer to that question came out of the researches of the 16th and 17th centuries into the laws which govern motion.

Newton's force law

Pre-Renaissance physics was largely classical Greek philosophy, mainly Aristotle's, interpreted by scholars like St Thomas Aquinas. Its central theme emphasized the static order of the Universe, with every object in its proper place according to the wisdom of the Creator. Motion was considered a temporary and possibly unnatural state. A stone fell in order to reach its rightful place, on the ground; flames rose in order to join the divine fires. What actually happened while these things were moving was left vague. Presumably an arrow in flight was pushed by the air in some way. Certainly the planets could only move in their orbits if they were pulled along by angels (when they weren't dancing on the heads of pins).

So firmly entrenched was this view of how things worked that the poet Dante, writing in *The Divine Comedy* in about 1300, was able to use it as a metaphor for the soul seeking God, knowing that his readers would certainly understand:

> "And as fire mounts, urged upward by the pure
> Impulsion of its form, which must aspire
> Towards its own matter, where 'twill best endure,
>
> So the enamoured soul falls to desire—
> A motion spiritual—nor rest can find
> Till its loved object it enjoy entire."

Galileo changed all that. By drawing attention to the simple observation that things fell more slowly through water than through air, he argued that a body didn't stop moving because of a sense of divine rightness, but because something else stopped it. If you could imagine getting right away from the air and the Earth, then a body that was moving would continue to move, perhaps forever. In other words, if a body was stationary it took effort to get it moving, and if it was moving it took effort to stop it. He summarized this by saying that all bodies had something called **inertia**, which was simply a measure of how difficult it was to change the body's state of motion. Today this quantity is called **mass**; and the brilliant piece of insight, that a moving body will continue to move in a straight line unless acted on by some external force, is known as **Galileo's law of inertia**.

The next, and most crucial step was taken by the English scientist, mathematician, misogynist and Master of the Royal Mint, Isaac Newton who was born in the year that Galileo died and of whom you will hear much more later. He proposed that a body would only change the way it was moving if it was acted on by a force. What is more, he regarded this as a universal law of nature, obeyed by heavenly as well as earthly bodies. A planet did not need angels to push it along. It was bound to the Sun by the force of gravity—the same force that made apples fall on the heads of unwary scholars—and behaved like a stone being whirled around on the end of a string.

To explore the consequences of this law, Newton invented a new branch of mathematics, called calculus, which we needn't go into here. However I will need to talk about forces, and I will use the technical word 'force' as though it had its ordinary meaning. When you lift a heavy weight you can feel the force of gravity. When you pull against a strong spring you can feel the force with which it pulls back on you. That is what a force is, and that is what Newton said causes change of motion.

The first problem to which Newton applied his new laws, was that of the movement of the planets round the Sun. He published the results in one of the most influential books in the history of Western thought, *The Mathematical Principles of Natural Philosophy*—usually known by its abbreviated Latin title, the *Principia.* In it he showed that Kepler's three laws were logical consequences of the existence of a universal force of gravitation; so that, at long last, science could explain *why* the Solar System behaved as it did. It was like a giant piece of clockwork which was presumably wound up by the Creator at the beginning of time but which proceeded forever after under the influence of the same laws that apply here on Earth.

The consequences of Newton's laws are almost limitless, and are still being explored today. But for our present purposes, there is just one particular kind of motion we are interested in—vibration. Again however, it is more sensible to discuss this in modern terms; and so once more I will break the historical sequence to talk about the theory of vibrations, but first I will have to introduce the concept of **energy**.

Energy

A good place to start is by thinking about a **pile driver**. From medieval times, engineers have created the foundations of their buildings by such devices. A heavy weight is hoisted up to the top of the crane and then released. It falls and hits a post at the bottom, driving it into the ground. How far the post goes in, depends on how heavy the weight is (its mass) and how fast it is moving when it hits (its velocity). So the effect of the force of gravity acting on the weight is to give it something which drives the post this distance into the ground. This something used to be called *vis viva* (literally 'living power').

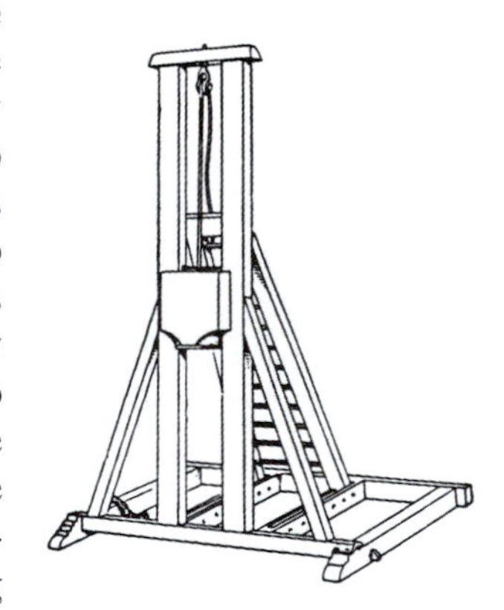

Now think about a medieval crossbow. You primed it by pulling back the string hard against the natural springiness of the cross-bar, and then loaded a bolt. When you released the catch the bolt was thrown forward with lots of *vis viva.* It was well known by soldiers that a heavy bolt would be thrown out with a slower speed than a light one. So *vis viva* was clearly some combination of mass and velocity.

Since then, experiments have shown that the quantity we are looking for is what is now called **kinetic energy**, defined by the formula,

$$\text{kinetic energy} = 1/2 \text{ mass} \times \text{velocity}^2.$$

The form of the law of motion I will use says that the effect of a force is to change a body's kinetic energy. (I should be more careful here and point out that this is only strictly true if the force acts in the same direction as the body is moving; but I will ignore that complication).

But now think about it in another way. Potentially, the ability to drive the pile into the ground was there from the time the weight was lifted up, before it was actually moving. Similarly, priming the crossbow gave it the potential to throw the bolt. The effort you made to lift the weight or prime the bow was somehow stored for later use. This was a new idea, and a new concept was invented to express it, **potential energy**. When a body was lifted up it was said to gain gravitational potential energy, and the higher it was lifted the more it gained. Similarly a spring could store elastic potential energy, the amount depending on how strong the spring was and how much it was compressed.

This new idea turned out to be very useful in describing the changes that occurred when forces were acting. When a weight was dropped its potential energy changed into kinetic; but if it fell onto a trampoline (say), it would reverse its velocity on contact and climb again to its original height (almost), changing kinetic energy back into potential. Both Galileo and Newton noticed a similar thing on examining Kepler's planetary motions. When its orbit took Mars a bit closer to the Sun (lowering its potential energy) it speeded up (increasing its kinetic). Later when it moved out again its speed went back to what it was before. The conclusion was drawn that both energies were really the same kind of thing, given the general name **mechanical energy**; and under ideal conditions, the total amount of this energy did not seem to change. That was the first hint of what is probably the most important result in all of physics, the law of conservation of energy.

It was soon realized that there were other kinds of energy. Cannons were first used in Europe at the fall of Constantinople, right at the start of the Renaissance. A charge of gunpowder obviously had more potential energy than the spring of a crossbow, but it was of a different kind, which we now call **chemical energy**. It would be over a century before scientists realized that **heat** was also a form of energy, and that whenever mechanical energy seemed to be lost (as when a bouncing ball did not quite get back to its original height), it was usually being changed into heat. I will talk a lot more about this later, but for now let me quote the **law of conservation of energy** in its modern form:

- Energy cannot be created or destroyed, it can only be changed from one form to another.

With this concept in our minds we can return to answer the question I started this chapter with: why do things vibrate as they do?

The theory of vibrations

The first point to appreciate is that vibrations are very common. Trees sway in the breeze, boats bob up and down on the harbour, tides ebb and flow, flags flutter, cradles rock, hearts throb, lights flicker. Sometimes a vibration lasts a long time, like a pendulum clock; sometimes it is over quickly, like a bouncing ball. So in order to talk about vibrations generally, we had better agree on some more technical terms.

- If some motion continually repeats itself, with the same amplitude, after the same period of time, it is a **periodic vibration**.
- If the motion repeats itself but the amplitude decreases, the vibration is said to **decay**. The time it takes for the amplitude to fall to about half its original value is called the **decay time**.

Let us go back and think about pendulums again. Imagine I pull the bob to one side and let go. It swings back and forth. Now I want to ask the question: why does it vibrate like this?

It often happens in scientific enquiry that the simplest and most naïve answer is also the most revealing. The pendulum swings because I pulled it sideways. If I hadn't disturbed it, it wouldn't be swinging now. I obviously asked the wrong question. Perhaps it should have been: when I don't do anything to the pendulum, why does it stay exactly where it is? What's so special about the rest position?

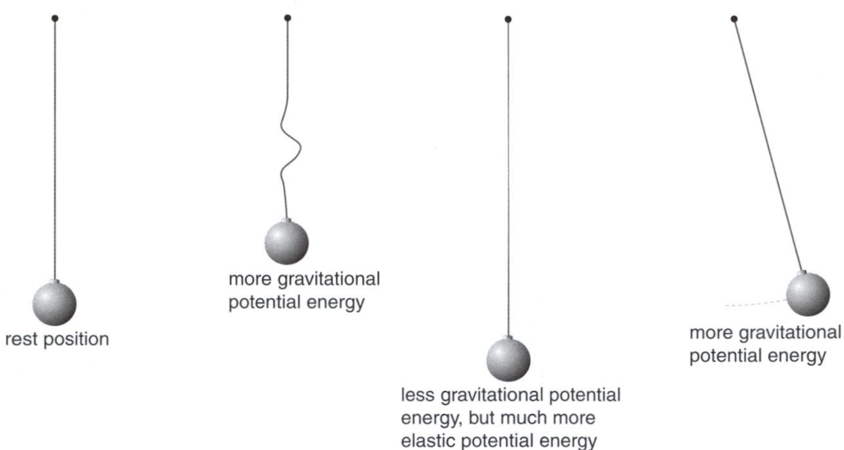

When the bob is at rest, it is a certain distance above the ground. It has a certain amount of gravitational potential energy. If it weren't at that exact spot, but were (say) a little higher, then it would have a little more potential energy. If it were lower it would have a little less gravitational potential energy, but the string would have to be stretched. That would take a lot of effort because strings do not stretch easily, so it would have a large amount of elastic potential energy. Likewise, if it were pulled to either side it would be higher than its rest position, since it can only be moved in an arc without stretching the string. Again more potential energy.

In other words, its rest position is special because that is the place where the total potential energy is less than for any other place nearby.

Now consider its kinetic energy. As it hangs there without moving, it has none. If it were moving at all in any direction, it would have a positive amount, since there is no such thing as negative kinetic energy. So, at equilibrium, its kinetic energy is minimum also. I can combine these two conclusions and make a very general statement (since the kind of argument I went through doesn't just hold for pendulums):

- **Equilibrium** corresponds to a minimum value of the total mechanical energy.

Now think what happens when I deliberately disturb this pendulum. If I pull the bob slowly to one side I increase its potential energy only. On the other hand, I could disturb it by hitting it, which is the same as giving it kinetic energy. No matter what I do I increase its total energy; it has excess energy above what it had at equilibrium.

You should be able to see what will happen. The pendulum will try to return to equilibrium. But it has too much energy, and we know that energy cannot just disappear. Therefore if none of it gets changed into non-mechanical forms, all the bob can do is to juggle this unwanted energy like a hot potato. It swings up until it comes to rest and the energy changes into potential; but gravity makes if fall again and the energy changes back into kinetic. Since there are only two forms the energy can take, it must repeat itself. It is now a vibration.

The motion will stop eventually. Some energy will turn into heat through friction at the point of support; and some will go into setting the air moving. In the end all the excess energy will be dissipated. Often these processes take a long time and the vibration decays very slowly. But it is still possible to keep the vibration going if you somehow keep feeding in just as much energy as is being dissipated.

I have only talked about pendulums, but the same ideas apply much more widely. In any vibration at all, energy is put in, it changes backwards and forwards between various forms, and it gradually trickles away. The vibration of a bouncing ball decays quickly because there is only a limited amount of energy and a lot is changed into heat each time it hits the floor. During each swing of a pendulum clock a small spring gives the mechanism a kick, so it will continue to vibrate as long as the spring is kept wound. A cradle continues to rock so long as someone keeps giving it tiny pushes. Your heart will beat so long as electrical nervous impulses replace the energy used to pump blood through your arteries. The tides will continue so long as the moon pulls on the oceans each day; but note that energy is dissipated by water sloshing over the sea floor and this will eventually stop the moon revolving round the earth. (Luckily the decay time of that vibration is around hundreds of millions of years).

Frequency of vibration

The test of whether we really understand all this will be if we can work out something quantitative, for example what the frequency of any vibration should be. Let us consider a weight bobbing up and down on the end of a strong spring. Its energy transformations are: potential energy (the spring is fully compressed), to kinetic (the bob is moving past its rest position), to potential (the spring is fully stretched), to kinetic, and repeat. The period of vibration, and hence its frequency, depends on the time it takes for these transformations to occur. This in turn depends on the velocity of the bob as it moves from one position to another. There are three obvious quantities that affect this velocity.

- **The strength of the spring.**

  If the spring is very stiff, when it is compressed or stretched it will exert a large force on the bob. The bob will start moving quickly in response. I would expect the time between transformations to be short and the frequency to be high. But if the spring is loose the weak forces it exerts will make the bob respond slowly. The period should be long and the frequency low. This conclusion can be checked by experiment. For a mass on the end of a spring the frequency increases as the spring is made stronger.

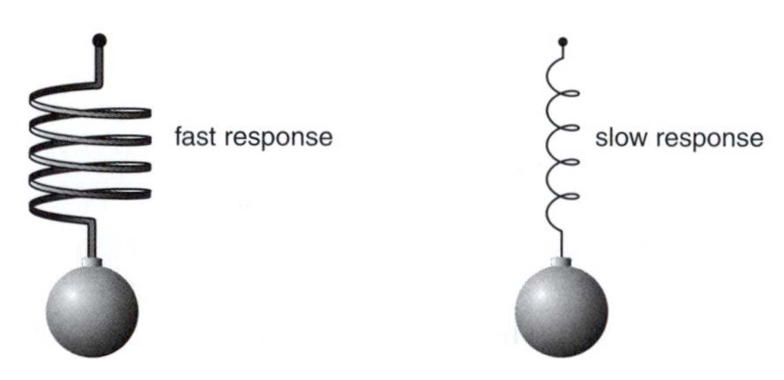

Since all vibrations are similar, the same effect should be noticeable elsewhere. Hold a stiff wooden ruler projecting over the edge of a table and twang the end. It will vibrate, at maybe 10 or 20 Hz. Do the same thing with a flexible plastic ruler and the frequency is markedly lower. Or think about the difference between a gymnast bouncing on a taut trampoline (higher frequency) and on a slack one (lower).

- **The mass of the bob.** A heavy bob has a large inertia and will respond sluggishly to the force of the spring. Everything will happen slowly and it should vibrate with a relatively low frequency. But a light bob will respond more quickly to the same force, and its frequency of vibration should be higher. Check it experimentally. The frequency of vibration decreases as the bob is made more massive.

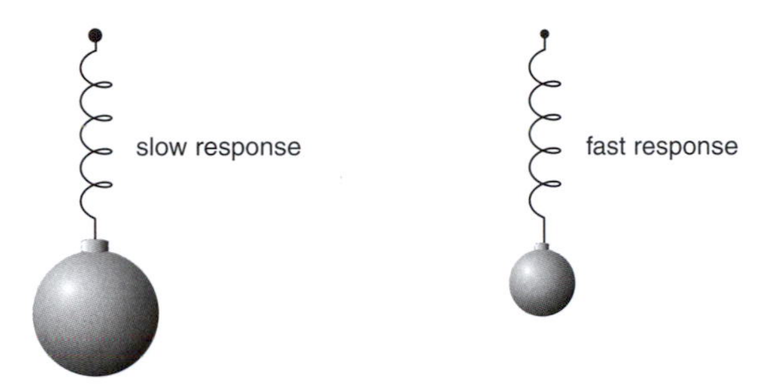

Again this conclusion applies elsewhere. Put a lump of plasticine on the flexible ruler and its frequency gets even lower. Or think about how a small child bounces on the trampoline with higher frequency than a large adult.

The arguments I have just given show in a general way how frequency depends on spring strength and bob mass. A more thorough analysis gives this numerical formula:

$$\text{frequency of spring} = \frac{1}{2\pi}\sqrt{\frac{\text{strength of spring}}{\text{mass of bob}}}$$

In this formula, the $1/2\pi$ is just a number. It actually tells you that you're measuring time for a complete cycle of the motion. Numbers like this keep cropping up, and I don't think there's much point in worrying about where they come from. However the square root sign needs some comment. Remember that the effect of a force is to change the bob's kinetic energy, which involves velocity squared. So when I work out the frequency by calculating the velocity first, I have to take the square root.

There is a similar formula for the frequency of a pendulum:

$$\text{frequency of pendulum} = \frac{1}{2\pi}\sqrt{\frac{\text{acceleration due to gravity}}{\text{length of string}}}$$

One difference here however, is that the term on the top line, representing the force of gravity which causes the bob to move, also depends on the mass of the bob. Therefore the mass cancels out and doesn't appear in the final formula. Also, the **acceleration due to gravity** with which all bodies fall to earth is 9.8 m s^{-2}. So it is easy to see, for example, that if the length of the string is 1 m, the frequency will be very close 0.5 Hz, a fact which I have already mentioned in Chapter 2.

- **The size of the system.**

  The most important result I drew attention to earlier was the relationship between frequency and length for at least two different vibrating systems. Now a weight on a spring, or a pendulum, is a simple system in that most of its inertia is located in one small part, the bob. It has only two simple kinds of energy and there are only two possible transformations. The repetitive nature of its vibration is straightforward. But if the vibrating body is extended and its inertia is spread out, different parts of it can have either potential or kinetic energy at different times. In general its motion might take a long time before it repeats itself exactly. So it seems plausible that, the more bits you can imagine the system being divided into, the more energy transformations will be possible, and therefore the longer should be the period of repetition. That must be why the frequency decreases as the size of the system increases.

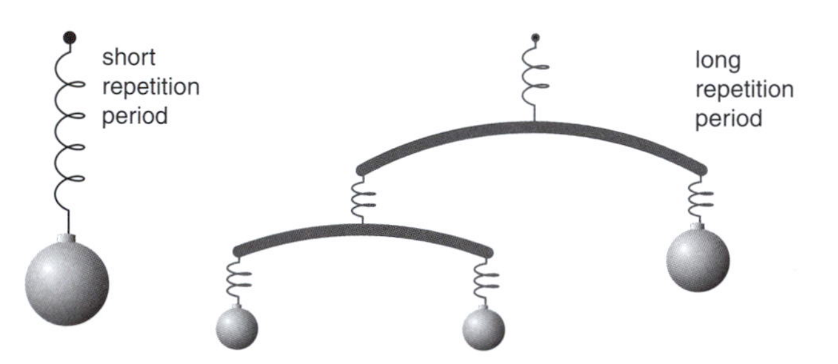

It is certainly a general result. As you shorten the length of the ruler overhanging the table, its twanging frequency gets higher. A small trampoline bounces more rapidly than a large one.

For the two particular extended vibrators that are important in music—a narrow column of air vibrating in its fundamental mode and a stretched string—the formulas are:

$$\text{frequency of air column} = \frac{0.59}{\text{length}}\sqrt{\frac{\text{pressure of air outside}}{\text{density of air inside}}}$$

$$\text{frequency of string} = \frac{0.50}{\text{length}}\sqrt{\frac{\text{tension on string}}{\text{line density}}}$$

In this formula, the term **line density** means the mass of one metre of the string. Again, don't worry about the numbers which appear—especially the 0.59. That comes from internal details about the air molecules, but won't add anything to your intuitive understanding of what's going on. Again also, it is easy to verify the figures I quoted in Interlude 1 for the frequencies of various lengths of air columns. In the appropriate S.I. units, normal air pressure is 101 000 and the density of air is 1.20, so an air column 150 cm long should have a fundamental frequency of 110 Hz.

Anyone interested in musical instruments should be familiar with these formulas, though it is not necessary to memorize them. They are the basis of the design of most instruments. But for now, the main point is that all vibrators are alike and they all have similar formulas. Under a square root sign there is a term on the top line indicating how strong the force is, and a term representing inertia on the bottom line; and all this is divided by a term representing the size of the system (if that is relevant). If you understand one, you understand them all. This appreciation that the laws of nature are universal is perhaps the most important legacy left to us by the scholars of the Renaissance.

Universal harmony

The last formula is a concise statement of three different experimental facts that were first discovered (or at least written down) in the early 17th century. They are usually called the **laws of stretched strings**:

- The vibration frequency of a stretched string is inversely proportional to its length.
- It is directly proportional to the square root of the tension.
- It is inversely proportional to the square root of the mass per unit length.

These laws are more or less common knowledge to any player of a stringed instrument. Pressing the string against fingerboard or frets shortens it and raises the pitch. This is the first law. To change the tuning, the pitch is raised by tightening the adjusting screws—the second law. And the lower strings are always thicker than the higher ones, in accordance with the third law. Such instruments have been played for a very long time—there were lyres in ancient Mesopotamia; yet the laws were not known until comparatively recently. Refer back to the illustration on page 6, which you will recall was drawn in 1492. The weights hanging on the end of the strings will give the wrong frequencies, according to our formula. Why is it that these laws escaped understanding for so long?

There's probably no real answer to that. It's got something to do with the need for the right intellectual climate. My guess is that earlier scientists and musicians just wouldn't have considered these observations important. But certainly, as so often happens in science, when the climate was right, several people made the discovery at the same time. These laws appeared in Galileo's later writings, but also in the works of other scientists. By common consent, the man who probably deserves the credit was a French monk, whose name has cropped up already, Marin Mersenne.

Mersenne was born in 1588, died in 1648 and led a singularly uneventful life in between. At a time when Europe was suffering the religious wars of the Reformation and France was the turbulent place described in *The Three Musketeers*, he went to university at the Sorbonne, was ordained a Catholic priest at age 24, and spent most of the rest of his life in a monastery in Paris. He made only a few original contributions to knowledge. Mathematicians know his name from an obscure formula in the theory of prime numbers. Physicists remember him for his early measurements of audible frequencies and the speed of sound, and of course the laws of stretched strings. Musicologists still know his writings on musical theory and his invaluable descriptions of the musical instruments of his time.

What Mersenne did do was to realize that knowledge, like manure, is no good unless it's spread around (to use an expression coined by Francis Bacon when talking about money). Long before the time of learned journals and international conferences, he made himself a one-man clearing house for knowledge in Europe. He wrote letters. He kept up an immense correspondence with scholars from as far away as Constantinople: theologians, scientists, philosophers, composers. He informed them of one another's work, made suggestions based on what he knew was being done elsewhere and urged them to publicize their results. He made sure that the works of Galileo were read outside Italy and translated some of them himself. At one time he was the only connection with the great philosopher Descartes who was bumbling around Holland in the middle of the Lutheran wars. His cell in Paris became a gathering place for intellectuals, and the weekly meetings there soon became famous. Out of these eventually developed the French Academy of Sciences. .

For me, Mersenne's place in history is proof that it is not only the specialist who can do something useful and be remembered. The generalist is also important to the advance of knowledge—which many of our present day scientists and musicians seem to forget. He stands for the universality of knowledge and the power of reason to uncover the grand design. Fittingly, his most important book on musical theory and acoustics was entitled *Harmonie Universelle*.

The music of the period

The period of history during which this revolution in scientific ideas occurred, from about 1587 when Galileo was supposed to have dropped the cannonballs from the leaning tower, to when Newton died in 1727, corresponds fairly closely to what musicians call the **Baroque period**. That name comes from a French word meaning 'bizarre' or 'in bad taste', and is usually understood to imply that music, particularly towards the end of the period, had become overly florid and ornamented—though it's hard to see that it was any more so than at certain other periods. However there was one technical feature of that music which was uniquely characteristic of the time.

Baroque music was from the start dominated by Italian ideas. The most important composers at the beginning of the period were the likes of Monteverdi, Caccini, Viadana. It began as a reaction against the elaborate polyphony of the Renaissance, advocated in the 1580's (as I described earlier) by that group of Florentine scholars which included Vincenzo Galilei—the Camerata. Initially its characteristic sound was **recitative**—a solo melodic line imitating the accents of speech, with little more than a few chords as accompaniment. As time went on, the bass line became more fully developed, but it never lost its character as being primarily a support for the melody. Unlike the bass part of a polyphonic setting which was frequently interrupted, the new accompaniment tended to be played continuously, and came to be known as a **basso continuo** (or by the older name of **thorough bass**).

The new style caught on quickly and became immensely popular, and a lot of older music was rewritten with a continuo accompaniment in order to sound more 'modern'. This meant that accompanists suddenly became very important, but they also found that increasing technical demands were made on them. In time, to accommodate a wide range of expertise, a kind of musical shorthand developed, known as **figured bass**. The chords of the accompaniment weren't written out in full but only symbolized by figures above or below the corresponding notes on the stave. The actual notes to be played, the 'realization' of the harmony, was largely left to the discretion and the skill of the performer.

These are the first few bars of the madrigal "*Svogava con le stelle*", written by Caccini in 1602. Compare the bass line that the composer supplied (bottom stave) with what a good continuo player would have been expected to play (bottom two staves).

This new way of organizing the music meant that players had to think in terms of chords rather than separate lines, and the job of continuo players in those days must have been onerous indeed. Not only did they have to fill out and decorate the chords to realize what the composer had intended, they also had to listen to what the soloists or other instrumentalists were doing and tailor their dynamics (the louds and softs) to suit the overall texture of the music. In short, the continuo player was often called upon to be a kind of performing composer. But it is the kind of music they produced, with that thick underfilling of harmony that comes from the continuo, that is so characteristic of the Baroque period.

Because so many players needed this kind of expertise, the knowledge of how to construct chords had to be widely available. There were floods of books published on how to play the continuo, but before they could be written the actual rules had to be worked out. So, though the same sorts of harmonies were used as had been employed by pre-Baroque composers, the understanding of what was being done was much more formalized—as was only appropriate for a form of music which tended to be planned out as an integrated harmonic structure from the start.

The Baroque period lasted well into the 18th century, reaching its highest peak with Johann Sebastian Bach and George Frideric Handel. By their time the system of harmonic relationships had become so firmly established that there was no need to make it explicit and the basso continuo gradually disappeared. However, almost as though in compensation, much of the music became more and more florid—and baroque—until the inevitable reaction set in. The new age, the **Classical period**, was beginning.

We know that architecture, painting and sculpture shared some of the characteristics of the music of the period. A fascinating question is whether science did also. It is not obvious to me that the physics of the time could really be described as 'baroque', even in the non-pejorative sense of the word. Yet there is one point of similarity. Today, when school students learn physics, they *start* with Galileo's law of inertia and Newton's laws of motion. In a very real sense these represent the beginnings of modern physics. In the same way, when modern school students learn the subject we call **musicianship**, they begin studying harmonization with the same rules which were codified and written down by the musicians who lived between the times of Monteverdi and Bach.

The new instruments

There is another connection between the science and the music of the time, and that is through the instruments which were developing. In the early days of recitative, as the composer Caccini said, the best instrument to play the accompaniment was the lute—particularly the bass version, the **theorbo**. It was delicate and able to follow the nuances of the music, and above all it could play the thick rich chords which were such an integral part of the continuo. But for larger public performances—most especially as the new operas gained popularity—it was almost universal to have at least one **keyboard instrument** to play the accompaniment, because not only were they also ideally suited to playing chords but they were also much louder. However, what strikes me as interesting is that the construction of these relied on (at least) some familiarity with the kind of scientific observations I have been discussing.

The earliest keyboard instrument was the organ; with one pipe for each note, controlled by an array of keys. Now before you can make an organ, the single most important item you must know is how long each pipe has to be. Organs had been built as early as 200 B.C., so the makers obviously knew these lengths. Once the Church had permitted instrumental music, nearly every big church throughout Europe had one; they were the natural instruments to accompany sacred music. But for secular music during the Baroque, keyboard *stringed* instruments were more widely used, and the construction of these is more complicated, at least in theory.

In order to design keyboard stringed instruments, there are three quantities you must know for each string, and it must be very difficult if you don't know how to work them out. I am not saying that it is impossible to build them by trial and error. Hurdy-gurdies with four strings were made in the 12th century, and clavichords with twenty-two in the 15th. But it seems that, for instruments with many strings and many more moving parts, it was just not economical and reliable to build them without being able to work out in advance what length each string had to be, how much it should be tightened and what it should be made of.

However, from the end of the 16th century keyboard stringed instruments proliferated and became cheap enough to get into private houses. Clavichords, spinets, virginals, harpsichords, and eventually pianofortes. And the question I find intriguing about all this is what caused what? I should be careful about being too speculative because clearly there were many factors at work. The old medieval guilds were breaking down and the new economic structure of society enabled specialist instrument-making firms to get going. But it is plausible that the new scientific insight stimulated the development of these instruments and this caused music itself to change. Equally well it could be that the changes already taking place in music produced a demand for the new instruments and this turned the attention of scientists in that direction. Probably the truth lies in-between: both music and science developed together in response to the new intellectual climate of confidence in human ability to know and influence the harmonies of the natural world.

Equal temperament

There was an even more fundamental consequence of the ubiquity of keyboard instruments in the 17th century. It raised the question of what was the most sensible *scale* for the new music; for truly the just scale had started running into serious problems. Composers wanted to put variety into their music by modulating—that is by changing in the middle of a piece from one diatonic scale to another (one which used a different note as tonic). A simple example should explain why this raised difficulties.

> Refer back to the table of just frequencies on page 36. If you had a keyboard instrument with only those notes; you could play melodies in the key of C major. To play in, say, the key of F (352 Hz) you would need extra notes. For example the subdominant (fourth), which should be called a B, should have the frequency:
>
> $$4/3 \times \text{frequency of F} = 469.3 \text{ Hz.}$$
>
> This is much flatter than the B you already have (495 Hz), so your instrument needs a new note, which you will call B♭.

You might also want to modulate to the key of B. This needs a leading note (seventh), called an A, whose frequency is:

$$15/8 \times \text{frequency of low B} = 464.1 \text{ Hz}.$$

This is clearly higher than Concert A, so it would be called A♯. Now, this A♯ is close to the B♭ you had before, but it doesn't take a very good ear to tell that they are definitely not the same. So there are two notes between A and B. Clearly to modulate into any key at all you will need a lot of extra notes.

On an instrument where the player has complete control over the pitch, like a violin or the human voice, it is possible (at least in principle) to modulate into any key and still use a just scale. But when the pitch of the notes is fixed, by frets or finger-holes, and most particularly by a keyboard, there are only two choices. Either don't modulate into remote keys, or change the scales. It was typical of the intellectual climate of the time that the second option was chosen. The tuning of each note was altered slightly—it was called **tempering**—and the problems disappeared (which is not to say that new ones didn't arise).

Let me explain what they did. You will recall that, in the just scale, there were two different whole tones and one semitone, which were not simple ratios of one another. The trick now was to abolish the difference between the major and minor tones; and also to make all the semitones the same, each equal to half a standard tone. This meant adding five more notes to the octave, to cut each of the five tones in half. These were the black notes on the keyboard. The octave then consisted of twelve equal semitones. The octave ratio was left at exactly 2; so the semitone ratio had to be that number which gave 2 when multiplied by itself twelve times. Check it for yourself. The answer is, to sufficient accuracy:

$$\text{semitone ratio} = 1.0595$$

Notice that this is close to, but not the same as, either the Pythagorean semitone (1.0535) or the just semitone (1.0667). The difference between these three can just be detected by a good ear.

It is worth noting here that the usual way of measuring very small differences in pitch is in terms of an interval called the **cent**, which is one hundredth part of a standard semitone. I have given the strict definition in Appendix 3. For now, it is sufficient to note that the smallest interval which even a good ear can discern accurately is somewhat less than 10 cents, or 1/10 of a standard semitone. The differences between the equal tempered (standard) semitone and the Pythagorean and just ones are 10 and 12 cents respectively.

Anyhow, if you now take the standard value for concert A of 440 Hz, and use a pocket calculator, it is easy to work out the frequency for any note at all. For example, the fourth octave on the piano keyboard is:

NOTES	EQUAL TEMPERED FREQUENCIES	JUST SCALE FREQUENCIES
A	440.00 Hz	440.00 Hz
A♯	466.16	
B	493.88	495.00
C	523.25	
C♯	554.37	550.00
D	587.33	586.67
D♯	622.25	
E	659.26	660.00
F	698.46	
F♯	739.99	733.33
G	783.99	
G♯	830.61	825.00
A	880.00	880.00

If you want to work out the corresponding frequencies in any other octave range, you only need to multiply or divide by 2 the appropriate number of times.

This is the **equal tempered scale** that we use today. I hope it is clear that there is no longer any problem with modulation. The notes of the scale are constructed by choosing a starting frequency and then successively multiplying by the magic number 1.0595; the actual scale is just a selection of notes from the set generated in this manner. Then when you modulate to a new equal tempered scale you simply choose a different selection from the notes you have already.

Although the intervals of this scale are different from those of the just scale, they are very close. That is why it works so well. Look at the fifth for example. The tempered E (659.26 Hz) is lower than the just E (660 Hz) by less than 1 Hz or 2 cents. Very few musicians, even with exceptionally acute ears, can detect that it is flat. The difference is too small to be heard by most ears (except under special laboratory conditions). The interval sounds perfectly consonant with either E. It is this physiological fact that is in the end responsible for the rich harmonic modulation that is such a feature of western music today.

The Well Tempered musical world

The changeover to the equal tempered scale took about 400 years. Long before theorists had worked out what should be done, players and instrument makers were experimenting with their own tuning. There are paintings dating from the late 15th century which show Spanish guitars and Italian lutes with frets arranged, so far as you can tell, for something very close to equal temperament. By Mersenne's time there were several different systems of tuning in common use.

The first suggestion that the whole musical community should standardize on equal temperament seems to have been made in 1550 by Adriaan Willaert. Zarlino and Vincenzo (when they were still friends) had worked out very good approximations to the required ratios. So, in 1636, did Mersenne who was always an eloquent champion for equal temperament. But, like metric conversion in our own day, it was a long time coming.

It was Johann Sebastian Bach who finally forced the issue. Just after he had moved to the position as Director of Music at the court of Cöthen he became interested in writing teaching pieces. Though he didn't usually approve of mathematicians interfering in matters musical, most scholars agree that he tuned his own domestic clavichords to equal temperament. In 1722 he wrote and published his first set of 24 preludes and fugues to include all twelve major and all twelve minor keys. By naming them *The Well Tempered Clavier* he was in effect making a public manifesto, rather like Martin Luther nailing his 95 propositions to the church door in Wittenburg.

Not everyone adopted equal temperament immediately, particularly the makers of church organs. And at least one famous English piano-making firm only changed over in 1856. Nevertheless I think it is fair to say that, with the *Well Tempered Clavier*, Bach dragged the musical world, kicking and screaming, into the Age of Reason.

Interlude 2

The piano

If you look inside a piano it is not immediately obvious which are the important parts so far as the sound is concerned. There is an array of steel strings stretched between pins on an iron frame, and in front of them a row of felt tipped hammers. Press a key and one hammer flies forward and hits a string (for some hammers it is 2 or 3 strings). That much is simple. Yet the sound is surprisingly loud; and careful investigation shows that it doesn't actually come from the string, but that it comes from all over the back of the instrument, from the large sounding board to which the strings are attached at one point. Obviously a vibration of the string makes the soundboard vibrate and that's what you hear.

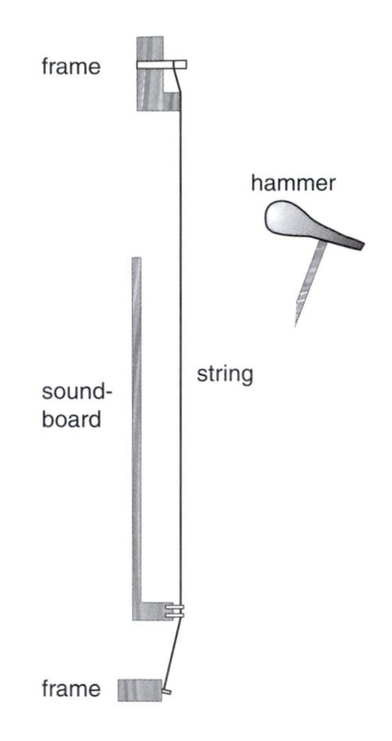

In terms of the general features of all musical instruments, the hammer hitting the string is the **generator**, the soundboard is the **radiator** and the job of the **resonator** is done by both the string and the board.

At this stage there is not much I can say about the soundboard. You can guess why it's there. The natural sound that a string makes is very soft. If you are to hear any sound, the air has to be set vibrating; but a string cannot do that efficiently because of its small size. On the other hand, a soundboard has a large area and can get a lot of air moving. The board acts as a sort of amplifier but doesn't really affect the pitch or the timbre of the note. It is obviously similar in that function to the bell of a brass instrument. I'll leave till later the whole question of how the vibration is communicated, both string to board and board to air—a question which was not understood by scientists until the late 19th century. For now I'll concentrate on the strings only.

Some general properties of strings

Let me go back to Mersenne's formula for the vibration frequency of a stretched string. I know that arithmetic can often be a bit of a bore, but there is information in that formula that I want to draw attention to.

1. The forces involved

Before we can do any calculations at all we need to know some line densities. These depend on what the strings are made of and how thick they are. Piano wires are steel, but other instruments have strings made of gut. Here are a few typical values:

MATERIAL	DIAMETER	LINE DENSITY
steel	1.0 mm	.0059 kg/m
steel	0.5 mm	.0015 kg/m
gut	0.6 mm	.0003 kg/m

These numbers will help you get a feel for the sort of tension that piano wires are under. The Middle C string on my piano is about 80 cm long and 1 mm thick. It vibrates at 262 Hz. So, using the formula on page 65, I calculate that its tension must be around 1030 newtons. (The **newton** is the standard unit in which force is measured in the S.I. system.)

The highest C, (C_8, 4192 Hz) is only 8 cm and 0.5 mm wide; the lowest C (C_1, 32 Hz) is about 1.4 m and has a double winding of copper wire round it, making its diameter effectively about 4 mm. For both of these I calculate their tension is around 700 N. Just for comparison, the E string on my guitar (gut, 65 cm long, 0.6 mm diameter) is roughly 60 N.

To appreciate what these numbers mean, I find it easiest to think of the force of 1 newton as being about the weight of a medium sized apple (actually 102 grams). So each string on the piano feels as though it is holding up a weight of 70 to 100 kg. There are 88 keys on the piano, with a total of some 230 strings. Therefore the total force is the same as if it had a 20 tonne truck sitting on top of it.

This is the first general point I want to make. *The tension forces on a piano are enormous.*

2. Questions of energy

The loudness of the sound you hear obviously depends on the energy of vibration of the air. That energy can only come, through the soundboard, from the vibration of the string. Therefore to get a good loud sound you must have lots of energy in the string.

Now think about how this energy of vibration depends on the length of the string. Recall the formula for kinetic energy. It involves mass and velocity squared. The mass of the string clearly depends on its length (for a given thickness and density). The velocity with which the string moves also depends on its length, as I think you will understand from the following argument:

> Velocity is distance moved divided by time taken. The time the string takes to swing from side to side and back again is always the same (1/262 seconds for Middle C). So the velocity depends only on how far it swings.
>
> This distance depends on what you do to start the vibration going—how far you pull it, or how hard you hit it. But there is a limitation on how great it can be. In order for the string to vibrate cleanly the wire shouldn't *bend* too much, else the forces of rigidity in the metal itself start to interfere with its motion (I'll return to this point shortly). Simple geometry will tell you that a longer string can move further sideways than a shorter one before it runs into this difficulty. Therefore the longer string can move with the higher velocity.

Putting these conclusions together means that the energy of vibration depends on the length of the string several times over. This is my second general point. *Even a modest increase in length of the string will give a big increase in possible loudness.*

3. How the vibration is excited

A piano string gets the energy for its vibration when it is hit by the hammer. It gets no more energy after that, so the note will decay in its own good time. If you want to, you can hasten the decay by pressing felt pads, or **dampers**, against the string; but there is nothing you can do to keep it going longer. The so-called sustaining pedal is misnamed. All it does is to keep these dampers away from the strings. What I've just said is also true for a harpsichord where the vibration is started by mechanically plucking the strings with a plectrum.

Another point to note is that, once the hammer or the plectrum has done its job, it should get out of the way quickly. It mustn't touch the string again once it has started vibrating. There is only a very short time for it to move in (1/262 seconds for Middle C) and no human player can be relied on to respond as fast as that. This will prove to be critical in the mechanical design of these instruments.

Lastly, the way the vibration is excited, and where on the string it happens, both effect the timbre of the note. But that's another of those questions I'll have to talk about later when we have covered the properties of overtones.

The development of keyboard stringed instruments

I suppose it would be fair to say that the guiding philosophy behind the development of these instruments is this. If the instrument maker could take away all the difficulty of actually producing the notes, then you, the player, would have more time to devote to putting the notes together and making music. In crude terms this comes down to two principles. First, each note has its own string, carefully tuned so that you don't have to worry about getting the pitch right. Second, the action is mechanical so you just press a lever and the note is made for you. In these terms we can identify the ancestors of the modern piano.

Two very early instruments were the **psaltery** and the **dulcimer**. They were very similar to one another, as you can see from these rather charming carvings on the ceiling of Manchester Cathedral.

Each consisted of a small soundboard or shallow sound-box, with a dozen or so strings run from side to side between two bridges on top. The difference was that the psaltery was played by plucking the strings and the dulcimer by hitting them with small wooden hammers. Both emerged in the 12th century, probably brought to Europe from the Near East during the crusades.

A different kind of stringed instrument, also common in the 12th century, was the **hurdy-gurdy**. In many ways it was rather like a mechanized viol. The earliest ones needed two players. One turned a crank handle which rubbed against the three or four strings, keeping them in continuous vibration. The other player worked the keys, bringing various bridges into contact with the strings, thus changing the notes. Because of the continuous nature of its sound, it probably sounded rather like bagpipes. Later it was streamlined so that it only needed one person to play it. It is interesting to note that the hurdy-gurdy was very popular as a street instrument, and when it was superseded by the street piano (which is also played by turning a handle) the name passed with it.

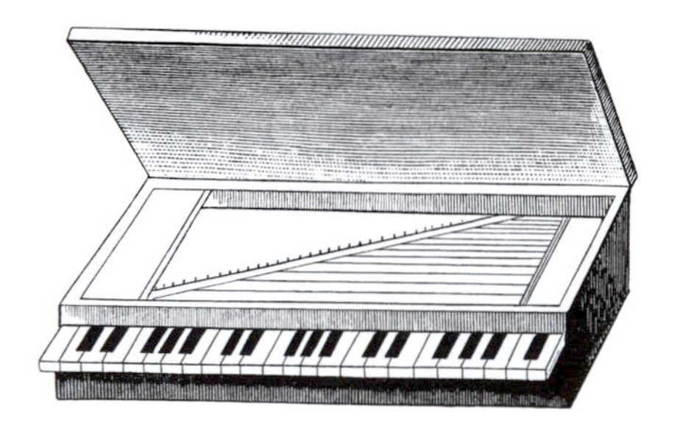

Both of these principles first came together in the 15th century with the **clavichord**. This grew out of our old favourite, the Pythagorean monochord, which was still being used to teach musical theory. However instead of a single string it had about twenty; and the job of placing the bridges in the right positions was done by means of keys (the Latin word *clavis* means 'key'). The whole thing was put inside a wooden box, about the size of a large suitcase, and that was a clavichord.

The strings were arranged to run parallel to the keyboard, and this meant that each one could be used to produce more than one note, simply by striking them at different points. In order to prevent the unwanted length of the string from sounding, it had cloth dampers to stop any vibration. All this added to its compactness.

There were also several other features which carry over to modern instruments. It had metal strings; and there was an independent soundboard in the bottom of the box which didn't double as a frame for the strings. Over the next 300 years the clavichord continually improved, but it had one limitation it could never overcome.

The problem was in its **action**, that is the way it made its notes. In simple outline, each key was connected by a lever mechanism to a small piece of metal (which is called a **tangent**). When you pressed the key the lever brought this piece of metal into contact with the string and held it there.

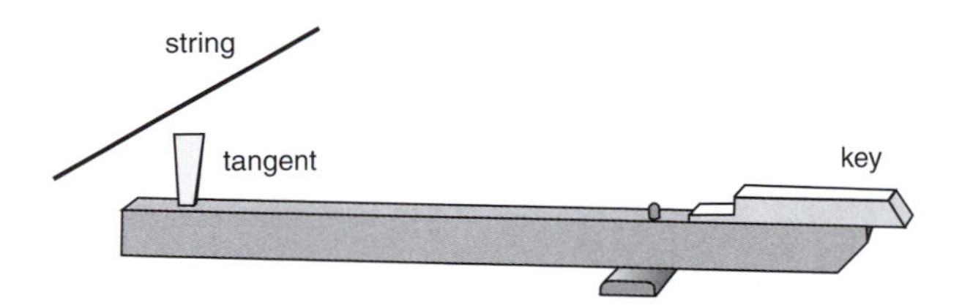

The tangent acted as both a bridge for determining the pitch of the note and a hammer for starting the vibration. Ingenious yes, but limited. This mechanism can only make a very soft sound. It's almost the same as pressing down a guitar string suddenly against one of the frets without plucking it. You can get nearly as loud a tone by just blowing over the strings. Therefore, though clavichords were popular domestic instruments, used a lot by organists for practising, they weren't much good for public performance.

An idea arose in the late 15th century, to sound the tone by mechanically plucking the strings. Each key operated a thin wooden rod, or **jack**, which carried a small plectrum made of leather or quill—not attached to the jack directly, but to a thin sliver, the **tongue**.

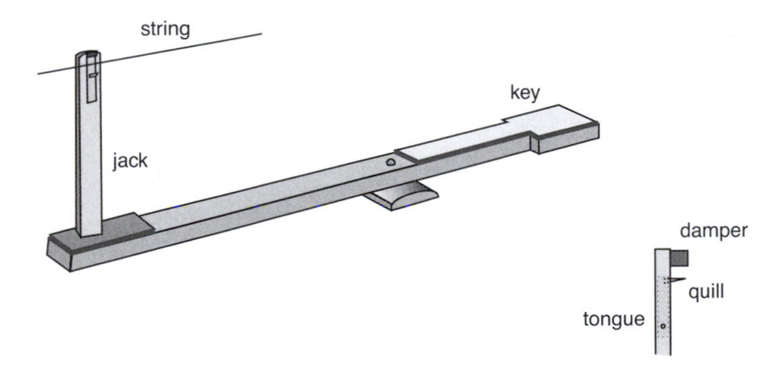

When you pressed the key the jack rose and plucked the string. When you released the key the jack fell back, and by an ingenious spring mechanism the tongue pivoted back so that the plectrum didn't pluck the string on its return. When the jack had fallen right back, a felt damper stopped the string vibrating.

The **virginals** were the first to use this new device. They had longer strings than a clavichord, to produce louder sounds, but otherwise looked rather similar from the outside. They were obviously still intended for domestic use. Why they were so named is not clear, probably because they were so often represented as being played by young women, as in this illustration from the cover of the first English publication of music for virginals in 1612.

The quest for louder sounds led very soon to even longer strings, larger soundboards and the invention of the **harpsichord** (and a cheaper version of the same thing, the **spinet**). With it came many features that we recognize in grand pianos today. The strings were so long that it was more sensible to run them perpendicular to the keyboard away from the player, rather than parallel to it. The case was no longer rectangular but followed the variation in length of the strings to give the familiar wing shape. For even more loudness, as well as better tone quality, two strings were used for each note. There was a 'forte' pedal which lifted the dampers from all the strings to allow sustained tones.

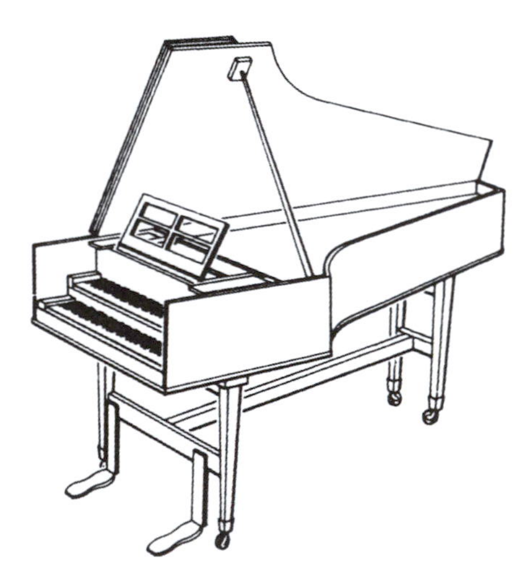

The harpsichord was enormously popular in the 17th and 18th centuries and much prized by composers for its distinctive tone quality; but it declined in popularity as its rival, the piano, improved. In the 19th century it became almost extinct, though it has made a comeback in the late 20th.

The reasons for its decline were again inherent in its design. It still wasn't loud enough. Attempts were made to increase the volume by using heavier strings, but the quill plectrum couldn't excite a very massive string and the wooden cases weren't strong enough to bear the increased tension.

But there was a more serious defect. It was very difficult to play a harpsichord with **dynamics**—that is, softly or loudly. And this, again, was because of its action. Whether the key was hit hard or softly, the plectrum always pulled the string to the same distance before it let go. So the note was always the same loudness. Good players could force some variation, but not much. This meant that it was impossible to accentuate, to change the stress on different notes as much as they wanted, especially compared with the piano. So even though the harpsichord has many musical qualities that its rival does not have, it lost that particular evolutionary battle.

The pianoforte

In 1709, the Italian harpsichord maker Bartolommeo Christofori built the first instrument with hammers instead of plectrums. It was essentially a mechanized dulcimer, in the same way that a harpsichord could be considered a mechanized psaltery. But because it gave the player complete control over how hard the hammers hit the strings, the difficulty about dynamics was solved. With an unerring instinct that it pays to advertise, Christofori named his new instrument simply **piano-forte**.

Its strings were heavier and its case was stronger than a harpsichord's. In 1720 he brought out an improved model with his own specially invented action. Remember the point I made about how important it was to get the hammer out of the way quickly? Well his device had a cunning catapult action which threw the free-swinging hammer at the string and caught it again as it bounced back. For 150 years inventors worked to improve the new instrument, particularly its revolutionary action. Constant striving for ever more volume led to thicker and heavier strings. Eventually the wooden frames of the earlier pianos couldn't take the increased tension and in 1855 the American manufacturer Henry Steinway brought out a grand piano with a cast iron frame. All later pianos are built along the same general lines. Small refinements are constantly being introduced, but there has been no fundamental change in the design or construction of pianos since then.

It is worth keeping in mind that, during all this time while the piano was changing drastically, a lot of music was written for it. But in many ways the piano of the earlier years was a different instrument from that of the later period. In the last few decades of the 20th century there has been renewed interest in hearing music on the original instruments for which it was written, so it has become fashionable to distinguish the earlier types of pianos from the later.

Originally the names 'piano-forte' and 'forte-piano' were used interchangeably. It was well into the 19th century before the agreed name had settled down to the first of these. Modern scholars have revived the second name, using it to denote the earlier instruments. Therefore when you hear people talking about a **forte-piano** they are referring to the instruments of the late 18th century for which the likes of Mozart, Beethoven and Schubert wrote much of their keyboard music.

Some technical features of the piano

Firstly I should say a little more about the action of the modern piano. You can appreciate how ingenious it is by looking at this simplified sketch of Christofori's original device.

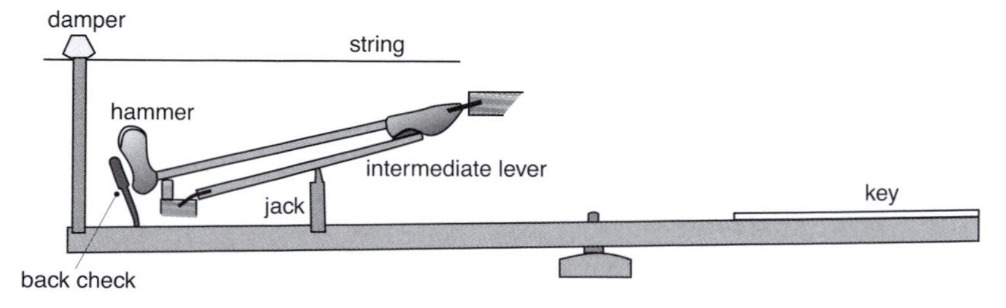

When you press the key, the jack pushes up against the intermediate lever. The end of this lever moves fast enough to kick the stem of the hammer and throw it against the string. When it falls back, its lower end is caught by the back check, because you don't want it to rebound and hit the string again. If your finger is still on the key, the intermediate lever is still raised so the hammer will not fall all the way back, but will be ready to be thrown again. This allows for rapid repetition of notes. There is a damper for each string which is raised when you press the key, but it falls again and stops the vibration as soon as you take your finger off.

The action on modern pianos is a more complex version of the same thing, with many more moving parts. It is slightly different in uprights and in grand pianos. In the former, you rely on the hammer to bounce back from the string, but in the latter, because the hammer is underneath, gravity assists the process. That is probably why the action on a grand is faster and smoother than on an upright. In a grand piano also, the whole mechanism is arranged so that it can be moved a short distance to one side if you press the 'soft' pedal. This shifts the hammers so that they hit two strings instead of three. In an upright, all the 'soft' pedal does is to move the hammers closer to the strings.

There are many more technical features about pianos which are outside the scope of this book, but there is one further point I want to mention. The strings on the modern piano are made of steel and are relatively thick. Thick steel wires are stiff and do not bend easily. This **stiffness** affects the way they vibrate.

If you think about this carefully you can predict just what this effect should be. Use this line of reasoning:

> Compare two strings of the same length, mass and tension; one made of steel and the other of something much more flexible, like gut. (I don't know what material could possibly have the same density as steel and the flexibility of gut; but this is only a 'thought experiment' to help us isolate in our minds the effect of stiffness).
>
> Now imagine starting them vibrating, by pulling them both to the side an equal distance. The gut string has potential energy because you have pulled it against its tension. Ditto for the steel string. But there is something else. It takes effort to bend a steel wire even if it isn't under tension. So the steel wire has a small amount of elastic potential energy that the other does not. It therefore has a little more energy in total.
>
> Now let the two strings go. All their energy gets converted to kinetic. The steel wire had more energy at the start, so it has more kinetic energy now. It must be moving a bit faster. It will complete its swing in a shorter time. Its frequency will be a little bit higher than the gut string.

Therefore the steel wire, because of its stiffness, will produce a slightly sharper note. Stiffness is more noticeable in shorter strings, so this effect should be more important for higher notes. Of course your piano is tuned to the tuner's ear and not to some mathematical formula, so all the strings will give out the fundamental frequencies that the tuner intends. But don't forget, all real notes have overtones in them. It is these which will be affected by the stiffness of the piano wire. Compared with a violin, say, the notes on a piano have overtones which are a little sharp. In physics speak, they are **anharmonic**, that is, *not* integer multiples of the fundamental.

This has a particularly serious consequence. Piano tuners usually do their job in the following sequence. They tune one string (most probably Middle C) with reference to an external standard—a tuning fork or electronic generator. Then they tune all the octaves of this string by matching the fundamental of the higher to the first overtone of the lower. After that, they work on the other strings, relying on their ear to set the appropriate intervals. Now I have already said that the overtones of a stretched steel wire are a little sharp, and therefore this kind of tuning cannot be accurate. And indeed if you measure the frequencies of two notes an octave apart on a well tuned piano, you will find their ratio is a little greater than 2. (It is known in the trade as a **stretched octave**). But it sounds right musically; either that or we have just got used to it. Attempts to tune pianos absolutely usually result in them being judged 'out of tune'.

Whether this is a good or a bad thing is not for me to say. It is certainly one of the features that gives a piano its distinctive tone quality; and over the last 300 years the public has shown it likes the sound of the piano. But it probably means that the piano can't change much more. The strings can't get much longer or the instruments won't fit into ordinary rooms; and they can't get much thicker or they won't sound harmonious. I wouldn't be surprised if the evolution of the piano has come to a natural end.

Touch

We all know pianists who bang out everything they play. We say they have a 'heavy' touch. Good players who get lots of light and shade into their music by varying the loudness are said to have a 'light' touch. That sort of 'touch' describes dynamics. But there is a much more contentious question which comes in here. Is it possible for different players to get different tone qualities out of the same piano? Can they vary the timbre of the notes they play?

Many pianists—I would even guess most pianists—would answer yes. Piano teachers commonly advocate stroking the keys, using supple fingers for this effect and stiff fingers for that, and so on. Some editions of Beethoven sonatas have explicit instructions about what timbre to aim for, with markings like *quasi clarinet*, *quasi horn* etc. There was a book published in 1911, by one Tobias Matthay, which lists 42 different ways to play a single note.

On the other hand, the physicist looks at the action of the piano and notices that, in the fraction of a second before the hammer hits the string, it has been thrown clear of the mechanism. The player has completely lost control over it. It has been given some kinetic energy and that determines how loud the note is. How much energy it has depends only on its velocity and it doesn't matter how it got it. There seems no possible way any player could produce another note of the same loudness with a different timbre, no matter what is done to the key (leaving the dampers alone of course).

So who is right? The sensible course is to check it experimentally. In the last sixty or seventy years, following the improvement in sound measuring techniques that occurred during the First World War, this experiment has been done many times. Usually the procedure is to get a good player to sound single notes of different loudness, and a bad player (or a mechanical device) to do the same. Then listeners, who cannot see how the notes are produced, are tested to find out if they can hear any difference. The results are always that they cannot.

In actual fact, there is probably a sensible resolution to this controversy. Some careful experiments done in the 1930s showed that, although notes produced with the same loudness on the same piano always sound exactly the same, notes played with different loudness can have very different timbres. In general, loud notes have more high overtones than soft ones. Now a skilled player is quite capable of striking the different notes that make up a *chord* with different degrees of force. So it is entirely plausible that you could hear differences in the timbre of chords played by a pianist with a good touch. And of course, skilful use of the dampers will also add another factor.

Whatever the rights and wrongs of this controversy, and I suspect we haven't heard the last of it, there is a message about the place of science in music. Scientists are slowly increasing our knowledge about some facets of music-making, and musicians should not ignore that knowledge just because it might conflict with some deeply held belief. It seems wrong for example, that some piano teachers should continue as though the experiments I talked about had never been done. It is difficult enough to learn the right things to do to play a piano well, without learning wrong things too. But, equally well, scientists must be careful about being arrogant. They may (or may not) know a lot, but their knowledge is of a special kind. They always try to *simplify* things, to understand a whole by knowing all about the parts that make it up. On the other hand, experienced musicians may not understand all the little pieces, but they certainly know how they fit together. It will be a long time, if ever, before science can match that kind of understanding of music.

Chapter 4

Overtones of enlightenment

In the history of science, 1666 was *annus mirabilis*—a miraculous year. The plague, which had reached London the preceding spring, was spreading throughout the south of England, and was only brought under control by the Great Fire twelve months later. In the meantime, Cambridge University had closed its doors and sent all its students home, among them the young Isaac Newton. At his mother's farm at Woolsthorpe he spent a year of enforced idleness—as legend would have it, watching apples fall. But in that year he invented differential and integral calculus, he created his theory of universal gravitation and he proved experimentally that white light is made up of many colours. When asked how he managed to do all these in so short a time, his unhelpful reply was: "By thinking about them."

Born in 1642, his childhood had been ill-starred. His father died before he was born and when his mother remarried he was sent to live with his grandmother. He was always sickly, and showed no particular intellectual promise at school. He entered University with the intention of studying theology, until almost by chance he stumbled on a copy of Euclid's *Elements of Geometry*. And then there came that year of 'idleness'.

Within five years of his return to Cambridge, the importance of what he had achieved persuaded the professor of mathematics to step down in his favour, and Isaac filled that position for the next three decades. In that time he extended the Calculus, he formulated the laws of motion which I have already talked about, he perfected his model of the Solar System and recalculated all the planetary orbits, he investigated the way light was broken up into colours by a prism, and he invented a new kind of telescope—one which used mirrors rather than lenses and which is the basis for most modern instruments.

It wasn't until 1687, at the insistence of his friend Edmond Halley, that the *Principia* was published. The prediction, by this same Halley, that a particular comet which had been seen in 1682 would return every 76 years was one of the first and most spectacular successes of Newtonian astronomy; and in 1985, when Halley's comet reappeared on schedule, it gave the modern world a small reminder of the magnitude of Newton's achievement.

There can be no doubt that Isaac Newton was one of the greatest intellects who ever lived, if not *the* greatest; yet he was a curiously mixed person. He never married, or showed any interest in women. He was ridiculously absent-minded, neurotically worried about his health, pathologically sensitive to criticism and petulant in response to it. At one time or another he fought with most of the famous scientists of his time. He lacked judgement even in academic matters—writing millions of useless words about alchemy and mystical speculation. Yet, with all that, he remained modest and unassuming, and the world still remembers that he said:

> "I do not know what I may appear to the world, but to myself I seem to have been only like a boy playing on the sea-shore, and diverting myself now and then finding a smoother pebble or prettier shell than ordinary, whilst the great ocean of truth lay all undiscovered before me."

On the misguided advice of friends he entered parliament but never made a speech while there. He was appointed Master of the Mint in 1699 and, though he did an excellent job by all accounts, it was a waste of his talents. He did no more science except to write up his work on optics and to prepare an English edition of the *Principia*. He died in 1727.

Modern historians of science sum up the importance of his achievements when they speak of the 'Newtonian revolution'. But his scholarly influence was not limited to science. He was one of the key thinkers who shaped that chapter in the intellectual life of Europe, which we now call the **Enlightenment**. The poet Alexander Pope wrote this epitaph:

> "Nature and Nature's laws lay hid in night:
> God said, Let Newton be! and all was light."

The problem of overtones

After Newton's death, science and mathematics went through a period of consolidation, largely dominated by the learned societies and the journals they published. The most prestigious of these were the French **Académie des Sciences** which had grown out of those meetings in Mersenne's monastery cell, and the English **Royal Society**, of which Newton had been president for the last twenty years of his life. Their activities gave cohesiveness to the international scientific community, in spite of disputes which broke out from time to time. There was, for example, one particularly bitter dispute about whether Newton was the real originator of the Calculus, or if the credit should go to the German, Gottfried Leibnitz, a brilliant but rather devious man who was the original model for Dr Pangloss in Voltaire's *Candide*.

Perhaps even more noteworthy was an eagerness for knowledge on the part of society itself. Semi-technical books, with names like *Mathematics Made Easy* and *Outlines of Philosophy*, proliferated. The leaders of fashion seriously attended lectures in order to join the scientific conversations in their elegant salons. Madame du Pompadour had her portrait painted with mathematical instruments and a telescope at her feet.

The same period produced the great encyclopedias. In 1751, in Paris, Denis Diderot published the first volume of his immensely influential *Encyclopédie* which aimed to bring together all the scientific views of the age in one place. The even more popular *Britannica* was first released in 1771. All in all, the spirit of the age was one of belief in the natural order of things and confidence in human (rather than divine) understanding—the spirit to which we give the name **classicism**.

It was in this climate that the mathematicians of Europe worked to change the Calculus from a forbidding, obscure discipline which only a few minds could grasp, into the simple elegant tool which scientists everywhere could use to solve problems, and which school students would be expected to master. They did this by setting puzzles for one another, in competitions sponsored by the societies. And there was one particular problem which occupied them for most of the century.

The list of people who helped solve this problem reads like a 'who's-who' of 18th century mathematics: the Englishman Brook Taylor, a friend of Newton and Halley, who was also a gifted artist and musician; the Swiss Daniel Bernoulli, just one of an amazing family that was as rich in mathematicians as their contemporaries, the Bachs, were in musicians; the Frenchman Jean le Rond d'Alembert, who was Diderot's main collaborator on the *Encyclopédie*; another Swiss, Leonhard Euler, who was the leading light of the St Petersburg Academy under the patronage of Catherine the Great; and the Italian, Joseph Louis Lagrange, who despite being a friend of Marie Antoinette, managed to survive the French Revolution and helped draw up a new system of weights and measures which later became the *Système Internationale*.

A collection of luminaries indeed; and the problem which they all worked on was this: how is it that a single stretched string can vibrate with many different frequencies at the same time?

The fact that a vibrating string will sound not only a fundamental note, but also a number of overtones is one of those things which musicians must always have known, yet which no one seems to have drawn attention to until Mersenne. Then in 1704 the Frenchman, Joseph Sauveur, who had been a deaf-mute for the first eleven years of his life, published his momentous treatise *On Harmonic Sounds*. In it he collected all that was known at the time: that a musical note from any natural vibrator has overtones in it; that for a string and an air column these overtones form a perfect harmonic series; that bells and other resonant bodies have overtones also, but they don't form a harmonic series; and that a string sounding one of its overtones vibrates with nodes and antinodes (just as I described in Interlude 1). Physicists, as well as mathematicians, were interested in this problem.

The final solution was achieved in the first years of the next century. Jean-Baptiste-Joseph Fourier was born in 1768 and led a rather more exciting life than most mathematicians. Prevented from entering the priesthood by the start of the French Revolution, he put his sermonizing skills to good use in politics. He was lucky enough to catch the eye of the rapidly rising Napoleon Buonaparte, and was made Director of the new *École Polytechnique* established for the training of modern engineers. He accompanied Napoleon to Egypt on that disastrous expedition to bring French enlightenment to the barbarians, but was left holding the baby when his friend and master fled the country. When he returned to France some years later, he was rewarded with a regional governorship; and there he was secure so long as Napoleon was in power. But after Waterloo, no one wanted to know him. He was lucky to retain a poorly paid position as secretary of the *Académie*. He died in 1830, a forgotten relic of a turbulent era.

But science remembers him because, as early as 1807, he had tied the final loose ends on that century-old problem: and today, all students of physics and mathematics learn, as a standard part of their training, the **Fourier theory of harmonic analysis**.

How to visualize a vibration

Clearly at this point I am going to have to stop the historical narrative again to discuss harmonic analysis in modern terms. In this case it is doubly necessary because the whole subject is very difficult to talk about without the information that can be gained from certain modern scientific instruments.

The problem is that it is almost impossible to visualize exactly what is happening when an audible vibration occurs, because everything moves so fast. Unless you can see what is going on, how can you begin to understand it? But today there are two particular devices which get round this problem: the **microphone** and the **oscilloscope**.

Now it is not absolutely necessary for you to know how these work, in order to understand what they can show. But it seems silly to me not to have *some* idea of how the information you work with is obtained; so I will try to give a very brief, principles only, description of these instruments. If you're just not interested then you can skip the next four paragraphs.

First the **microphone**. If you strip away all the extraneous bits, its essential operation is like this:

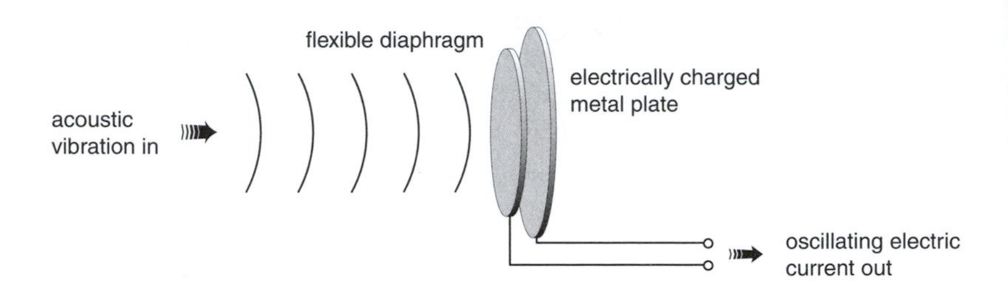

Pressure vibrations in the air, which constitute a sound, cause a thin flexible diaphragm to vibrate exactly in sympathy. This causes it to move closer to and further away from a charged metal plate, which, in turn, causes tiny electric currents to flow, either towards the diaphragm or away from it. (This same effect can be achieved in other ways, but the one I've described is conceptually the simplest.) The result is that an electric current, which is oscillating exactly in step with the original sound, flows out of the microphone into the connecting wires.

The second instrument is a device which makes this oscillation visible; hence its name—an **oscilloscope**. We all own television sets, so the operation of this instrument is easier to understand than it otherwise might be. The picture on a TV screen is produced, one line at a time, by a single spot of light moving across the screen. It starts at the upper left-hand corner, and sweeps across in about 1/15,000 of a second, leaving a glowing trail of variable brightness behind it. It then flicks back and draws the next line down. After about 600 lines the screen is full, and the spot goes back to the upper left corner and starts again. It does this 25 times each second. The important point to have clear in your mind is that this spot can move extremely quickly, and is capable of independent horizontal and vertical movement.

Now, an oscilloscope is a simpler version of the same thing. The spot on the screen is produced by a very narrow beam of fast-moving, electrically charged, atomic particles. These used to be called **cathode rays**, which explains why this instrument is technically still known as a **cathode ray oscilloscope** (or **CRO** for short). The trick here is to deflect the vertical path of this beam by means of a pair of electrically charged, metal plates.

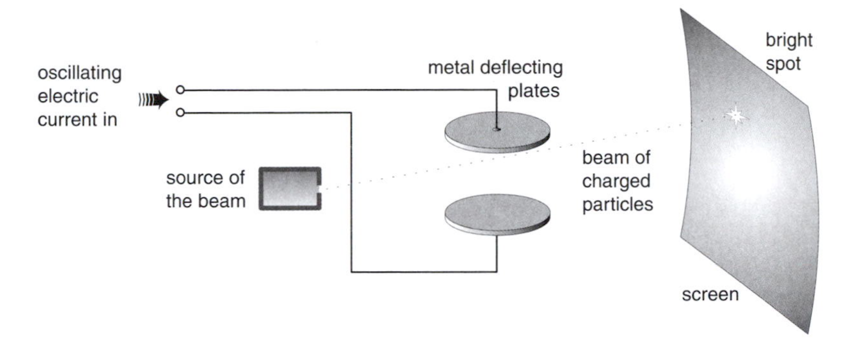

An oscillating electric current (from the microphone) changes the distribution of charge on these plates—positive charges on the top one and negative on the bottom, or vice versa, depending on which way the current is flowing at that instant. When the top plate is positive the beam is pulled up, and the spot rises; when it is negative the beam is pushed down and the spot falls.

While this is happening the spot is moving horizontally very rapidly and completely independently, hundreds of times a second. So what you see on the front of the screen is a curve which shows how large the current is (indicated by its vertical position) at different instants of time (indicated by its horizontal position). Then if the oscillating current is periodic (as it would be if it originated from a musical note) this same trace will be redrawn every time the spot sweeps across the screen, and you will see it as a perfectly steady pattern.

These then are the instruments which have revolutionized the study of musical sounds. Today they are cheap enough that most school laboratories have them as a matter of course. With them you can *see* exactly how the air pressure is varying when you hear a musical note; and I now want to describe what they show.

Start off by sounding a note at Concert A, but make it a note which is usually described as 'pure' or 'thin'—from a tuning fork or a softly blown recorder. This is the pattern the oscilloscope will trace out:

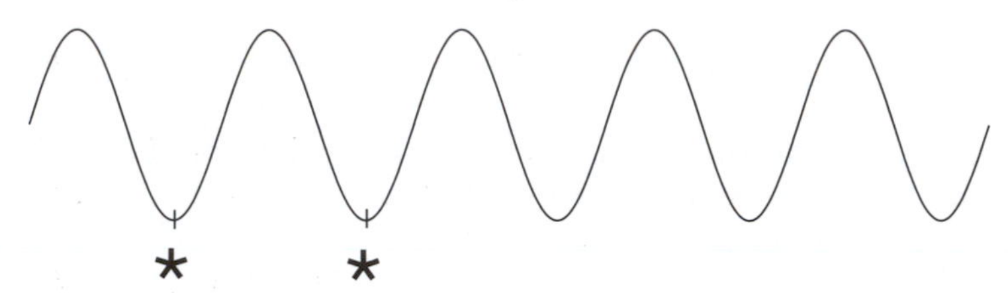

The distance between the two points marked ⋆ corresponds to a time of 1/440 seconds (remember there are 440 complete vibrations in one second). That distance represents the **period** of the oscillation.

Likewise an A5 tuning fork (880 Hz) gives this trace which, because its period is half that of the former, has its peaks half the distance apart:

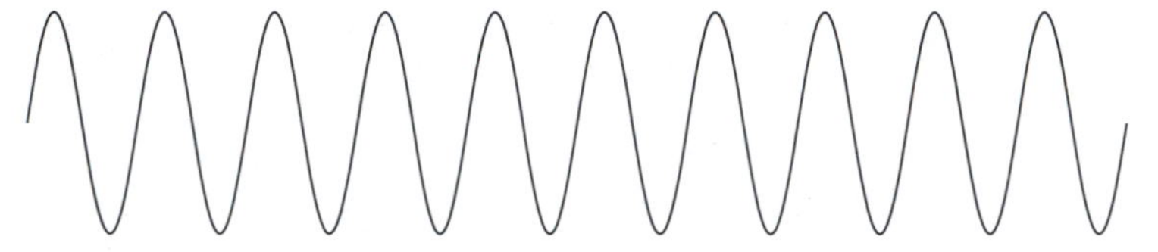

If you make it louder, or simply move the source closer to the microphone, you will get the same curve, but expanded in the vertical direction.

This trace, incidentally, is exactly the same as you would get from a swinging pendulum (only that would be much slower of course). It therefore represents a very basic kind of periodic motion, which is characteristic not only of pendulums and tuning forks, but also of many other vibrators. In Chapter 3, I gave this the name **simple harmonic motion**; and the curve the oscilloscope traced out is called a **sine curve** (for reasons that we needn't go into here).

However when you start feeding other musical notes into the oscilloscope you begin to get more interesting information back. Let me give you just a few traces from different musical instruments playing the same pitch—Concert A. You won't get exactly the same traces every time you sound these notes, the actual pattern will fluctuate a bit with time, but nonetheless these are representative of what you will see, on average.

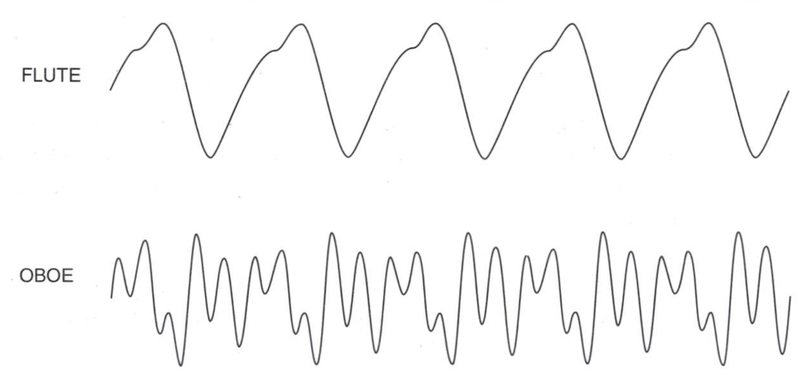

You can see now, I hope, why I said that these scientific instruments revolutionized the study of musical sounds. All these traces repeat themselves after the same time (1/440 seconds), so they do have the same basic pitch. But it is now possible to *see* something which has always been easy to *hear*: that steady notes of the same pitch played on various instruments are quite different from one another. We now have an simple description of what this difference is. They all have different **oscillation traces**. At long last we have found a characteristic of physical vibrations which corresponds to the musical attribute of **timbre**.

Combinations of oscillations

One important result of having found a way to visualize vibrations is that it allows us to see what happens when we start to combine them (and remember that in the end we want to understand how musical notes sound together). It also immediately explains two well-known musical effects.

The first is what we call **beats**. If two notes, whose frequencies are very close but not quite the same, are sounded together, you hear the sound pulsate. These pulsations disappear only if the two frequencies are made exactly equal. It is a reasonably common effect and, for example, piano tuners regularly use it when tuning piano strings.

Now it is easy to explain what causes this, with the aid of the diagrams I have just been talking about. Here I have drawn two simple harmonic oscillations with slightly different frequencies, the period of the second being just a little longer than the first.

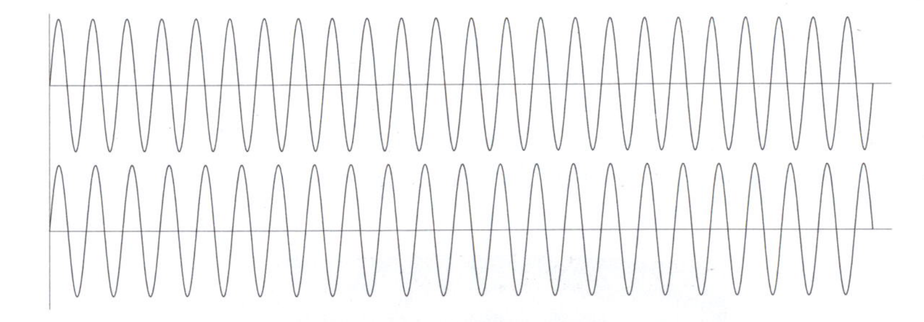

(Remember that the horizontal axis on these graphs represents time, and the vertical axis represents pressure of the air, or whatever else it is which is vibrating.)

Now when you sound the two notes together, the pressure changes will follow both vibrations simultaneously, and to represent the resulting motion I would need to **add** these two sine curves. By 'adding' I mean constructing a new trace by combining the heights at various times—on an expanded horizontal scale like this:

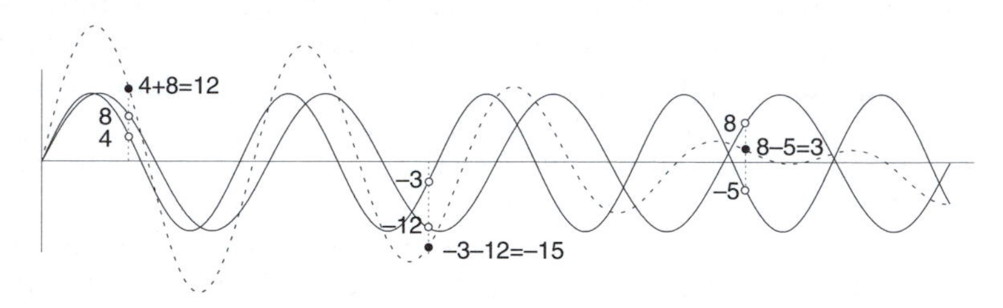

Let me do this for the two oscillations above, at all possible times along the horizontal axis. At some points the two curves reinforce one another, and at others they completely cancel. The result looks like this:

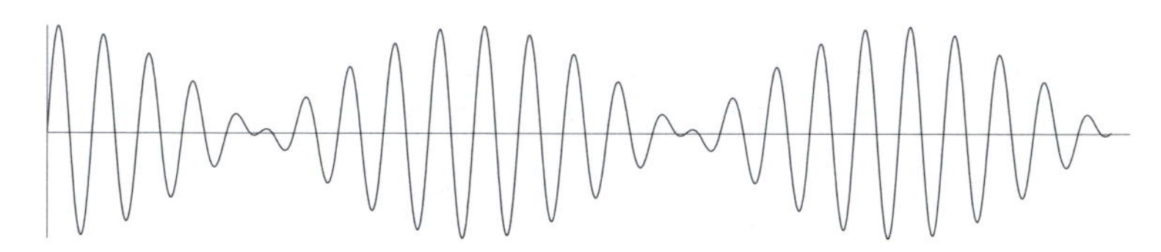

And that is what you hear, isn't it? A tone whose loudness periodically increases and decreases with time.

If you want to think it through carefully, you should be able to work out that the number of pulsations you will hear in one second is equal to the difference in frequency of the two notes. So this is an example of how the visual way of describing sounds is useful. But be clear, it doesn't really tell you anything you couldn't have worked out in other ways. The explanation of this effect didn't have to await the invention of the oscilloscope. Sauveur understood it perfectly in 1707, and used it in an ingenious measurement of the frequency of musical notes.

Likewise, it's not difficult to find other, very simple, situations where you can observe the same effect, and where what is happening is obvious—for example, between two pocket combs whose teeth spacing is almost, but not quite, the same. (Do try this for yourself. It's very easy to do).

There is a second effect related to this. You hear beats only when the frequencies of the two notes are very close. If you widen the interval between them, the pulsations that you can see in the diagram become too fast for for you to hear as individual beats. The question is: should you be able to hear them as a *tone*?

You can't answer that question without knowing a lot more about how your ear works. I'll talk about the ear in Chapter 7, but let me just say, at this point, that under certain conditions, you can hear this tone, and its frequency is equal to the difference in the frequencies of the two original notes. But only sometimes, most obviously when the original notes are quite loud. These are usually called **difference tones**, though in earlier books you may find them referred to as **grave harmonics** or **Tartini tones**, after the Italian composer and violinist who was the first to describe them in 1754.

They are clearly important in music. You will remember that the frequencies of any two consonant notes are related to one another by a simple ratio. Therefore the difference between them should give another simple ratio. As an example, if the two notes sounded were A_4 (440 Hz) and E_4 (660 Hz), then the difference tone will have a frequency of 220 Hz. In other words you will sometimes hear the note A_3.

Let us move on now to look at what happens when I add together simple harmonic oscillations whose frequencies are numerically related. Firstly two which are an octave apart—one frequency twice the other.

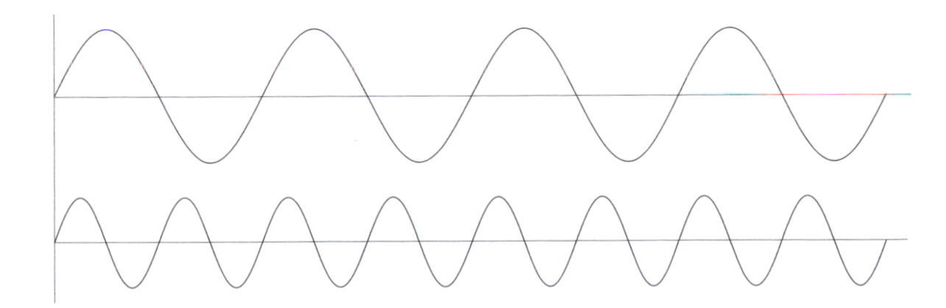

When these are added together you get this (solid) curve:

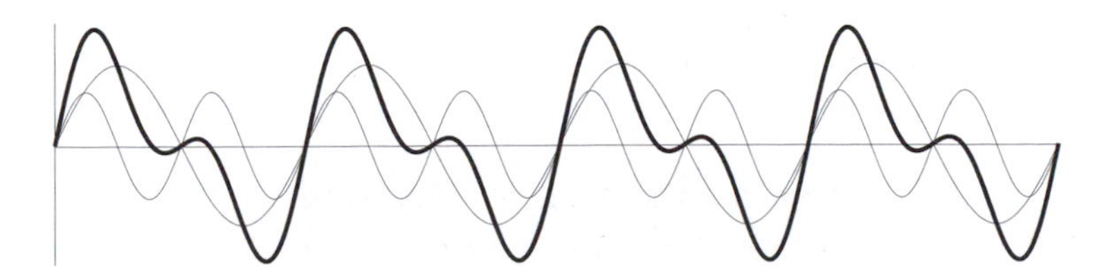

The detailed shape of this combination curve is not important of itself. I could alter it by changing the amplitude of either of the two sine curves, or by starting one of them at a different time in its cycle (called changing the **phase**). But there is one point about this diagram which I hope is easy to see. The combination curve is perfectly **periodic**—it will go on repeating itself exactly, for as long as I care to keep drawing it. And its period is the same as that of the first sine curve.

Now this observation is going to be most important, and it is vital to be clear about why it is so. It comes about because two complete cycles of the second curve fit *exactly* into one period of the first. It has to be exact because, it they were just a little different, then the combination curve wouldn't be quite the same from one cycle to the next. Instead it would look quite strange. The point at which the two curves add up to be a maximum would change from one cycle to the next, and you'd get a kind of progressive beat like this:

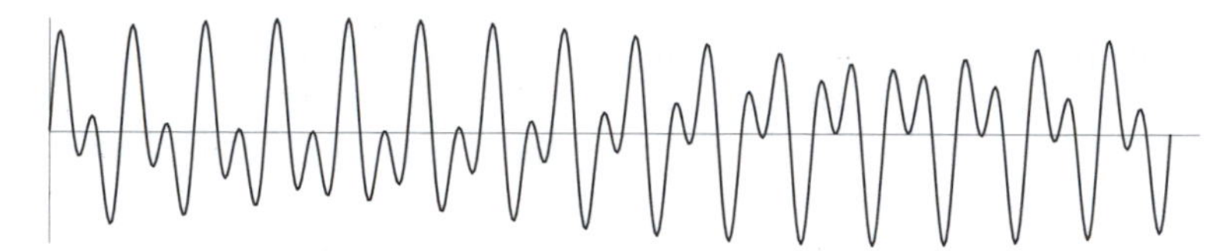

Only when the ratio of the two frequencies is *exactly* equal to 2 will the combination curve be *perfectly periodic*.

Let us press on. If I add a third simple sine curve of three times (exactly) the frequency of the first, I get an even more complex combination.

Again I have perfect periodicity, and for the same basic reason, because three complete cycles of the third curve fit exactly into one period of the first.

Clearly there is a trend emerging. I will be able to produce a combination curve of increasingly complicated shape, but still perfectly periodic, provided I add only sine curves whose frequencies are *whole number* multiples of the first. And furthermore, this procedure will generate an enormous variety of outcomes. There are so many variables I could change, there seems to be no limit to the number of different shapes I could produce.

Here are a couple of examples (which happen to be important in electronics).

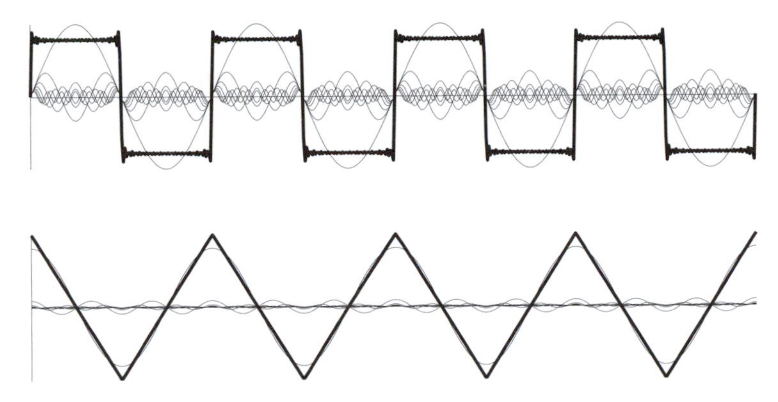

There is one small point I would like you to notice here, for future reference. The general overall shape of these combination curves is determined by the low frequency components. But to get sharp corners, I had to add in a lot of high frequencies. We'll come back to that observation in later chapters.

Anyhow, the impression I want to leave you with is that, with enough patience (and computer-assisted drawing facilities), I could construct the shape of *any* periodic oscillation at all, by adding together simple sine curves of the appropriate amplitudes. I have tried to make this plausible. It was Fourier who first *proved* it.

Harmonic analysis

The **Fourier theorem**is a formal statement of this result, and it can be expressed as follows:

- Any periodic oscillation curve, with frequency f, can be broken up, or **analysed**, into a set of simple sine curves with frequencies f, $2f$, $3f$, $4f$, $5f$ ...etc, each with its own amplitude.

  Conversely, adding together this set of sine curves will reconstruct, or **synthesize**, the original oscillation curve.

In musical terms, what I have just been doing is synthesizing complex musical sounds from pure tones. This is exactly what is done by the new breed of modern electronic instruments, and that is why they are called **synthesizers**. But now I want to look at the converse procedure—**analysis**.

The component sine curves, into which the original curve can be analysed, are called a set of **harmonics**. The first is the **fundamental** (also called the **first harmonic**), the next is the **second harmonic**, and so on. You will appreciate the reason for this name if you look at the ratios between the frequencies of successive members of the following table.

COMPONENT	FREQUENCY	RATIO	INTERVAL
Fundamental	f		
		2/1	octave
2nd harmonic	$2f$		
		3/2	perfect fifth
3rd harmonic	$3f$		
		4/3	perfect fourth
4th harmonic	$4f$		
		5/4	major third
5th harmonic	$5f$		
		6/5	minor third
6th harmonic	$6f$		
		7/6	(3– semitones)
7th harmonic	$7f$		
		8/7	(2+ semitones)
8th harmonic	$8f$		
		9/8	major tone
9th harmonic	$9f$		
		10/9	minor tone
10th harmonic	$10f$		
			etc. . .

In other words, this set of components forms a perfect **harmonic series**. You might recall that I made essentially this same point when I was talking about the notes a horn can play in Interlude 1. Therefore in this context the word ‘harmonic’ is exclusively used to mean a frequency which is a whole number multiple of the fundamental.

Clearly I am talking about musical overtones now. But before I get into that there is a small point of nomenclature here that is worth being fussy about. Which particular overtones an instrument will play depends on many things—how it is made and what the player does to it. The lowest frequency (the pitch you hear) is called the fundamental; the next the *first* overtone, and so on. So the words ‘overtone’ and ‘harmonic’ don’t have the same meaning. Back in Interlude 1, I showed that, for a horn (or any conical tube of air) the first overtone is the *second* harmonic, the second overtone the *third* harmonic, and so on. But for a cylinder closed at one end, the first overtone was the third harmonic, the second the fifth, etc. For a didjeridu the first overtone (which you recall was usually a tenth up) is therefore not a harmonic at all.

Please be clear about this last point. If a note can be kept perfectly steady (by blowing or bowing) so that it is genuinely periodic, then the Fourier theorem says that its overtones must be harmonic (whole number multiples of the fundamental). It cannot be otherwise. But when the note is not steady (when it is plucked or hit, or when it increases, decays or otherwise varies) this need not be true. I have already pointed out that the overtones of a piano string are a little sharp, therefore not quite harmonic. Those of a drum or a bell are not even nearly so.

Incidentally, in many books on theoretical music, you will come across the word 'partial'. More often than not it means what I mean by 'overtone'; but in some cases the author seems to mean 'harmonic' instead. Therefore I think it's best to avoid that word altogether.

The importance of all this to music is that we can now describe with some sort of precision, what the *timbre* of a steady note means. I showed you previously the oscillation traces for a flute, an oboe and a violin, all playing Concert A. But it is difficult to interpret such pictures; and I have already said that they aren't completely reproducible. However, any such shape can be analysed into a set of harmonic components.

The easiest way to exhibit the information about what overtones are present and what their amplitudes are, would be on a kind of *bar graph*. In the following diagrams I will represent the amplitude, or loudness, of each of these components by the height of vertical lines drawn against the corresponding frequency. (Actually, for reasons which I discuss in Appendix 4, it is sensible to use a *logarithmic scale* for this.)

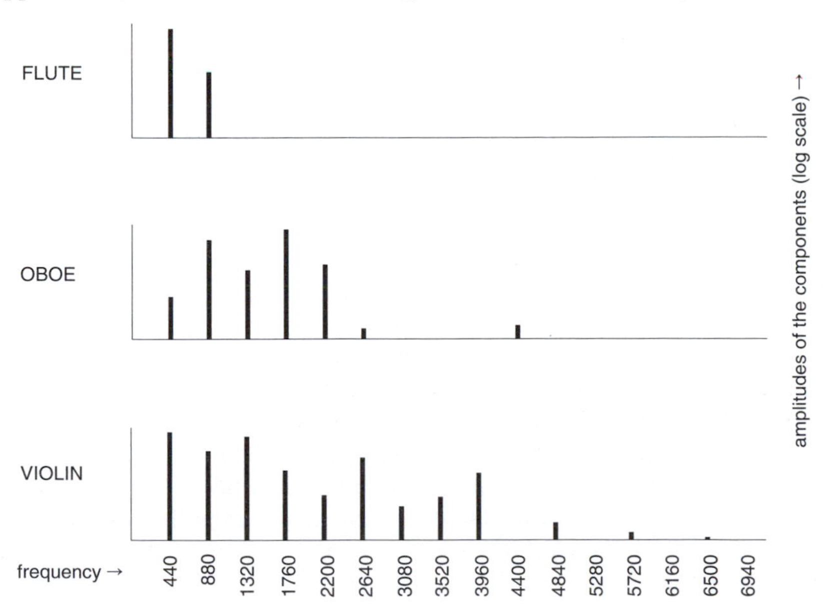

Such diagrams are technically known as **harmonic spectrums**; and the important thing about them is that they are reasonably reproducible by players playing their instruments in the same kind of way.

This gives us a completely new method for visualizing musical notes. And furthermore it is easy to interpret such diagrams. You can see at a glance that the flute has only one significant overtone—the second harmonic. The oboe has more, with the second and fourth harmonics particularly prominent. The violin on the other hand has a very rich spectrum with strong harmonics up to the fifteenth. What the players are doing to achieve this I can't say, but these are the same terms in which a trained musician would describe the tones. A less precise listener might simply say that the flute's tone is very 'thin', the oboe's 'bright' and the violin's 'rich'.

What I have been discussing for most of this chapter represents the spectacular advance in the understanding of musical sounds which slowly accumulated during the 18th century. But it wasn't necessary to wait until Fourier had tied the final theoretical loose ends before this knowledge could be used by practical musicians. For, while all this was in the air in the first half of that century, Jean-Phillipe Rameau made a gigantic step forward in the theory of musical harmony.

Rameau and the science of music

The foremost French musician of his time, Rameau had a career unlike that of any other major composer in history. Born in 1683, and by all accounts musically precocious as a youngster, he held various posts in different parts of France, but remained essentially unknown until the age of forty. Then in 1722 he published his famous *Treatise on Harmony*, so that when he moved to Paris in the following year, he had a reputation as a theorist rather than as a composer.

That proved something of a handicap, for the public seemed to believe that his airs, dances and musical comedies, being the work of a philosopher, were not supposed to be enjoyed. It wasn't until 1733, when he was fifty, that he started producing the operas and ballets (twenty-four in all) for which he became famous. At first he was considered daring and innovative, but as time went on his popularity grew and he ended up being regarded as typifying all that was good in French music.

In the closing years of his life he returned to his theoretical writings and reworked his early ideas into what he liked to call the 'science of music'. He died in 1764, a perfectionist to the end; and it is a fondly told story that, on his deathbed, he chided the priest who had come to give the last rites for his poor chanting.

His theoretical work is what is of most interest to us, because it has had such a profound influence on Western music ever since. But, despite the reputation of his music for clarity and grace, I must say that I find his writings are very hard to follow. Part of the difficulty comes because his early work is described in terms of the semi-mystical theories of numbers that Zarlino had passed on from Pythagoras. It was only in his later books, after he had finally discovered Sauveur's work, that he could (to his great delight) present a 'natural' basis for his ideas: and it is in this light that I will try to describe them.

But there is one thing to keep in mind. A work of musical theory is not the same as a scientific treatise. It doesn't deal only with 'objective' matters, but also contains a strong element of aesthetic judgement. I will not attempt to comment on the latter at all, and for the former, I will try to describe what was new in Rameau's work in the kind of terms in which I understand them, using the concepts I have just been developing (which musicians may find rather idiosyncratic).

The first key feature is the idea that the **chord** is the basic element of music. Now any real note is always accompanied by overtones, and so you shouldn't think of the note in isolation from its harmonics. The same must be true for a chord, but its spectrum will be a very complicated affair. For example, the chord consisting of a tonic plus a fifth will look like this:

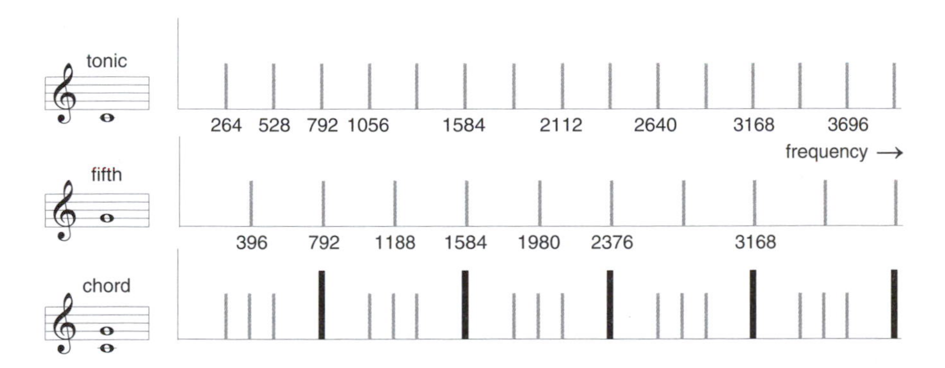

A word of caution. What I am drawing here are not 'real' spectrums, in the sense that they do not indicate the actual amplitudes of the overtones—that would depend on what particular instrument is playing the note. Here I am merely trying to show which overtones are present. But the important feature to notice is that some of the harmonics overlap and reinforce each other.

Now the ear, like the eye, is very good at noticing features which are different from those around them. Consider how hard it is to see a candle flame if you hold it near a strong electric light. Yet if you move it away slightly, so that the background is not so bright, your eye picks it up easily.

Well it seems plausible that, whatever is the mechanism in your ear which responds to overtones, it should notice particularly those which reinforce one another and are therefore stronger than their neighbours. So when you hear the tonic and the fifth sounded together,your ear will probably register the pattern of overlapping.

If I now add a major third, I get the most important chord of them all, the **major triad**:

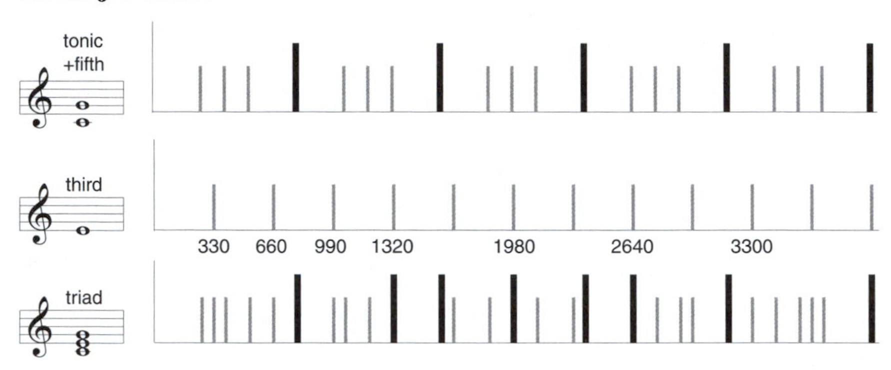

The spectrum is getting very complicated, but there is a definite pattern in the way the notes of the chords 'fit together' through the overlapping of their harmonics. What is very obvious here is that the major triad is particularly rich in such overlaps, and presumably this is somehow related to the fact that it is such a satisfying chord to listen to.

The next key feature concerns what is called the **inversion of chords**. If you construct triads with the same three notes, but in a different order, you will get these patterns of overtones:

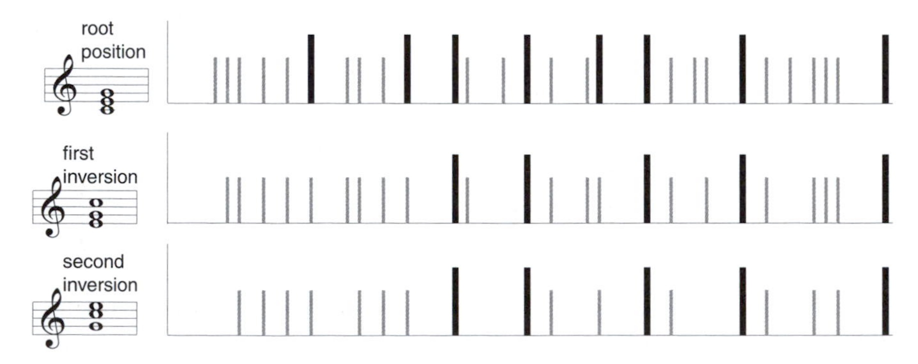

Clearly, these chords are not the same; yet the interesting point is that a lot of the pattern is common to all three. So, if indeed your ear responds to this kind of regularity, it ought to recognize this. Rameau proposed (and later musicians have concurred) that these three should be considered, in some fundamental way, the 'same' chord. Furthermore it would seem useful to talk about these chords as though they all had the C as their lowest note, even though two of them do not. Rameau called this note the **fundamental bass**.

The next logical step is to consider the relation of different chords to one another. Of particular interest are the three triads from which the major scale is constructed: those based on the tonic, the fourth and the fifth (as I outlined in Chapter 1). From these are derived the most important chords in the theory of harmony:

They are called, in order left to right, the **Tonic chord**, the **Subdominant chord** and the **Dominant seventh** (which is just the triad on the fifth with the seventh of the scale added). The tonic chord is repeated to round off the sound.

Let me draw the spectrums for these chords, but this time I will expand the vertical scale, and butt them up against one another so that you can see the overlap in their overtones.

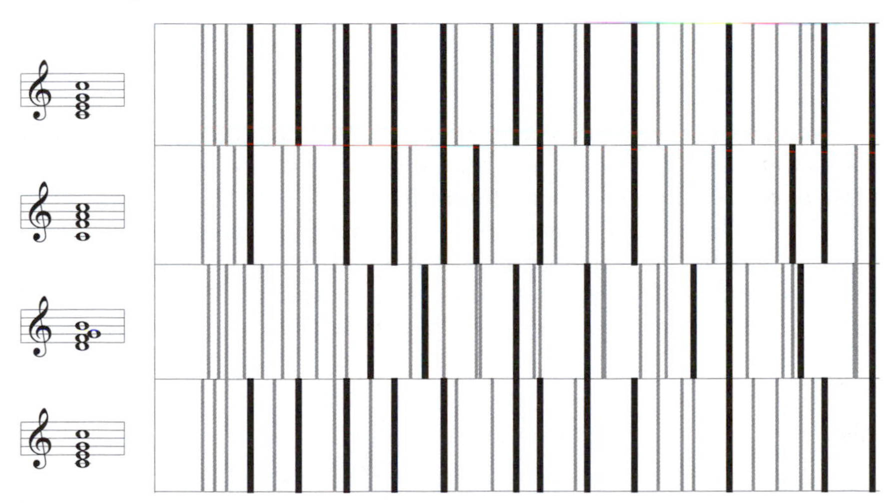

Again there is still a high degree of overlap. Some overtones are common to all of these chords, many more are shared between different pairs. This raises an intriguing conjecture. If your ear does respond to such patterns, then it will also recognize that when one of these chords is played immediately *after* another, there is a pattern to the change—again because of the overlap.

A visual analogy of what I am saying can be found in those fascinating drawings of the late Maurice Escher, like this one.

Here the transition from one figure to the other is made interesting (pleasing?) to your eye by the artist's drawing attention to the lines they have in common. Perhaps your ear might enjoy hearing some particular pair of chords in succession because of a similar overlap of fine detail.

Now this is a controversial idea. It implies that *harmony is the basis of melody*. Rameau believed this, but other musicians do not agree. I certainly wouldn't venture an opinion, but nevertheless it is true that large sections of the works of composers as late as Haydn, Mozart and Beethoven can be harmonized using just the three chords I wrote down above.

Now consider the question of **modulation**. In what I have been saying, the major triad G–B–D was considered as belonging to the key of C because it consists of notes from that scale. But for the same reason it belongs to other keys; and in particular the key of G, in which it is the tonic chord. So the pattern which I have suggested exists between related chords might be somewhat similar to the pattern your ear detects during modulation of the music between different keys.

This parallels another concept which Rameau introduced into musical theory—that certain chords perform particular jobs in the structure of a piece of music. He called it **functional harmony**. So, for example, when one chord, which could belong to either of two keys, occurs in the middle of a modulation between those keys, it was called a **pivot chord**.

Observe these few bars taken from *The Fitzwilliam Sonatas, No.1*, by Handel, from which I am quoting the piano part only.

You will notice that it starts in the key of B♭ major (the key signature has two flats), and ends up in C major on the tonic chord. In the third bar the triad F-A-C is the dominant chord of the initial key, but is also the subdominant chord of the final key. So it is the pivot on which the modulation turns.

Let me go back to the visual metaphor of the Escher drawings. Here is another small segment from the same work I used before (which incidentally is a lithograph entitled *Metamorphose*).

Notice that, in this drawing, the castle has a definite function: it is the feature on which the transition hinges. Just like a pivot chord. Perhaps the interest your ear finds in the transition from one key to another is similar to the delight your eye finds in looking at pictures like this. .

I can't possibly discuss all of the innovations that Rameau introduced, but I should at least mention the difficulties he found with minor harmonies. If I construct a **minor triad**, consisting of a perfect fifth and a minor third (the ratio 6:5 remember), these are the resulting overtones:

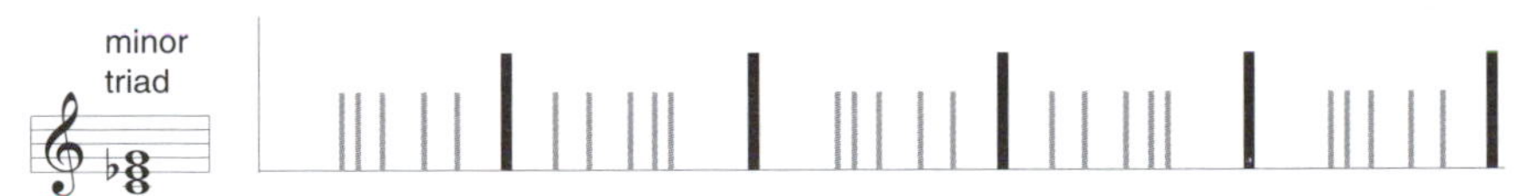

Again there is an obvious pattern. So the fact that this chord is considered harmonious is consistent with everything I have said so far. Furthermore it seems reasonable to me that your ear might react differently to this chord and to the major triad because this pattern has a quite different kind of regularity. But Rameau's difficulty lay in deciding why it should be harmonious at all, mainly because of his beliefs of how harmony was to be 'explained'.

Today, we are content to talk in a quite abstract way about 'pattern recognition'—based largely on our familiarity with the idea that it is something computers can be programmed to do. But Rameau was limited to less flexible concepts, like the numerical divisions of a string, or (later) the intervals that occurred naturally between harmonic overtones. Because the minor third did not occur early in the harmonic series, whereas the major third did, he could find no satisfactory 'explanation' of its harmoniousness. So he tended to relegate all minor harmonies, and minor scales, to a subsidiary position. This prejudice earned him a lot of criticism, but you can still see its influence in our music.

The legacy of the Enlightenment

It is often said that the Age of Reason ended in 1778 with the death of Voltaire, that celebrated author, historian, dramatist and poet, whose caustic wit earned him world wide fame as well as several sojourns in prison; yet who found the time, and the enthusiasm, to champion Newtonian science in France, and to himself write a commentary on the *Principia*. He was the living embodiment of the ideal that it was proper for a person of humanistic culture to understand and admire science, and for a scientist to love the humanities. The great social upheavals which occurred in the following decades changed that intellectual climate completely. It is from that period, if any, that the great split developed between what C.P. Snow has called the 'two cultures'.

For one thing, science expanded enormously and became more and more specialized. In earlier times it was possible for a single individual to understand all of natural philosophy and to contribute to many areas. Not any more. Chemistry, after shaking off the trappings of medieval alchemy under the guidance of Antoine Lavoisier, became a blossoming science on its own. The Swede, Carl Linneus, who died in 1778, was responsible for the classification scheme which systematized the study of biology; and the Scotsman James Hutton, who died in 1797, virtually invented geology single-handed. But even more importantly, that other revolution, the Industrial, had got under way in 1769, with the production of James Watt's steam engine. Science and technology became wedded to commerce. So, as scientific knowledge grew and grew, and as scientific education turned into vocational training, the Enlightenment ideal of the well balanced scholar faded.

Nowhere was this more obvious than in the divorce of science and music. Sauveur himself advocated that physics should concern itself with all sound, and not just that which was musically pleasing; so the field of study which had previously been called 'harmonics', came to be known as 'acoustics'. The quadrivium had ceased to be a formal part of university courses and with it the idea that musical theory should be studied as a branch of a wider intellectual discipline. Musical education started to be seen as mainly professional training and was relegated to its own specialist institutions. The Paris Conservatoire was founded in 1784 and the London Royal Academy of Music in 1822. Perhaps two of the last people to combine notable professional careers in both music and science were the Herschels, William and his sister Caroline, he an organist and she a singer, who in 1777, aged 38 and 27 respectively, took up the physical sciences to become the most distinguished astronomers in England—and, I might add, to discover a new planet, Uranus, the first for over 3000 years. The only other example I can think of is the Russian composer Alexander Borodin, who earned his living as a professor of chemistry.

So, by the beginning of the 19th century things had changed irreversibly. Natural philosophy was now a collection of different sciences; and music, together with the other arts, left the classical traditions behind and entered the Romantic period. The transition points may be difficult to tie down but nonetheless the changes were real. An era had ended. For me, all that was wonderful in that era, as well as many of its shortcomings and the abruptness of its passing, are symbolized in two figures whose careers had a number of curious parallels: Antoine Laurent Lavoisier and Wolfgang Amadeus Mozart.

Both were born to unusually intelligent and devoted parents and showed extraordinary ability in childhood. Mozart, born 1756, was an infant prodigy who from the age of six regularly toured the courts of Europe, under the relentless tutelage of his father.

In maturity he produced work of such magnitude that it is difficult to know what to say about it. His symphonies, sonatas, chamber music and especially his magnificent operas are an integral part of our musical culture. Even today, surveys carried out by researchers of popular taste usually show that his music is performed on modern concert platforms more frequently than that of any other composer.

Lavoisier, born 13 years earlier, was originally trained for the law but crowned an outstanding career as a student by choosing to specialize in chemistry. The works of his maturity are of a similar status to those of Mozart. It was he who changed Chemistry from a descriptive, cook-book kind of undertaking into a quantitative science. Chemical knowledge has increased a lot since his time of course, yet the basic laws of how substances combine with one another and in what proportions, are all attributable to him. He is rightly called the 'father of modern chemistry'.

Both of these brilliant careers were tragically cut short by causes stemming from the need for money and naïvety in dealing with others. Very early, Lavoisier decided that he didn't want the teaching and administrative duties that went with a university post, so he chose to finance his researches privately. He bought into a tax collecting enterprise which, in pre-Revolutionary France, was carried out (very profitably) by private businessmen. They were known as 'tax farmers', and were understandably despised. Given the political climate of the time, it would seem to have been an ill-judged move; but he made an even worse mistake in 1770. A journalist who fancied himself as a scientist, one Jean-Paul Marat, applied for membership of the Académie; and Lavoisier black-balled him because he considered his scientific writings worthless. After the revolution, when Marat came to power, he proved he had a long memory.

In Mozart's case, the sorry story of his poverty is well known. As his music matured it became more difficult and his popular appeal waned. He slipped further into debt. It has always been believed that had he been more astute in his dealings with members of the Viennese court, particularly the court composer, Antonio Salieri, he might have secured a position that would have supported him adequately. The idea that Salieri positively hounded him out of any such appointment has been voiced, in Rimsky-Korsakov's opera *Mozart and Salieri* (1898), and more recently in Peter Schaffer's play *Amadeus*. Whether or not it is true, it is plausible that things could have been easier had Mozart been more accommodating to the wishes of his employers; but then, had he been a different person, he might not have written the music he did.

The end to both these stories was sudden and tragic. In 1791 Mozart, desperately overworked to support the demands of his family and hopelessly in debt, contracted uremia and died within a few days, leaving unfinished his sublime *Requiem*. He was buried in an unmarked pauper's grave. Less than three years later the Reign of Terror caught up with the tax farmers—including Lavoisier. His chief prosecutor was Marat himself and he had no chance. He died on the guillotine. The arresting officer was reputed to have said:

> "The Revolution has no need of scientists."

Interlude 3

The violin

The musical possibilities of stringed instruments must have been realized from the day when the first caveman twanged the string on the first bow and arrow. From the earliest recorded times these strings have been cut in long thin strips from the intestines of slaughtered animals, hence the label 'gut'. They were very flexible, remarkably strong and relatively easy to make uniform; and when stretched near to breaking between two supports on top of a sound-board they could be made to vibrate at an accurately controlled frequency. In time it was found that many metals could be drawn out into wires which were even stronger and more uniform and which could be used for the same purpose.

Mersenne's laws were, of course, not known quantitatively until the 17th century, but their musical consequences had been recognized much earlier; and over the centuries a bewildering array of instruments evolved. So, while it can be said that, in principle, it is easy to understand how these instruments work, in practice, there is a vast amount of fine detail which science cannot explain fully to this day.

The development of stringed instruments

The easiest way to get some sort of overview of the different kinds of stringed instruments that are around today is to consider how they developed. But, because there are so many of them, I will group together the historical instruments according to how their strings are sounded, and we will be able to follow several different threads in this development.

(1) Instruments with open strings

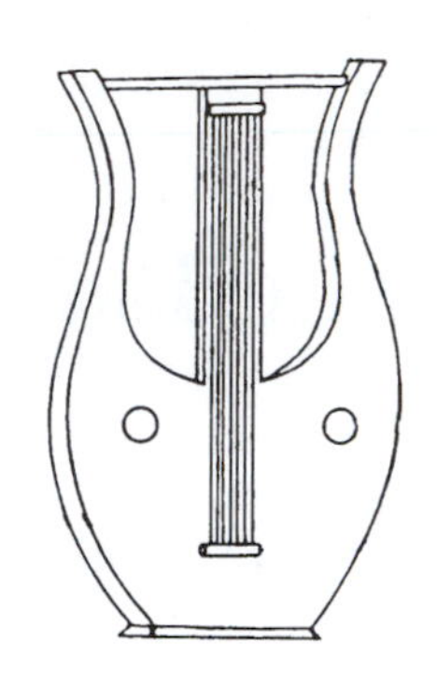

The very earliest instrument, known to have existed in ancient Sumeria and Egypt, was the **lyre**. It had between four and six strings, each tuned to a different note, stretched across some kind of sound-board onto a yoke. It was mostly used to accompany singing and was usually played by strumming with a plectrum held in the right hand. The instrument itself was held in the left hand in such a way that the fingers can be used to silence any strings not required to sound. In Classical Greece lyres had only four strings, which were all that were necessary to play the small number of acceptable chords.

It can't have been very long before adventurous musicians rebelled against the severely limited range of the lyre. The addition of more strings increased the number of notes that could be played and produced the **cithara** or **harp**, the biblical instrument beloved of King David. By early medieval times, small portable harps with about 25 strings and the characteristic sloping soundbox were very popular, particularly in Ireland and Wales.

But harps were still restricted to a diatonic scale and as modulation increased in late Renaissance music, they became less and less popular. They faced competition from similar instruments from Eastern traditions, the psaltery and the dulcimer, and when these eventually developed keyboards (as I talked about in Interlude 2) the harp was completely overshadowed.

Nevertheless it has survived as a specialist instrument. The modern concert harp has up to 47 strings and an elaborate system of pedals for changing the tuning of each one. It doesn't have a very extensive solo repertoire but it plays a prominent role in the modern symphony orchestra. Its sound is characterized by the long decay time of individual notes (because the strings are so lightly damped) and the player's ability to play chords spanning many octaves in spectacular *arpeggios*.

(2) Fretted Instruments

No one knows just when musicians realized that you could *stop* a string at any point with your finger, thus shortening the vibrating length and (*viva* Mersenne!) raising the pitch. Very early in medieval Christendom there appeared instruments with long **finger-boards** which had narrow raised ridges against which the strings could be pressed down. These **frets**, as they were called, were often no more than pieces of gut tied across the fingerboard perpendicular to the strings, and served the same function as did the movable bridges on the old monochord. But it is important to realize that they do more than merely tell you where to press down. The pad of your finger, being soft, will absorb energy from a vibrating string, so by pressing down *behind* the fret you ensure that the string has a hard surface to vibrate against and will not be so severely damped.

In the middle ages in Europe there were a great number of fretted instruments. They all had a relatively small number of strings; all were played by plucking with a plectrum; and all had capacious hollow sound boxes with elaborately carved holes to communicate the sound to the outside air. There were many individual differences, but three main lines of development can be discerned.

The first line was characterized by an instrument called the **gittern**, which appeared in the 13th century. It usually had four strings and movable gut frets, and its sound-box was flat on top and bottom with incurving sides—whether for acoustic reasons or to make it easy to carry we do not know. It was played with a small quill plectrum, usually of goose feather. It was used primarily as an accompaniment to singing, and, although it appeared every now and then as late as the 17th century, it disappeared after the Restoration.

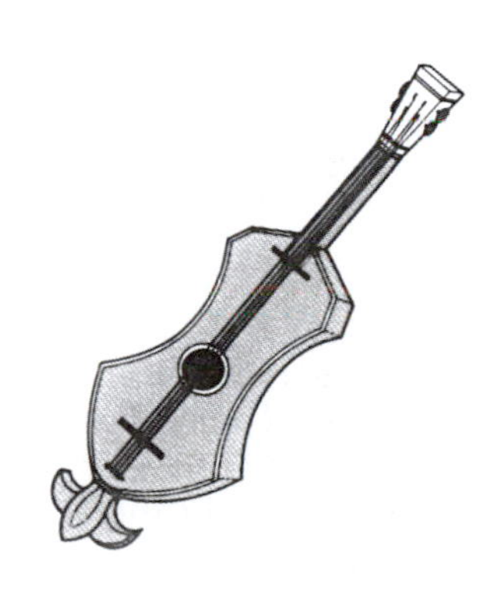

By the 16th century a small instrument with four pairs of gut strings was popular in southern Europe, particularly in Spain, called the **vihuela** (sometimes referred to as the **guittara**). Traditionally they were strummed, again with a quill plectrum, in complicated rhythms for dancing. For some time it was the most important solo instrument in the Spanish court, and a lot of the music written for it survives. But the instrument itself did not.

As time went on the number of strings settled down to six, and the frets became fixed and made of metal. In the 19th century more and more composers started writing for these instruments, which meant that increased loudness was called for and the sound-box became greatly enlarged. The result is the well-known modern **guitar** which, because of its association with pop music, is arguably the most widely played instrument in the Western world today. Nor has its evolution stopped there; but the **electric guitar** is a modification which I will talk about later.

The second line of development can be picked up in Renaissance Italy with one of the most popular instruments of its day, the **cittern**. It was entirely strung with metal wires, usually six pairs. It had fixed metal frets and was again plucked with a plectrum. Since the vibrations of metal strings are relatively energetic, the sound-box, which was traditionally pear-shaped, could afford to be very shallow without sacrificing loudness. The cittern was therefore robust and inexpensive; and as players could get many interesting sounds from the metal strings, it was very popular, particularly among working class musicians. In fact it seems to me that it shares many features with the modern electric guitar. Nevertheless it faded from use in the 18th century, and you have to look for its descendents among various folk instruments like the Rumanian **bandura** and the Austrian **zither**.

The third line came from the Moorish occupation of Spain—the instrument which the Arabs called *al 'ud* and which we know as the **lute**. For nearly 400 years, until the 18th century, it held the unquestioned place of honour in music, second only to the human voice. It was without equal as a court instrument and had an enormous repertoire written for it, only a tiny fraction of which is heard today. It reached its most sublime flowering in Elizabethan England where lived the greatest lutenist composer, John Dowland.

The lute usually had six double gut strings with movable gut frets. But the thing that gave it its special character was the particularly deep sound-

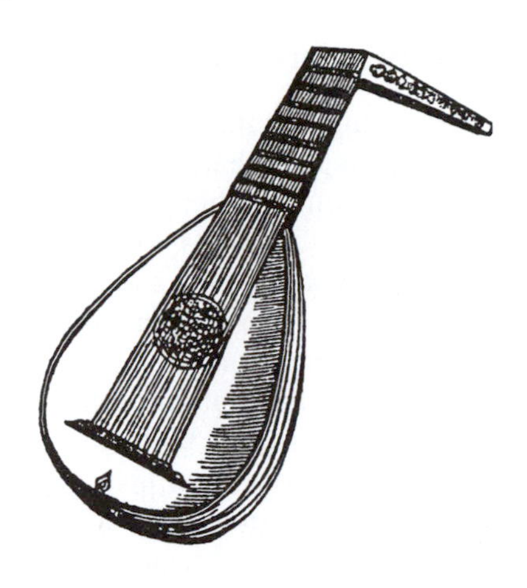

box with its unmistakably rounded back and elaborately carved rose-hole in the front. Its construction was incredibly light, the back being built up of a series of paper-thin ribs, bent and glued together edge to edge. Such a delicate structure couldn't withstand too much strain, so the dozen or so strings were quite loosely tensioned. This meant that they had to be relatively thin to compensate, and the highest string in particular was so fine that it was in constant danger of breaking.

When first introduced the lute had been a sturdier affair, primarily an ensemble instrument. But as it developed, its soft delicate sound made it a natural accompaniment for the human voice. Then in the 16th century the new *basso continuo* became popular and a bass version of the lute, the **theorbo**, was developed which could give stronger harmonic support to the voice, especially the bass line. So lutes became chordal instruments. But the new kind of music required more than simple strumming. Chords had to be carefully selected from the appropriate strings; so the plectrum was discarded and players used their fingertips only. But this meant that extra care had to be taken to avoid all unnecessary damping—no grand sweep of fingers across the strings here. In the end, all of these things combined to produce a unique sound, of great sweetness and delicacy, of which the poet Richard Barnfield wrote:

> "Dowland to thee is dear; whose heavenly touch,
> Upon the lute doth ravish human sense"

Why the lute fell from popularity so suddenly in the early 18th century is not clear. Perhaps it is just that "all things have their day and cease to be." Nevertheless it did fall, and much of the music that had been written for it was lost. However interest in the lute has been revived in recent decades, and those of us who regret its passing can take some comfort from that.

(3) Bowed instruments

Considering that the hunting bow was the prototype for this whole class of instruments, it is hard to understand why it took so long for someone to think of using another bow to sound the strings. The idea seems to have arisen in Muslim lands. Possibly it developed from the plectrum itself. You can easily imagine that a long quill might have been used to produce some special effect on occasions by scraping instead of plucking the string: and it is a short step from there to equipping it with horsehair, or some such, to improve its frictional properties.

Anyhow in about 900 AD, Ibn Khaldun, in his *Introduction to History*, has a description of an instrument called the **rabab**. It was bowed by a string rubbed with resin attached to a bent shaft, the left hand stopping the strings and the right hand working the bow. It was played upright on the lap with the instrument facing away from the player. By the 11th century it had found its way into Byzantium and Spain, and, after the crusades, into medieval Europe.

Once there the bow caught on very quickly. At first it was used on existing instruments which had up to then been plucked—harps and lyres. It took longer for new types which were specially designed to be bowed to emerge. By about the 13th century there were a confusing array, of which the **fiddle** could be considered representative. It had a waisted box, with (usually) 3–5 strings. The bow (or **fiddle-stick**) varied a lot, but usually maintained the traditional hunting bow shape.

Others from that period, whose names you might come across, are the **rebec** and the **lira da braccio**. Although they differ in detail, they're pretty much the same in essential outlines.

Of all the bowed instruments of that period there is one that especially takes my fancy. It was called the **tromba marina** and consisted of a long triangular box fitted with a single string. It was, in fact, little more than a bowed monochord.

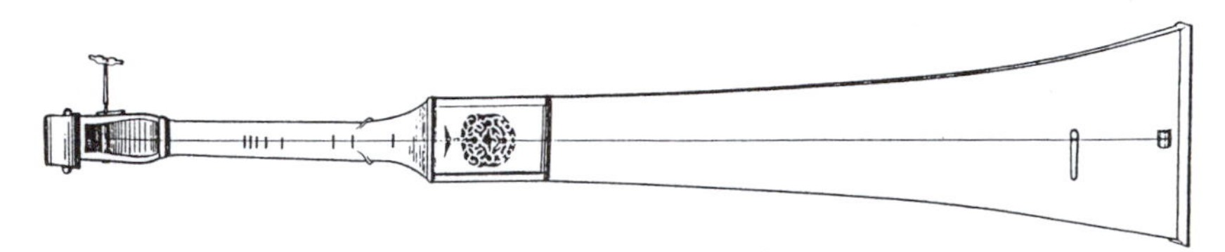

Because it was so long (often nearly 2 m) you couldn't bow it in the normal way. Instead you touched the string lightly with your thumb and drew the bow across the *shorter* length. For a loud note to result, *both* parts of the string must vibrate, but this can only happen at one of the harmonics of the open string (when the point you touch is a node). So the tromba marina played only a harmonic series, and this explains the 'tromba' part of its name, which means 'trumpet'—though there is doubt about what the 'marina' stands for. However, although it did develop a repertoire of its own, it was never terribly popular, being mainly of interest to theorists (like me).

There seemed to be a long period in which musicians experimented with ways of getting a comfortable position for holding the bow. Eventually two distinct solutions were found.

First was the *da gamba* position (in Italian *gamba* means 'leg'). The body of the instrument was held upright on, or between, the knees, and the bow was pushed horizontally across the strings using an underhand grip.

Then there was the *da braccio* ('on the arm') position. The instrument was held horizontally to the shoulder and the bow pulled almost vertically down with an overhand grip, which allowed the player to exert more force on the strings.

Towards the end of the Renaissance, two different families of instruments emerged which gained precedence over all the others—one to each of these solutions.

The first was the **viol** family. These developed from the Spanish vihuela and kept the same waisted, flat-bottomed sound-box. They had six strings, tuned a fourth apart, which gave them a very wide range of notes they could play. So wide was this range that for a long time only three members of the family were considered necessary—bass, tenor and treble. Later a kind of double bass, the **violone**, was added. They were all played in the *da gamba* position, sitting down.

By the beginning of the 17th century viols had become very fashionable. Together they made a perfect instrumental combination—sonorous and sustained—just the thing for Renaissance polyphony. Their wide fretted fingerboard reduced the technical demands on players, so they were ideal for social music-making. All in all, they were suitable instruments for ladies and gentlemen—for amateurs—and viol playing became a standard part of a good education.

At the same time a second instrument emerged, using *da braccio* bowing. It didn't develop out of the viol, as many believe, but in parallel with it, coming from, if anywhere, the medieval fiddle. It was initially a lower class affair, used mainly for dancing; and above all it was a professional instrument.

Loudness was most important, so the strings were tightened, and to compensate for the increased strain, their number was reduced to four, tuned in fifths. It had to be portable so the box was made smaller, but the top and bottom were arched to compensate. Playing a melodic line was more important than chordal accompaniment, so the fingerboard and bridge were curved to make it easier to play the individual strings. For flexibility of tuning frets were dispensed with; but the extra damping from the player's fingers was more than made up for by the extra pressure that could be exerted because the bow was held on top.

Thus evolved the **violin**, and as everyone knows it had the most incredible success story in the whole history of music. It reached its final form in about 1580 and within fifty years it was being played in church, chamber and opera house; it had become the foundation member of the modern string orchestra; and it had had written for it a wide and prodigiously difficult repertoire such as made it the envy of all other instruments.

The violin

What this instrument looks like is probably familiar enough, but since I want to talk about it in some detail, it would be wise to get the names of the various parts right.

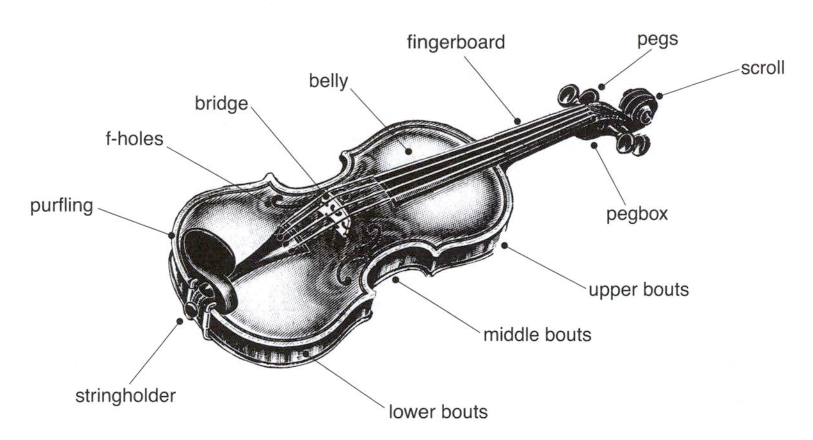

The sound-box is the item about which there is a lot of mythology. The top plate (or **belly**) is usually made of a soft wood like spruce or pine, carefully cut so that its grain runs lengthwise from peg-box to tail-piece. This is the main vibrating surface. The rest of the box—the ribs and the back—are made of a harder wood like maple; as are the neck, peg-box and scroll. The fingerboard is usually ebony. The arched shape of the plates is maintained by a long, specially shaped **bass bar** glued underneath the belly and a **sound-post**wedged upright between the two.

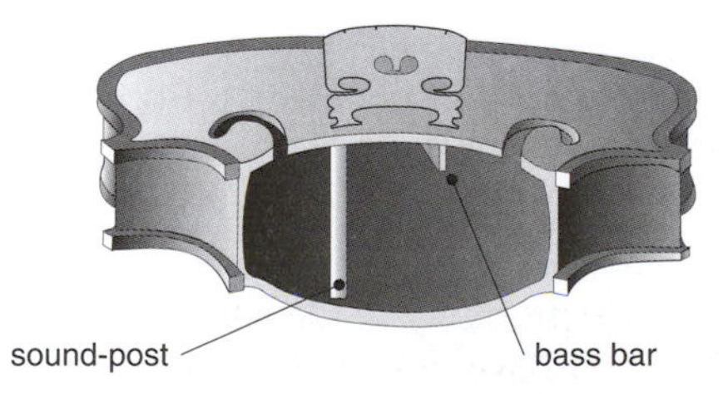

The whole box is varnished and, especially in the finest specimens, can be an exquisitely beautiful piece of craftsmanship. But the big debate among theorists has always been: how much of all this serves an acoustic function, and how much is purely decorative?

There are those who argue that the shape of the waist came about simply to allow the bow to move over the strings freely; but we have already seen that this shape actually predated bowing. So I think we must accept that violin makers decided on the present geometry of the box after much experimentation and deliberation, to give exactly the sound they wanted.

In the early 16th century craftsmen in the north of Italy were already making instruments with what we would recognize as real violin characteristics. In 1560 the king of France ordered 38 stringed instruments from one Andrea Amati from the town of Cremona, just 80 km from Milan. This order evidently enabled his designs to reach positions of great musical influence and ensured the fame and profitability of his family's business for the next four generations. The Amatis were responsible for much of the classical violin shape—flattening the body, deepening the middle bouts, sharpening the corners and turning the sound-holes into a more elegant shape. Their designs reached their highest peak of perfection during the time of Andrea's grandson, Niccolo.

Cremona became the acknowledged world centre of the industry. In the period from 1690 to 1750, one particular square in that town had eight different violin makers' workshops around it, whose names are to us today symbols of legendary genius and fabulous price-tags; but during that time a prospective buyer, in a single morning's shopping, might order an instrument from Amati, Stradivari, Guerneri, Bergonzi or Ruggeri. Eventually the art of violin making spread throughout Italy and beyond, but Cremona always remained its centre of gravity.

The figure of Antonio Stradivari is worth special attention. He began as Niccolo Amati's pupil in about 1660, and started signing his own violins in his twenties. Right from the start he experimented with each new instrument he made, constantly striving for perfection, of appearance as well as sound. His geometrically exact drawings survive to this day. He selected only the finest woods and worked them with exquisite care. Different parts of the plates he meticulously shaved to the exact thickness necessary for the box to vibrate equally for all the overtones he wanted. The resulting instruments had a tone that was full and mellow and a beautiful timbre that has never been surpassed.

Stradivari charged only modest prices for his instruments and therefore was never short of work. Notwithstanding the care he lavished on each one, he turned them out at an average rate of two per month, so that when he died at the ripe old age of 93 he was as famous in Cremona for his wealth as for his skill. He probably made about 1200 violins in his working life—of which about 550 have survived in playable condition, and fetch astronomical prices when they change hands today.

Now what I find interesting is the fact that, after Stradivari died, evolution of the violin stopped. Later makers ceased experimenting, spending all their time trying (unsuccessfully) to equal his standards. In the middle of the 19th century, there were a few minor changes. Necks were made a little longer and sloped backwards more, and bridges were made a fraction higher (and at the same time many old instruments were 'improved' in the same way); but apart from that there has been no real change in the violin in 250 years. That is truly remarkable. I can think of no other technological artifact of comparable complexity which is still in common use today and which has undergone so little modification since the early 18th century.

So we have to be circumspect when we set out to 'understand' how a violin works. Albert Einstein, arguably the most important scientist of the 20th century and a passable musician himself, admitted he knew of no way that mathematical formulae could express the principles involved in designing or constructing violins. Nevertheless I think it is possible, and useful, to try and get some kind of qualitative understanding.

The bow

Let us start by considering the bow itself, because what gives violin music its distinctive character—its sustained singing tone and its wide dynamic range—is the player's bowing technique. Rapid oscillations of the bow can produce a tremolo, a fast repetition of the same, or different, notes; up and down bows differ in attack; a phrase is shaped by playing a group of notes together with one pull of the bow, or by detaching one note and giving it a stroke by itself; the bow can hammer the string (*martellato*) or jump lightly on it (*saltato*) or turn over and play with the wood (*col legno*). Altogether the bow gives the player an astonishingly wide range of resources of great subtlety.

To this day horsehair is still used in stringing bows because no artificial fibre has been found to match its special frictional properties. Under a microscope, horsehair can be seen to be covered with tiny scales, all overlapping in the same direction. If you rub your finger against the hair, opposite to how these scales lie, you can feel an appreciable resistance; but if you rub in the other direction (that is with the scales) the resistance is noticeably less. This is true for most kinds of hair: pull one out of your own head and try for yourself. This frictional force can be increased by rubbing with **rosin**—a solid substance extracted from the gum of turpentine. It deposits microscopic sticky granules onto the hairs which strengthen the adhesion with any other surface pulled across it (in either direction).

Modern bows have a flat bundle of hairs stretched between their ends, half of them carefully aligned as they originally were in the tail of the animal, and the other half running the opposite way. Hence the bow can be used either forwards or backwards— the best bows are those in which the frictional force is the same in both directions.

The shape of the modern bow, which was changed during the 19th century, has the **stick** slightly arched *towards* the hair rather than away from it as in the traditional hunting bow.

The reason for this change was mainly mechanical. When you press down with the bow onto the string, the bundle of hairs itself bends; but this is counteracted by the tendency of the stick to *straighten*. In the older shape the tendency would be for both stick and hairs to bend even further. The virtue in keeping the hairs always as straight and taut as possible is simply to give the player the same control when playing loudly as when playing softly. There is therefore a great deal of skill in making bows. A good bow is prized no less than a good instrument and the names of great bow makers are equally renowned as those of great instrument makers.

Before I launch into a description of how the bow does its job, I must point out one important piece of information. There is a well-known law in the physics of friction which says:

- Moving friction is less than static friction.

This means that the frictional force between two surfaces is *less* when they are moving over one another than when they are not moving. You may not find this immediately obvious, but think about pushing a heavy box across the floor. It takes a lot of effort to get it moving, but once you've got it going you don't have to push so hard. The reason is that, on a microscopic level, friction is a kind of sticking process. Any kind of glue needs time to set; so if the surfaces are moving over one another, the adhesion doesn't have time to take effect properly.

On casual inspection, when the string vibrates, it seems to move in a long shallow curve, rather like the first mode of a skipping rope. But in fact the actual motion is more complicated. On careful inspection you can see that the string divides into two nearly straight-line segments, with a sharp bend between them. This bend races around the curved path that you see, making one complete round trip in each period of the vibration. I've tried to describe this in the diagram opposite.

It is not difficult to understand at least some of this motion. When you bow, the friction of the bow hair grabs the string and pulls it aside. But the further out the string is pulled, the greater is the restoring force pulling backwards. Eventually the friction can no longer maintain its hold. The string snaps back and starts to vibrate.

From then on a regular pattern is set up. As the string moves backwards it slides easily underneath the bow (because *moving* friction is low); but when it turns to swing forwards again, it finds itself moving at the same speed as the bow. *Static* friction takes over and the string is dragged forward until it breaks away again and the cycle repeats. It is behaving rather like a child being pushed on a swing. Energy is being fed in each cycle, which just balances the energy being lost. The vibration will be sustained as long as you keep the bow moving.

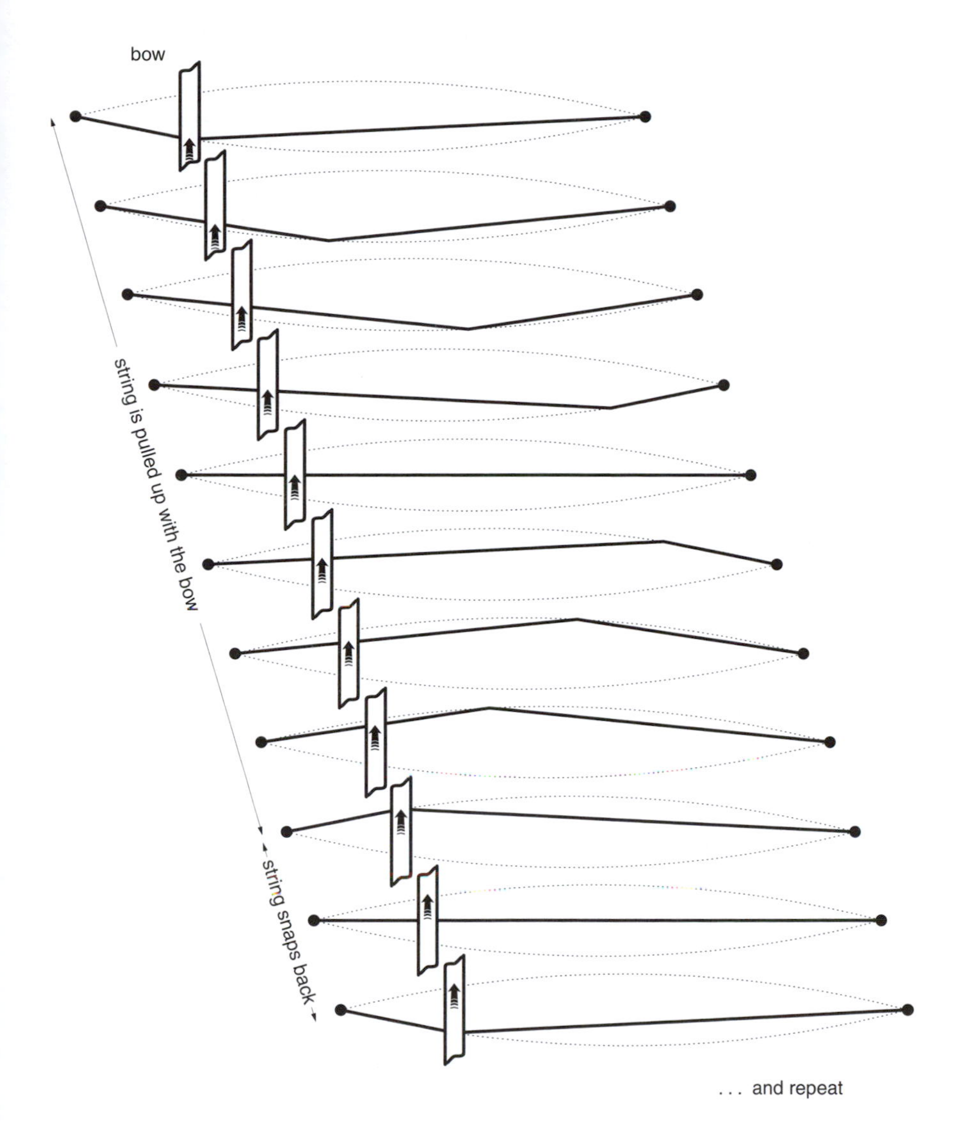

The motion of that part of the string in contact with the bow is perfectly periodic, but, as you can see from the diagram, it is certainly not simple harmonic. It doesn't behave like a pendulum—moving sideways, slowing down gradually, stopping, speeding up gradually backwards, and so on. On the contrary, it is pulled sideways *at a constant speed* (the speed of the bow), then it stops suddenly and snaps back very quickly. The actual speed with which it snaps back depends on how close the bowing point is to the end of the string: the closer the faster.

So the graph which describes its motion is nothing like a simple sine curve, but instead looks like this:

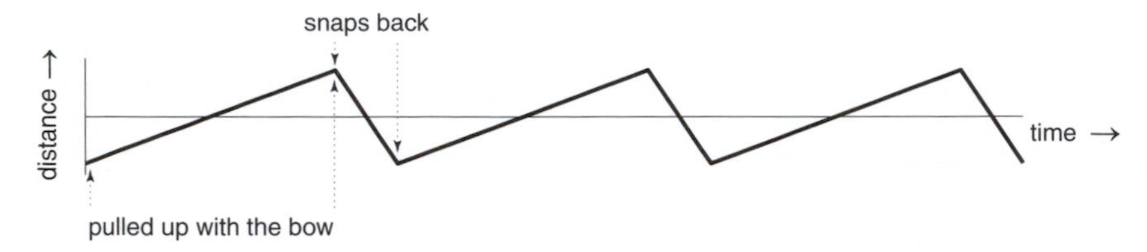

Anyhow, from my simple description of what happens, I think there are three basic deductions that can be made.

- By and large, you'll always get the same kind of (fundamental) vibration no matter *where* on the string you bow. Most players operate about halfway between the bridge and the end of the finger-board—that is about a fifth or a sixth of the way along the string.

- The position of bowing and the speed of the bow largely determine how loud the note is. Since the string is forced to move with the same speed as the bow, and since the time that elapses before it snaps back is fixed by the pitch of the note, that part of the string in contact with the bow moves the same distance each cycle, regardless of where you bow. Therefore if you bow towards the middle of the string the amplitude of vibration is limited to this distance. But when you bow closer to the bridge you get a much bigger overall vibration, and therefore a louder note.

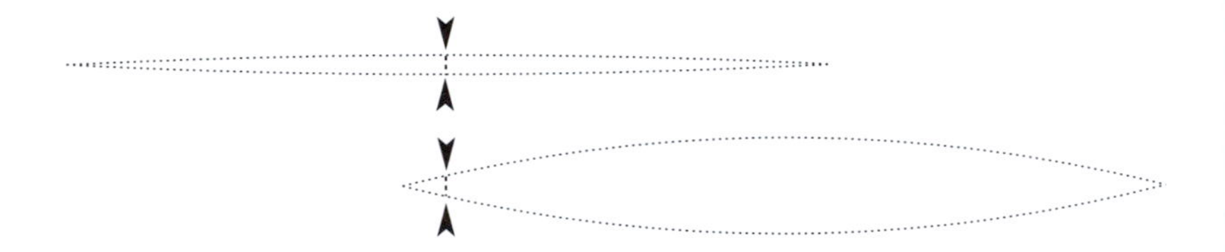

- The pressure you exert on the string is less critical. This is because the force of the bow mainly determines how much friction comes into play. If you press too softly the *static* friction won't be large enough to pull the string out the correct distance. So the vibration won't start properly. If you press too hard the *moving* friction won't let the string go when it should, and then all sorts of other unpredictable vibrations will ensue—you'll get squeals and squeaks. (Much the same will happen if you don't bow perpendicularly to the string; vibrations *along* the length of the string will be set up—more squeals and squeaks.) But in between you have a fair range of bow pressures to work with—except if you bow very close to the bridge when the pressure needed becomes greater and more critical.

The overtones of a bowed string

We are now in a position to deduce something about the timbre of a violin note. We know that the string moves in a periodic oscillation, so it should be just a question of using harmonic analysis. But before we start, please be clear that I am only talking about bowed strings; a plucked string would have quite a different pattern of motion and therefore a different timbre.

The key observation is that, no matter where you bow, the subsequent motion of the whole string looks approximately the same. But nevertheless any one part of the string moves in a different pattern from all the others. The part that we should concentrate on is that which is right next to the bridge, because it is the vibration of the bridge which is transferred to the body and directly produces the sound you hear. The motion of this part of the string is not the same as the diagram I drew on page 122 but rather, as you might guess, like this:

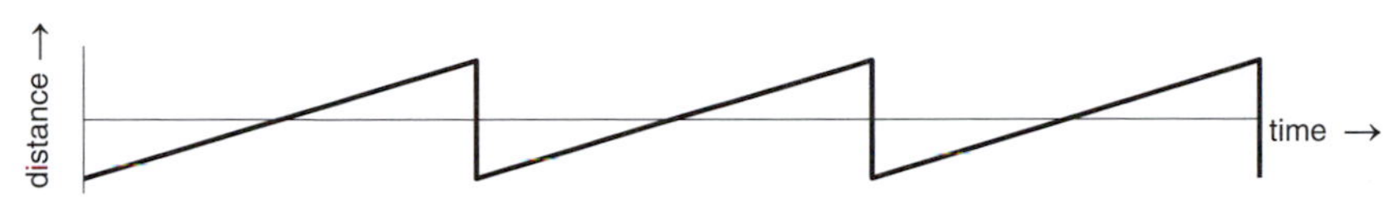

Now let us try to do a harmonic analysis of this shape. Since I haven't described the actual mathematical procedures, I will do it diagrammatically—step by step, adding one harmonic at a time.

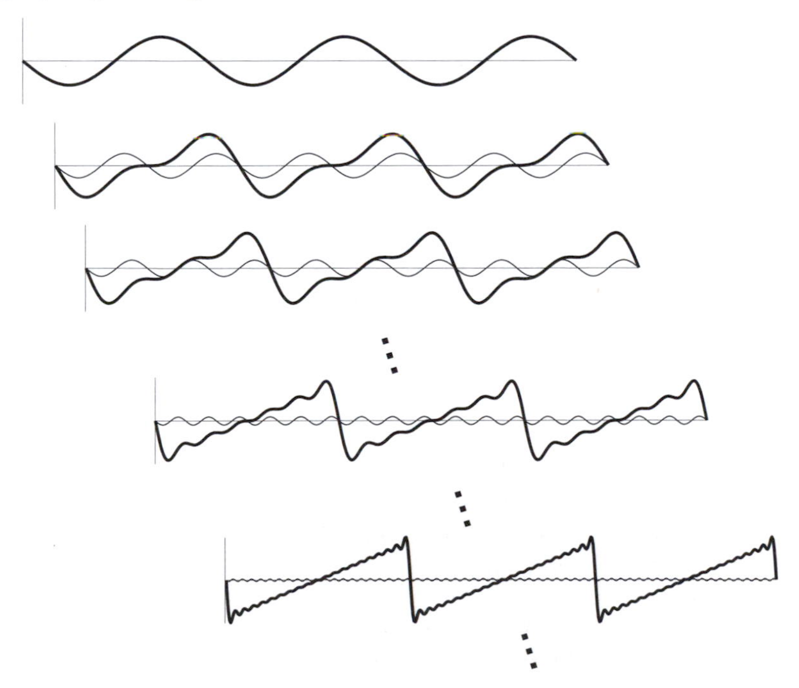

I would ask you to note especially that, broadly speaking, it is the low harmonics which are responsible for the general overall shape of the curve. The high harmonics chiefly contribute sharpness at the corners.

To determine what the amplitudes of the harmonics are, all you have to do is to measure the height of each component (with a ruler will do). You will find that they form a remarkably simple spectrum. If the amplitude of the fundamental is taken as 1 unit, then that of the second harmonic is 1/2; that of the third is 1/3; of the fourth, 1/4; and so on. In order to compare this spectrum with that of other instruments, it is conventional to plot the *logarithm* of these measurements against frequency. Why this is done relates to how your ear judges loudness, which is a subject I will take up later. For now it is enough to point out that the spectrums I showed in Chapter 4 were drawn by this rule, and the spectrum we are talking about here looks like this:

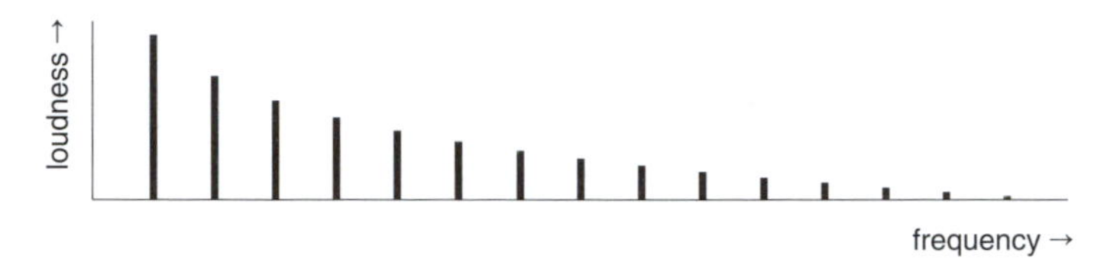

This discussion has been simplistic in one important respect. Look again at the diagram I drew on page 123, of the motion of the part of the string next to the bridge. It is a jagged curve, with very sharp corners. This shape implies that, when the bend in the string goes past you, the string changes its direction of motion extremely rapidly. Such behaviour is plausible, provided that the bow is operating right next to the part of the string you're looking at. But not otherwise. Inertia will always slow down rapid changes of motion in time. So if the part of the string you're looking at is not close to the bow, the bend will have had time to become less abrupt. Therefore this diagram should have sharp corners only when the bowing position is close to the bridge, and when it is further away the corners will be somewhat smoothed out or rounded off.

This has an important consequence. Since high harmonics are primarily responsible for sharpness in the vibration pattern, any smoothing out of corners must mean that the high harmonics are less prominent. So the spectrum of the vibration must be different depending on where, in relation to the bridge, you bow the string.

You must remember that this spectrum does not yet describe the sound you hear—only the vibration just before it goes into the sound-box. Even so, there are a few general statements you can make.

- The spectrum is rich in high harmonics. Even though their amplitudes decrease as they get higher, more of them remain appreciably strong than in the spectrum of other instruments (refer back to page 99).

- If you bow away from the bridge, towards the middle of the string, the vibration will suffer from the smoothing of corners effect, and the upper harmonics will tend to be lost. We also know that this bowing position gives a soft sound. Therefore if you bow towards the middle of the string, the timbre of the resulting note has a gentle character which composers often call for with the bowing instruction *sul tasto* (literally 'over the fingerboard').

- As you bow closer to the bridge the amplitudes of the high harmonics increase, but the pressure you have to apply becomes higher and more critical. Experienced players like this region of the string for its brilliance of tone, but beginners find it safer to play close to the fingerboard.

- Closer still to the bridge the pressure you have to apply becomes prohibitive and the whole pattern of string movement changes. The low harmonics become unstable and the sound turns into an eerie collection of high overtones which composers call for when they specify *sul pontecello* ('over the little bridge').

Incidentally, if you want to hear what *sul tasto* and *sul pontecello* bowing sound like, you can find them, as just one example that I happen to know, in the Debussy *Cello Sonata*.

The body

Once the string has been set in motion as the player wants, it is up to the body of the violin to use the energy of this vibration to get the surrounding air pressure oscillating so that we can hear the note—that is, it must act as an efficient **radiator**. (Some people call it an *amplifier*, but strictly speaking, that name should be reserved for a device which puts energy into the sound). Furthermore it must reproduce all aspects of the vibration as faithfully as possible, particularly the harmonic content. It should neither suppress overtones which the player has put into the note, nor add others the player doesn't want. Ideally the body should support all frequencies equally; but as you might guess, that is an unattainable ideal.

Consider the bridge for example. It is through here that the energy of vibration must pass. As the string swings from side to side, the bridge, being more or less pivoted on the sound post, oscillates in a small circular arc around that point. This forces the belly to vibrate in sympathy, in a vertical direction.

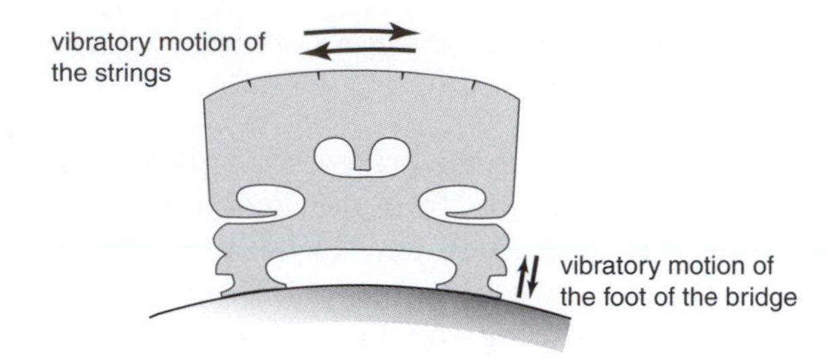

But if you think about it carefully, the bridge will do more than this simple oscillation. When you pull the string sideways, because its end is rigidly attached to the holder, the bridge will tend to be drawn very slightly parallel to the string, towards the neck. It will be drawn in the same direction when the string is pulled to the other side also. So as the string vibrates, the bridge will not only oscillate in a circular arc, but will also rock slightly in a direction parallel to the string, at *double* the frequency. This is an exceedingly small effect but nevertheless it means that the second harmonic of the string's vibration tends to be emphasized—already a small distortion.

When the energy reaches the box, the wooden surfaces are set vibrating, particularly the very flexible belly; and these efficiently produce pressure fluctuations in the surrounding atmosphere. The volume of air inside the box will also vibrate (just as inside a wind instrument) and this communicates to the outside through the **f-holes**. It is an interesting experiment to observe the reduction in loudness you get if you block these holes with tissue or cotton-wool. You should be able to observe that about as much of the sound comes from the f-holes as from the wood itself.

Any irregularly shaped volume of air or thin wooden plate will have its own natural frequencies at which it prefers to vibrate—for exactly the same sorts of reasons that a stretched string does. If you try to set them vibrating at one of these frequencies they will respond very vigourously. Technically they are said to **resonate**. But there is one important difference between these vibrators and a stretched string. They are much more complex in structure and the processes by which energy can move around inside them are almost infinitely varied. They tend not to resonate at clearly defined, single frequencies, but rather in a wide range (or **band**) around some average values.

Nevertheless each of these systems does have a 'fundamental' frequency band (where response is typically strongest) and higher, 'overtone' bands. As a result, the body of a violin will actually respond to vibrations of the bridge at any frequency at all, but in a very irregular fashion—sometimes strongly and sometimes weakly. And there are two frequency bands at which the response is particularly strong, which correspond to the fundamentals of the enclosed air and of the wooden surface.

The first of these, called the **air resonance**, , depends mainly on the volume of the box and the area of the exit holes. For all violins therefore it has much the same value, at around 280 Hz or just below D_4. The strings on a violin are tuned to G_3, D_4, A_4 and E_5. So the air resonance will tend to reinforce strongly the note of the open D string, and less so the succeeding notes.

The second, the **main wood resonance**, is much more variable (usually higher) and depends on all sorts of things about the wood and what is done to it. So where this resonance actually falls is an indication of the skill of the maker. In really good violins it is always found somewhere around 420 Hz, just below A_4. It therefore reinforces strongly the open A string. Furthermore it also reinforces the second harmonic of that frequency an octave lower, which means that it makes the open G string louder also. Therefore it seems that the secret of the great violin makers was in the compromises they made with the inescapable fact that no box can amplify *all* frequencies exactly equally. They experimented until they got the response to the high notes reasonably uniform (or at least as they wanted them to be), and for the low notes, where uniformity was unattainable, they ensured that the different *strings* had similar playing characteristics.

You can see this clearly if you measure the degree to which any particular violin box reinforces different vibration frequencies and represent the result graphically. (This is called a **frequency response curve** and I will have a lot more to say about these in Chapter 8.) Here are two actual response curves measured by the American researcher Carleen Maley Hutchins for a Stradivarius, and for a poorer violin of doubtful origin:

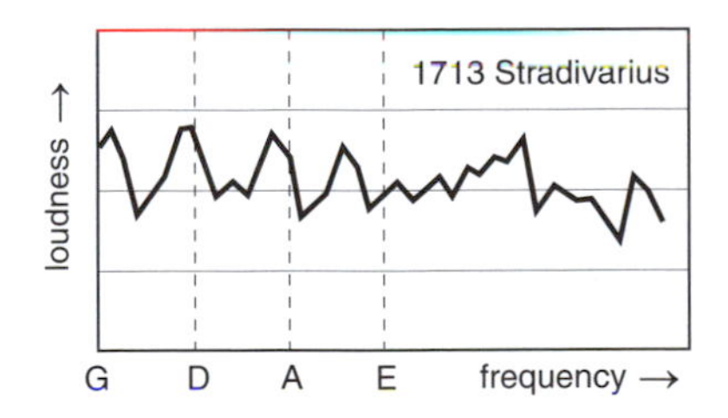

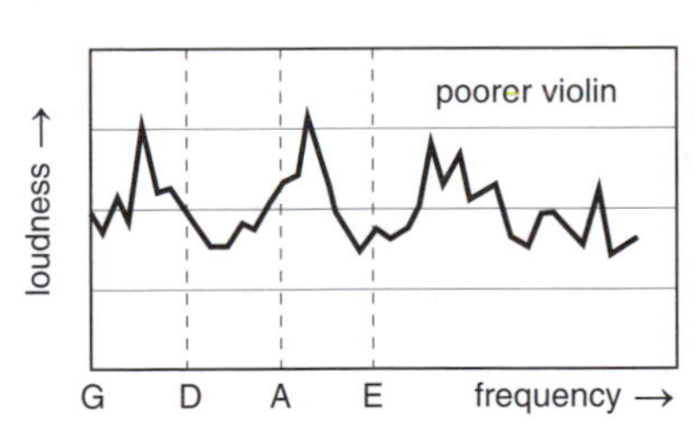

Notice that, whereas neither instrument reinforces all frequencies equally by any means—which is why a player has to spend a lot of time getting used to the instrument—the Strad is less extreme, with smaller variation between maximum and minimum response.

One last point. Most instruments of the violin family, including the very finest, often have one particular note at which there is an especially strong coupling between the strings and (usually) the main wood resonance. At this frequency the vibration of the string is very difficult to control, because energy moves backwards and forwards between string and box as each tries to make the other vibrate in sympathy with itself. As a result the note tends to warble unsteadily, often breaking into an octave higher, rather like the voice of an adolescent boy. This note is known as a **wolf tone**. Makers try to ensure that it falls exactly between two semitones on the normal scale so that it shouldn't ever be used; but tuning can change that, and then it takes an experienced player to tame it properly.

Other members of the family

Up till now I have talked exclusively about the violin, ignoring its close relatives—though of course the same physical principles apply to them all. The complete family, in order of increasing size, consists of:

- the **violin**, which is about 60 cm in overall length, with its four strings tuned, as I have already said, to G_3, D_4, A_4 and E_5;

- the **viola**, about 75 cm long, with four strings tuned a fifth lower to C_3, G_3, D_4 and A_4;

- the **cello** (more correctly the **violoncello**,), about 120 cm long, which plays an octave lower again, its strings tuned to C_2, G_2, D_3 and A_3;

- the **double bass** (sometimes called the **contrabass**), around 200 cm in length, its strings tuned (this time in fourths) to E_1, A_1, D_3 and G_2.

> The double bass is in fact closer to the viol family—as evidenced by its sloping shoulders and its being tuned in fourths. For that reason it is something of an outsider in this group; and when composers write for strings alone they commonly favour the 'complete' sound of the string quartet—in which the cello usually plays the bass line, the viola the tenor line and two violins take the soprano and alto parts.

We have seen many times that the most important characteristic of anything that vibrates in determining its frequency is usually its length. So you would think that, if you wanted to make a 'tenor violin' to play a fifth lower (like a viola), you should be able to do it simply by enlarging all the dimensions—string length, thickness of wood etc.—by a factor 3/2. This would make it about 90 cm long. Similarly a 'bass violin', which played music in the cello range (down another octave) would have to be twice as long as this, or 180 cm. The resulting instruments should, in principle, have very similar timbre to the violin itself. The trouble is that they wouldn't be easy to play since the human hands which have to play them cannot be enlarged similarly.

The construction of real instruments is a compromise. The viola is only 20–25% longer than the violin, and the cello is about twice as long; and the extra decrease in frequency required is attained by using thicker strings. However getting the tone right is not so simple. In particular, the main wood and air resonances are a problem. If the same body shape were used, these resonances would have much too high a value and wouldn't coincide with the notes of the open strings as they do in a violin. So in practice the shape is extensively modified.

Unfortunately the old Italian masters who perfected the violin never did enough work on the larger instruments to develop designs that would put these resonances in the right places. So the lower notes of the traditional viola and cello suffer somewhat through not being sufficiently reinforced. Viola players in particular seem to have developed something of an inferiority complex in regard to their instruments, and the profession abounds with derogatory "viola jokes". Just do a search on the Internet for 'viola jokes' and you'll see what I mean.

It is interesting that there is an active field of scientific research today, formulating **scaling** laws to design new instruments to overcome this drawback. The aim is to produce a whole family of different sized violins, each with exactly the same sound, perhaps with the thought of reviving much of the Renaissance music originally written for viols.

Many musicians would dispute my use of the word 'drawback' here. They would argue that the traditional viola and cello were never supposed to be simply scaled up clones of the violin, and composers have long recognized that they have individual characteristics. Mozart's quintets for two violas, for example, cannot be played completely satisfactorily on the new 'tenor violins'. One reason is that, in the ordinary viola, the air resonance falls between B and B♭ on the G string; and in the new instruments these strong tones don't occur where the players expect them, nor where Mozart counted on them to be.

So the timbres of the traditional violas and cellos are quite different from the violin and over the centuries they have developed their own definite personalities. Violas have always been considered less brilliant than the other two, and often used only to fill in harmonies. When composers bring them out of the ensemble it seems to be to suggest warmth and compassion rather than drama and excitement. It is typical that in Richard Strauss' tone poem *Don Quixote*, the viola plays the role of the solid and sensible Sancho Panza, while the more mercurial and romantic part of the Don is given to the cello. For the cello, though lower in pitch and normally the carrier of the main bass harmony in the orchestra, has always been recognized as a singer of melodies. It is frequently used in symphonic music to announce second subjects, and has an extensive concerto and solo repertoire. The double bass, on the other hand rarely steps out of the ensemble, being content most of the time to underscore the bass line; and only plays solo for extraordinary effects, as when John the Baptist loses his head in *Salome* (also Richard Strauss).

To sum up therefore, it should be obvious from my brief discussion of this family of instruments, that their manufacture is a mind-bogglingly complex business; and we can only marvel at the skill of the old masters. Nevertheless it must be remembered that no magic was involved. For many years there has been a mythology about the special properties of the Cremona varnishes whose secret recipes have since been lost. Properly controlled tests, however, have established, reasonably convincingly, that so long as it does its proper job of protecting the wood without interfering with its elastic properties, the varnish has little discernible effect on tone.

In the end what are important are the judgement and knowledge of the maker, which come from a lifetime of experience. Unfortunately such attributes are difficult to pass on to the next generation, as Stradivari's successors discovered. That is why I think it is most important for modern scientists to continue trying to understand the violin makers' art. After all, science is merely the knowledge and judgement of many lifetimes' experience, distilled into a form that can be passed on.

Chapter 5

Over the waves

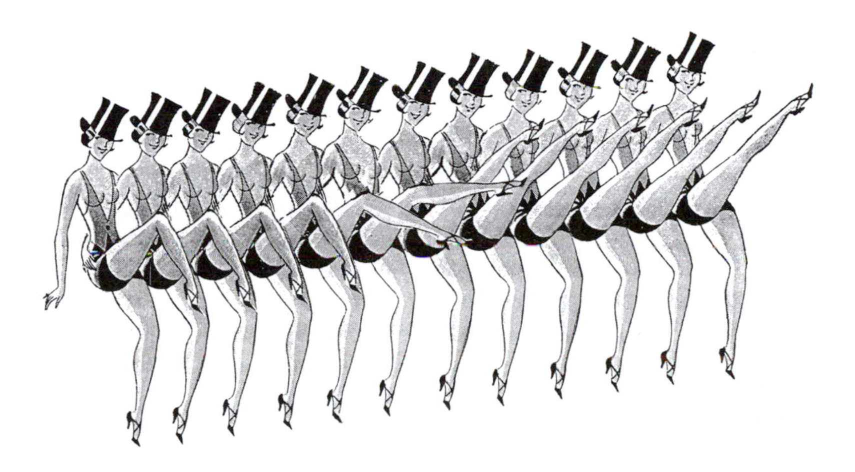

There is a theatrical manoeuvre much loved by choreographers in charge of well disciplined chorus lines. On the stage of the *Folies Bergères* or the Radio City Music Hall, or in the studios of a Busby Berkeley musical, the dancers arrange themselves in a long straight row. The one on the end starts off by performing some action—kicking her leg or doffing his top hat, it doesn't matter what. A fraction of a beat later the next one repeats the action, and after exactly the same time the next one does it. And the next, and the next, right to the end of the line. To the audience, a wave of legs (or top hats or whatever) seems to travel across the stage, even though it is perfectly obvious that the dancers stay exactly where they are.

Exactly the same thing can be seen in big sporting events. A group in the bleachers somewhere will stand up and cheer and wave their arms, or something. Then those nearby will copy them, and the next group, and the next. A wave of cheering and waving arms will travel around the stadium, keeping going as long as the audience is willing to participate. It is called a *Mexican wave*, having been given that name during the football World Cup held in Mexico in 1986. Again it is only the wave that travels round the stadium, not the football fans.

You can observe the same effect in the waves on a beach. Whichever way the tide is going, the waves come in and break on the shore. On every beach in every country in the world, throughout the hundreds of millions of years that the Earth has been in existence, it has been the same. The waves come in and never go out. Yet the oceans don't empty onto the land. Clearly, as with the chorus line or the football game, it is only the waves which travel and not the water itself.

The idea that sound must also travel in some such fashion goes back at least to Roman times. One of Julius Caesar's military engineers, Marcus Vitruvius Pollio, wrote extensively on the subject of theatrical acoustics. He seems to have been the first to suggest unambiguously that sound spreads out from any source like circular ripples from a stone thrown into still water. His works were lost for 1500 years (like so much else), but after their rediscovery Renaissance scholars soon verified his insight. They established that sound must have some medium to travel in (it can't go through a vacuum) but the medium doesn't travel with the sound—instead the only thing which propagates is a wave of localized disturbance.

In my account of the history of physics and music in earlier chapters, I have ignored this parallel scientific development, and concentrated on the growth of understanding of musical oscillations. So I will have to retrace my steps, through periods and people I have discussed already. But before I do that I want to talk more about waves themselves—in modern terms of course—and then it will be easier to see which parts of the historical story are important.

General features of waves

Waves occur in many different contexts and can take many different forms; but they all have some features in common. The best way to understand these is to think about the simplest medium in which you can get a wave to travel—a long rope (or garden hose). Lay it out straight along the ground, and then quickly jerk the end up and down once. A **pulse** will run along the rope.

If instead, you shake the end a few times, you will get a collection of such pulses going down the rope—which I will call an **extended wave**

packet. But if you smoothly vibrate the end from side to side, then you should be able to get a perfectly regular oscillating wave travelling continuously away from your hand—a **periodic wave train**. These three different classes of waves have straightforward analogues in acoustics—a click, a spoken word and a sung note.

The first thing to clarify is what the various parts of the rope must be doing to give the appearance that a wave is travelling along it. If you remember the chorus line, it should be obvious that each part is simply doing exactly what the previous one did a short time before. In the case of a periodic wave train, for example, each element of the rope is performing a simple oscillation, behaving exactly like a ball on the end of a spring attached to its original position. If I draw what each of these imaginary balls is doing at some particular time, then I get a clear picture of the shape of the wave you see.

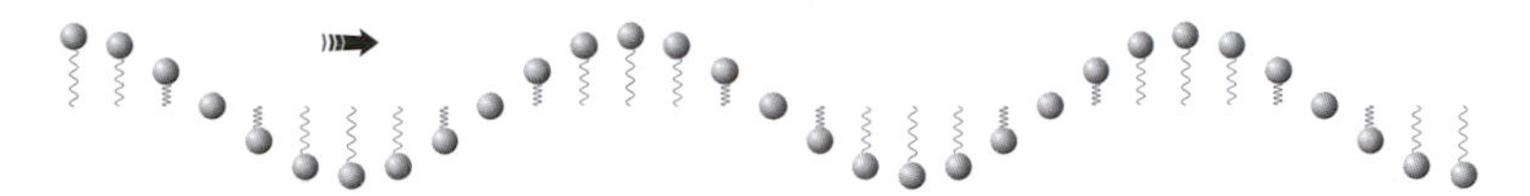

Since the motion of these elements is perpendicular to the direction in which the wave appears to travel, this kind of wave is called **transverse**.

Ripples on a still water surface are transverse waves, but deep water waves are different. In them the elements of water move in circles, behaving as though they were tied by inelastic strings to their original positions.

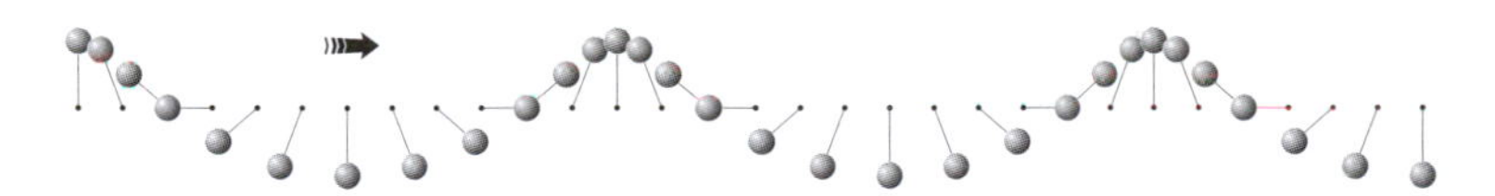

It is clear, I hope, that I have only drawn one row of water particles; whereas there should be many more, behind, underneath and in front of these. But at least you can see the crescent shape of the waves and appreciate why, when the bottom gets too shallow, the whole pattern trips over and the top crashes forward—the wave breaks.

There is yet another kind of wave, where the motion of the individual elements is again a simple oscillation, again as though each were attached by a spring to its original position; but this time they oscillate in the same direction as the wave travels.

This is called a **longitudinal wave**, and this is the way sound behaves. You will notice that at some places the elements cluster together densely.

Here the pressure is greatest. At other places the elements are dispersed: here the pressure is least. So you can think of a sound wave in two ways: as a travelling pattern of oscillatory movement *or* as a travelling pattern of pressure changes. I think the latter is easier.

These pictures I have drawn are no more than aids to visualization, as was the image of the chorus line. They simply say: if the movement of individual elements of the medium is like this, then you can understand that a wave shape will appear to travel down the medium. What they leave unexplained is *why*, in a real medium, the elements should be behaving in this regular kind of way in the first place. That is the question we must address next.

Why do waves travel?

There is an intriguing device, known to physicists as a **Newton's cradle**, which enjoys a vogue every now and again as an 'executive toy'. It consists of a row of steel balls supported on short strings like this:

You 'play' with this device by lifting up one of the end balls and letting it fall. It swings down and, when it hits the others, it stops dead and the ball on the other end flies out. In its turn this one falls back, hits the others and is brought to a stop, and the first ball is kicked out again. So the whole process repeats itself, and the end balls continue bouncing in and out accompanied by a series of sharp clicks with military precision—just the kind of noise, apparently, to soothe the jangled nerves of the average business executive.

For me however this device is a splendid example of wave propagation. If you simply let the first ball drop and *catch* the ball on the other end when it flies out, you have an archetypal example of a wave pulse. You disturbed one end of the medium (the row of balls), that disturbance travelled through and came out the other end; and after it had passed the medium was exactly as it was before. Furthermore it gives you an important clue to understanding. You put *energy* into one end of this medium and that same energy came out the other end. What actually travelled in this 'wave' was energy.

To follow exactly what is going on here, consider the behaviour of a Newton's cradle with only two balls.

You give one ball potential energy, which promptly gets converted into kinetic. When it hits the other ball it compresses it slightly, and very strong forces come into play. These forces slow the ball down; but the same forces act on the second ball and start it moving. By the time the first has come to rest completely the other has moved right out of the way and no further forces act—no more motion of the first ball results. Its energy has been completely transferred.

But why is this process so neat, so exact? That's a tricky question and we'll keep coming back to it. In the present case it's because the two balls are *identical*. If they weren't, things wouldn't happen as they do. If the second ball were much heavier than the first (say it were a cannon ball) then the first one, when it hit, would simply bounce back. On the other hand, if the second were much lighter (a ping-pong ball) then it would be belted out of the way unceremoniously by the first, which would hardly slow down at all. In either case, the first ball would still have energy after the collision. But here, because the two balls are absolutely identical, *all* the energy of the first gets transferred to the second.

Now go back to the long row of balls and you can see why the 'wave' propagates. Each ball passes on all its energy to the next one, so the packet of energy—which shows itself as a disturbance remember—travels down the row. It must be much the same for any sort of wave—say for a transverse pulse going down a rope. One bit of the rope moves sideways. In doing so it pulls on the next bit and starts it moving sideways. But in the same process the second bit drags back on the first, causing it to start moving backwards. Because the rope is uniform, all the 'bits' are identical and the energy of the (sideways) movement will pass cleanly from one to the other. The wave pulse will travel down the rope, and after it has passed the rope will be exactly as it was before.

Wave speed

If we think we understand *why* waves travel, the obvious question to ask is: how fast do they move? Well, think back to the chorus line. What determined the speed of that wave was the **time delay** between when neighbouring dancers repeated the basic action. In exactly the same way, the speed with which energy travels through the row of balls depends only on how long each one takes to pass its energy on to the next. And there are two quantities which control that.

- **Inertia**

  The mass of each ball is the first important item. If the balls are very heavy, they will respond sluggishly to the forces which try to slow them down or start them moving. The energy transfer will proceed slowly. But if the balls are light they will respond more quickly and the energy will be transferred in a shorter time. So:

  > *the speed of the wave decreases as the mass of the balls increases.*

  The same should be true for any sort of wave—in a string, say, or in water or air. But in such waves, it will not be the total mass of the string or whatever which is important. Rather it must be the mass of the 'little bits' of the medium which pass the energy on, one to another. That means it must be the **density** which comes into the formula.

- **Tension/pressure**

  The second consideration is how 'elastic' the balls are. As they are made of steel they do not compress easily, so that, when two hit, the slightest compression will cause enormous forces to act. The slowing down/speeding up takes place extremely quickly. But if the balls were (say) soft rubber, they could be compressed a lot while only bringing quite modest forces into play. Such forces would allow all the energy to be transferred eventually, but the whole process would take much longer. So:

  > *the speed of the wave increases as the internal forces increase.*

  For the row of balls I am clearly talking about the **rigidity** of the material they are made of: but in a string the equivalent thing would be the **tension** it is under, and in a column of air it would be the **pressure**.

You will remember that I went through a similar line of argument in Chapter 3 when I was talking about how things vibrate. What I was doing then, and what I am doing now, is trying to give you an intuitive feeling for how a proper analysis of the problem might be done, while sparing you the mathematical details. It is enough now (as then) to assure you that these same crude arguments can be applied to most wave systems, and the same kind of conclusions can always be drawn. So for the two waves of most importance to music, the following formulas can be found in most textbooks:

- For transverse waves in a stretched string,

$$\text{speed} = \sqrt{\frac{\text{tension}}{\text{line density}}}$$

- For longitudinal pressure waves in air,

$$\text{speed} = \sqrt{1.4 \times \frac{\text{pressure}}{\text{density}}}$$

  Note again the square root sign and the number 1.4. But, as before, don't let them worry you.

No doubt it is nice to be able to calculate the velocities of different kinds of waves, but the figure that will be of most importance to us is the speed of sound in air. The 'standard' value of atmospheric pressure, at a temperature of 20°C, is about 101 000 pascals (the **pascal** being of course the S.I. unit in which pressure is measured). This is probably not a number you have at your fingertips, unless you are interested in weather forecasting or look carefully at the pressure gauge whenever you put air in your car's tyres (and happen to live in a country, like Australia, which has made the changeover completely to S.I. units). Likewise the density of air comes as a surprise to many people. One cubic metre of air (at 20°C and standard pressure) has a mass of 1.20 kg! Putting these numbers into the above formula tells you that the speed of sound in air at standard conditions is about 343 m/s.

This number is certainly worth remembering. If you don't recognize it in those units, you might know it equivalently as (about) 1200 km/hr. This is the speed of a plane travelling at 'Mach 1': and planes flying faster than that are said to have 'broken the sound barrier'. In this age when we are all conscious of noise pollution, you should now be able to appreciate what happens when planes fly at such speeds. The energy of the disturbance they make in the air cannot radiate away in front of the plane by ordinary 'wave' processes. It simply collects into one big pulse of energy and arrives at your ears as a 'sonic boom' after the plane has passed.

So far as music is concerned, a speed of 343 m/s is not all that fast: a fact which often becomes obvious in musical performances, particularly of church music. Even a modest parish church might be 50 m long, and sound will take about 1/6 of a second to travel that distance. So if the organ is at the back of the church and the choir, or the congregation, is at the front, you have problems. After a note has been played, the organist won't hear the singers respond until nearly 1/3 of a second later. The temptation to slow down and let them catch up is hard to resist, and it takes great strength of mind to prevent the most joyous hymn turning into a funeral dirge.

This time delay can become important even over considerably shorter distances, like from the back to the front of a stage. In a large symphony orchestra there might be something like 10 m between the conductor and the back row of the brass, which corresponds to a time delay of 1/30 of a second. So if the players aren't absolutely precise, and take the beat from what they see the conductor doing, rather than what they hear from the other players, they can get seriously out of time. Even down the line of the violin section the beat can easily get lost. That is one reason (just one, of course) why conductors became so important in the 19th century when orchestras were getting much bigger.

Perhaps the most extreme example is the Berlioz *Te Deum*, first heard during the Paris Exhibition in 1885 in the cathedral of Saint-Eustache, a building close to 100 m long. It is performed by an orchestra and two choirs at the front of the nave, and the great organ at the back—it is popularly believed that it is supposed to be a dialogue between emperor and pope. There is also a children's choir, usually about half-way along. In order for these groups to keep in time with one another, three sub-conductors are usually required to communicate the beat to the performers. Berlioz was well aware of the difficulties, and in those parts where the ensemble might get difficult, the music has a certain directness about it, both rhythmically and in the simplicity of its harmonies. Even so, when conducting this piece, he always used an **electric metronome**—a device which he describes with great affection in his memoirs.

All this means that when music is performed in even bigger venues, it essentially has to be come out of loudspeakers. Part of the *Te Deum* was sung at the opening ceremony of the Sydney Olympic Games in 2000. That was in a stadium over 300 m from one end to the other. There's no way a choir can be heard directly by those sitting opposite, so of course it has to be amplified. But the sound can't come out of just one speaker, else it will be too loud in some places and too soft in others. It must come from many speakers distributed throughout the auditorium. But you have to be careful that listeners cannot hear more than the speakers closest to them, otherwise, over that kind of distance, the sound delays will scramble everything. So you can see that the sound engineers who run these audio systems are responsible for much of the success of the performance. It's a small step to their wanting complete control. In many of these events today, as was the case in the Sydney Olympics, the music is pre-recorded, and all the musicians do on the night is mime.

It seems to me that this has consequences for the performance of music in future that we have yet to come to terms with.

Properties of travelling waves

There are three well known effects which we associate with ordinary kinds of waves—**reflection**, **refraction** and **absorption**. I think the meaning of these terms is clear enough. Light rays (which are a kind of wave of course) *reflect* from mirrors; they bend, or *refract*, when passing from one medium to another—say from air to water; and they are *absorbed* by dark coloured materials. Similarly, water waves can be reflected from rocks or the wall of a breakwater; they can change direction as they enter a place where the water suddenly becomes shallower; and they can lose their energy and come to a stop by breaking on the beach.

Refraction is not very important in acoustics, at least as far as music is concerned, but the other two are. Sound bounces off walls or other hard surfaces, producing an **echo**. In a closed room, repeated reflections from many walls produce **reverberation**, and the sound takes some seconds to die away. Similarly sound can be absorbed by soft curtains and wall hangings, which decreases the time for which it lasts.

It is worthwhile being clear about what causes these effects. Think back to the Newton's cradle. You will recall that a pulse of energy travelled through the row like a wave because all of the balls were identical. But imagine what would happen if one of the balls were much more massive than the rest.

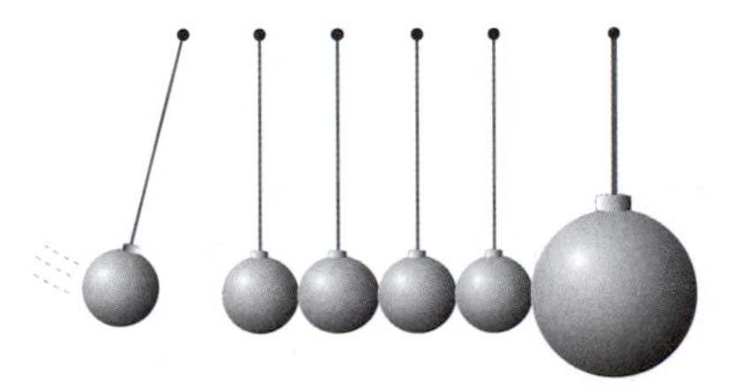

It's pretty obvious—isn't it?—that the energy pulse won't get past the big one. The ball that hits it will just bounce back; and then the energy will have to travel back along the row. This will cause the first ball to be thrown out again, and the pulse will have been *reflected*.

On the other hand, imagine you terminated the row with something soft, like a bag of sand.The ball that hits this won't bounce back, but neither will it move the sandbag much. Its energy will be swallowed up and lost from the wave—maybe the sand will be a bit hotter. The pulse will have been absorbed.

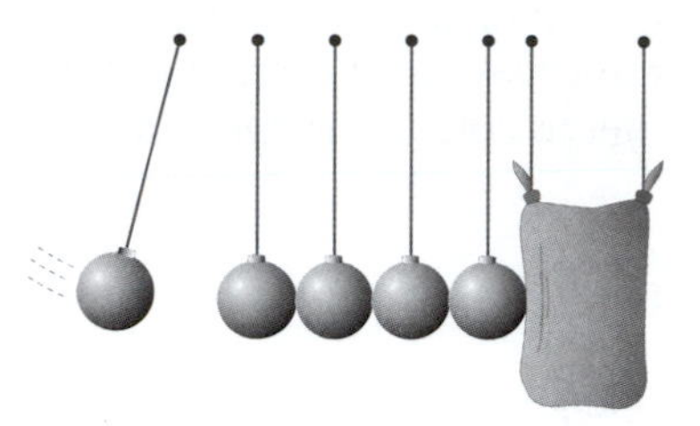

I will have occasion later to come back to this very simple metaphor for why waves propagate, because it shows the basic ideas so clearly. We will meet cases when waves do strange things, and you may find them difficult to understand. Just keep in mind that, before wave motion is possible at all, you've got to have special conditions established—in my metaphor all the balls of the row had to be absolutely identical. If these conditions aren't met, don't be surprised if the wave doesn't behave as you expect.

The three effects I have just talked about are shown by all waves—pulses or extended trains. But there is another effect, called **diffraction**, which really only occurs when the waves are **periodic**. Before I can talk about it, I must introduce another technical term—**wavelength**.

This is easiest to define by using my earlier diagram of a periodic transverse wave made up of imaginary balls on the ends of springs.

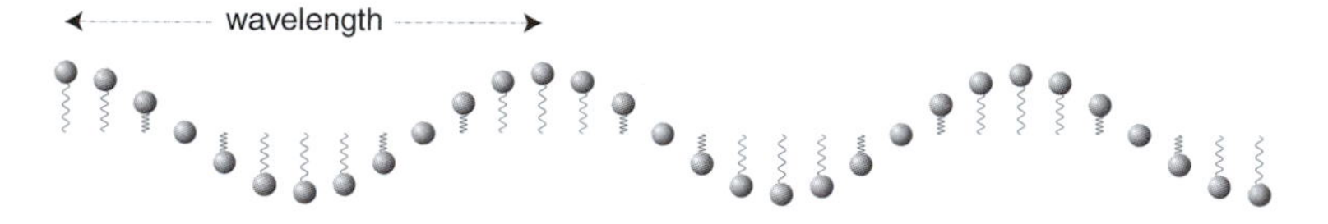

- In a periodic wave, the wavelength is the distance between two successive crests.

  Equivalently it can also be taken to be the distance between two troughs, or any other two points at the same stage in their oscillatory cycle.

There are then three quantities which specify all the important abstract features of a wave. And these apply to any sort of wave at all—transverse waves, longitudinal waves, water waves, sound waves, even light waves (and it isn't at all clear what is 'waving' in those). At each point there is an oscillation with some **amplitude** and **frequency**; and at nearby points there are oscillations of the same amplitude and frequency. But only at a distance of a **wavelength** away, are these oscillations 'in phase' (that is, at the same point in their cycle).

Now look at the diagram again—and, if it helps, think of the balls as chorus line dancers. As each one goes through its oscillation, the crest of the wave moves forward. In exactly one **period**, a new crest will arrive at the first ball, and the old crest will have moved forwards exactly one wavelength. So you can calculate the **speed** of the wave by dividing this distance by the time. Therefore, remembering that frequency is the reciprocal of the period,

$$\text{speed of the wave} = \text{wavelength} \times \text{frequency}$$

Now this *is* an important formula, especially as it applies to sound. We know the speed with which a sound wave travels through air (343 m/s under standard conditions) and we know the frequency of all the notes on our scales. So I can calculate a few representative wavelengths:

NOTE	FREQUENCY	WAVELENGTH in air
A_2	110 Hz	312 cm
A_3	220 Hz	156 cm
A_4	440 Hz	78 cm
A_5	880 Hz	39 cm
A_6	1760 Hz	19.5 cm

The important thing to observe here is how large these wavelengths are. When you consider that, more often than not, we listen to sounds in a particular sort of environment—namely a room—you can't help noticing that these wavelengths are of similar size to other things in that same environment—furniture, doorways, people. Which brings me at last to diffraction.

So far I have only talked about waves travelling along a straight line; but in two (or three) dimensions an extra complication arises—waves spread out. A vibrating source will send out energy in all directions equally; and if there is nothing in the way, it travels outwards in expanding circles (or spheres). As energy reaches each point in the medium, an oscillation starts there, and this in turn begins to send its energy out in all directions. But the complete wave-front keeps its circular shape because all the points in the medium oscillate in exactly the right phase.

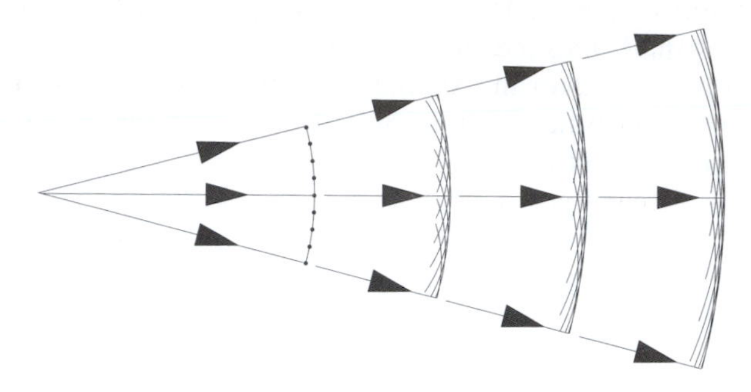

However, when the wave meets an obstacle, or tries to go through an opening, some of the wave is stopped. Where it can go forward the energy is beamed; and where it can't, a 'shadow' is cast. But energy always tends to spread sideways, so the edges of the beam or the shadow become fuzzy.

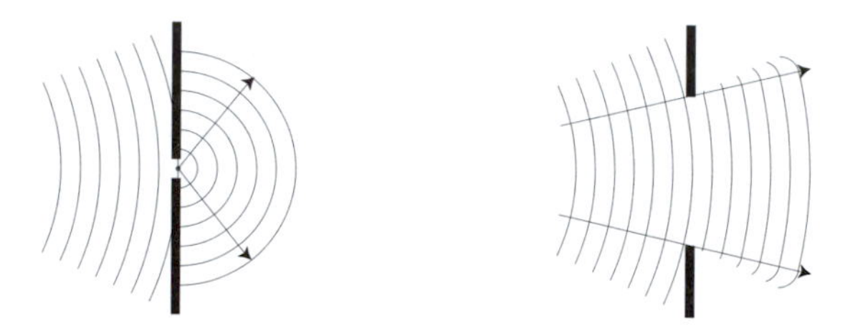

The degree of this fuzziness is largely determined by the wavelength—this being a measure of the distance any small oscillating element can send an appreciable fraction of its energy. Hence only if the width of the object is shorter than the wavelength will there be much spreading—or to give it its proper name—**diffraction**.

In the case of light, typical wavelengths are very small, less than a thousandth of a millimetre. So diffraction of light is not easy to observe, and ordinary sized objects cast very sharp shadows. But with sound, things are quite different. Its wavelength is often bigger than the obstacles it meets, so it *bends* around them, often hardly casting a shadow at all. And this effect is more pronounced for longer wavelengths (i.e. for lower frequencies). As a rule of thumb you can say that, if the wavelength is much larger than the size of an obstacle in its path, the wave will hardly notice it is there. This explains why bats, which hunt small insects by sending out pulses of sound and listening for the echo, have to use wavelengths of only a few millimetres. Such frequencies are much higher than can be heard by the human ear: they are called **ultrasonic**.

Clearly diffraction must be quite common in ordinary acoustics. It means, for example, that sound waves can travel round corners and you can hear a sound even when you're not in the line of sight of the source. The wavelength dependence is also easy to observe. If there are many obstacles in the path of a sound wave, then the long wavelengths (low frequencies), which bend round each one readily, will find it easier to get past all the obstacles than the short wavelengths (high frequencies). So, when you hear a marching band from a few streets away, or listen to a choir singing in another part of the building, you will notice that the high notes tend not to reach you, and all you hear is the bass line.

Though there are other effects which travelling waves show, only the three I have described—reflection, absorption and diffraction—are of much importance musically. If you understand those three, you can understand a great deal about the behaviour of sound inside theatres and concert halls, the area of study usually called **architectural acoustics**. It is a subject which was surprisingly late developing, being invented almost single-handedly by an American scientist, Wallace Clement Sabine, between 1895 and 1905; despite the fact that the important properties of sound waves had been understood for more than a century. However it is a rather specialized field and I'll defer talking about it till the next Interlude.

But there is another interesting point of contact between wave behaviour and music. Most composers obviously have a deep, intuitive understanding of sound and how it behaves—it is, after all, their stock in trade. Therefore, when a composer is using music to paint an aural picture, you can often identify that a particular wave property is being used.

On a very simple level, an echo can often be used to suggest the presence of enclosing walls. For example, in the witches' chorus in Purcell's *Dido and Aeneas*, the end of each line is repeated softly, contributing to the impression of being in a "deep vaulted cell". Similarly, the way sound reverberates inside a stone church is nowhere better suggested than in the chorus "This Little Babe", from Benjamin Britten's *A Ceremony of Carols*, in which the melody is sung in a three-part canon, the voices separated by only one beat.

There are more subtle examples than these. An orchestral piece which starts with a few bars of softly played low notes and then suddenly introduces a main theme on higher instruments—like for example the Schubert *Unfinished Symphony*—conveys the aural impression of opening a door into a room where music is being played. And conversely, when the cemetery scene in *Don Giovanni* ends with four bars of progressively decreasing pitch registers—first the baritone stops singing, then the bass, and last of all the cellos—the effect is similar to that of a television camera drawing away from the scene. Now you may find this a bit fanciful. After all it is highly unlikely that Mozart or Schubert knew anything about diffraction—though in fact, both were highly educated men. Nevertheless, the point I am making is that they clearly knew a lot about sound, and whether or not they knew the scientific jargon is irrelevant.

The history of waves

There are several historical threads we can follow through the story of the development of these ideas. The first is mainly experimental, concerning the measurement of the speed of sound. It is perfectly obvious to anyone who notices the time delay between a lightning stroke and the thunderclap which follows it, that sound does not travel instantaneously; yet there seems to be no record of anyone in the ancient world attempting to quantify this observation. It had to await the revolution in scientific thought, often attributed to Francis Bacon in the 16th century, that everything that can be measured, should be measured.

Although so much of Renaissance science was done in Italy, doubtless under the influence of Galileo, the speed of sound seems to have been a French preoccupation. The philosopher Pierre Gassendi, whose claims to fame include the fact that he was the teacher of the playwright Molière, made the first recorded estimate in 1635. By timing the delay between when the flash from a distant gun was seen and when it was heard, he got a value of 478 m/s—as you can see, almost 50% too high. He did however confirm accurately that sound-speed does not depend on the pitch of the note—which is what Aristotle thought to be the case. Mersenne repeated his measurements some years later, and got the (slightly, but not much) better value of 450 m/s. The main error of both these gentlemen seems to have been that they didn't allow for the effect of the speed of the wind.

Over the next century experiments continued to be refined, both in laboratories and out of doors—and many a scientist was regarded with deep suspicion as he travelled around the country with a keg of gunpowder in his baggage. In 1738, a team from the Paris *Académie des Sciences*, using a cannon as the source of sound and working only on perfectly still days, achieved the remarkably accurate figure (adjusted to a standard temperature of 20°C) of just under 344 m/s. You really must admire the care with which those Paris academicians did this work. Very few physical measurements have stood the test of time as well as theirs.

A second historical thread follows the development of the theory of wave motion. I mentioned earlier how Vitruvius had come up with the right idea before the time of Christ, but that his writings were lost for many centuries. Even when they were rediscovered, the climate of opinion was initially against him. Gassendi, for example, was quite convinced that, since you couldn't detect any motion of the air associated with a sound, it had to be propagated by a stream of very small, invisible particles shot out from the source. The German Jesuit, Athanasius Kircher, who wrote much and learnedly about music and architecture, was the first to try, in 1650, the experiment of ringing a bell inside a vacuum. He concluded that air was *not* necessary for the transmission of sound. Undoubtedly the trouble was that he didn't exclude the possibility of sound travelling through the walls of his containers. It wasn't until ten years later that the English scientist, Robert Boyle, repeated the experiment much more carefully; and showed conclusively that, in the absence of air or some other medium, sound definitely would not travel.

There now appears in the story another of those towering figures from the history of science, of whom there seem to have been so many in the 17th century. Christiaan Huygens was born in Holland in 1629, and in early life took up a career as an astronomer. At that time, the Dutch led the world in the manufacture of optical instruments, and the experience he gained in his brother's workshop enabled the young Christiaan to construct the most advanced telescope of his day, almost twice as long as Galileo's. With its aid he made a series of spectacular discoveries: the rings of Saturn, and its giant moon which he named Titan; the surface markings on Mars, which he mistakenly interpreted as seas and swamps; the strange celestial object which turned out to be a huge mass of gas and dust which we know as the Great Nebula in the constellation of Orion; and he made the first ever reasonable estimate of the distance to the nearest stars.

His observations demanded many specialized instruments, including something to measure time accurately. The best available devices were traditional water clocks, which marked off intervals of time as water flowed out of a container; or the more modern mechanical clocks, which used slowly falling weights for the same purpose. Neither was particularly accurate, so Huygens put Galileo's studies of pendulums to work—ten years after the old man's death—and produced the first ever pendulum clock. When he presented his 'grandfather's' clock to the Dutch government, the age of accurate time-keeping can be said to have begun.

Huygens was a theoretician as well as an experimenter. He was among the first, for example, to recognize the importance of kinetic energy. But his greatest interest was in wave motion. It was he who first used the principle of the Newton's cradle to explain how waves travel through material media, and it was he who invented those diagrams I reproduced earlier to explain how a collection of spherical wavelets can combine to produce a coherent travelling wavefront. Physicists today still use this geometrical method of analysing wave propagation, and it goes by the name of **Huygen's construction**. The diagram on the top of page 142 is an example of one of these.

It is here that the third historical thread can be taken up—a story which runs parallel to the understanding of the nature of sound, namely that of the nature of light. As Huygens' scientific achievements multiplied, his reputation in Europe grew. In 1665 he was, by invitation, a foundation member of the English Royal Society, and he spent many years as a scholar in the court of Louis XIV. Late in life he finally felt secure enough to do battle with the Grand Master himself.

Isaac Newton, as we have seen, had made the study of optics a major part of his life's work: and held that light and sound were fundamentally different in the way they travelled. Sound bent round corners, so was clearly a wave: but light did not. He firmly believed that only a stream of particles would travel in absolutely straight lines and throw sharp shadows, the way light did. Huygens spent his last years proving that, under certain circumstances, waves *can* appear to travel in straight lines and obey the optical laws of reflection and refraction; but he lost the battle, at least in his lifetime. He died in 1695, widely respected and admired, but in the minority of scientific opinion so far as the nature of light was concerned.

For nearly a century Newton's views reigned, but in 1803 an English physician named Thomas Young changed the picture. Young was reputedly a brilliant and cultured person, who played a wide variety of musical instruments, including the bagpipes; but he was not a success as a medical man. Those who knew him said he definitely lacked a bedside manner. However his researches into human sight and hearing paid dividends elsewhere.

He had become interested in the acoustical phenomenon of beats, which you will recall Sauveur had explained many years earlier. He wondered whether, if light were a wave, he might detect some similar effect. So he shone a light beam through two extremely narrow openings and observed that, when the two beams came together again, they produced a striped pattern of light and darkness—for all the world like the picture of two combs I reproduced on page 95. So with one simple experiment the question was settled. Light was a wave after all!

Unfortunately science often has a habit of never being quite so simple. Throughout the 19th century enormous effort was devoted to the study of light, seeking answers to such questions as: how fast did it travel, and did it need a medium to travel through? The outcome of that work a century later would overturn physics completely, but I'll leave that story till a later chapter. For now I want to focus on the understanding of wave motion that Young managed to uncover by his experiments.

The observation that two light rays (or any other waves) of the same wavelength travelling through the same medium in slightly different directions, will 'add up', sometimes to reinforce and sometimes to cancel one another, is given the name **interference**. The kind of experiment that Young did with light can be done with sound. It is possible to arrange for the same sound to come from two closely spaced sources, and to observe that, in some directions, the result is louder than in others. But in most rooms there are too many other reflections for this to be really noticeable, so this effect isn't of great relevance to music.

However there is one circumstance which is very important indeed, and that is when the two waves travel in exactly *opposite* directions. Then they 'add up' to give what are called **standing waves**; and the understanding of these threw a completely new light on that old problem I discussed in the last chapter—the modes of vibration of a stretched string.

Standing waves

To understand this new effect, I want to go back and think about how waves travel down a long rope; but this time I will examine what happens if one end of the rope is held fixed and not allowed to move. It is easy to observe, just by doing it, that if you send a pulse down such a rope, it will be reflected like this:

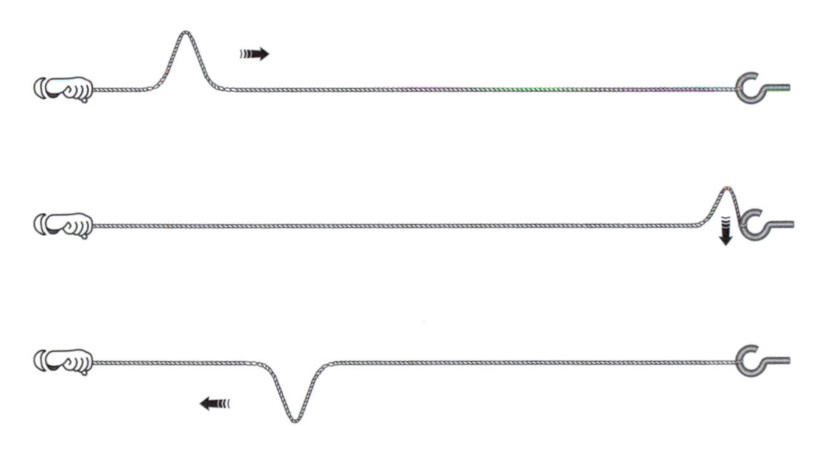

It isn't difficult to *understand* why the energy of the pulse is reflected; but it is extremely insightful to realize that you can *describe* exactly what happens here by imagining that two completely independent pulses travel along this rope. The first, an **initial wave**, simply moves down the rope to the right and disappears off the end when it gets there. The second, a **reflected wave**, which has exactly the same shape except that it is upside down and back-to-front, comes onto the rope at exactly the right time and the travels to the left. The trick is to choose the 'right time' so that, for the brief period when *both* pulses are visible, they exactly cancel one another at the very end point of the rope, where we know physically that no motion is allowed.

This bit of mental gymnastics is even more useful in visualizing what happens when a *periodic* wave is reflected from the fixed end. Again you describe the situation by drawing two independent waves, this time in the shape of sine curves. However, since an upside down and back-to-front sine curve looks just like another of the same, it is difficult to tell them apart, and you've got to keep careful track of which way each is travelling.

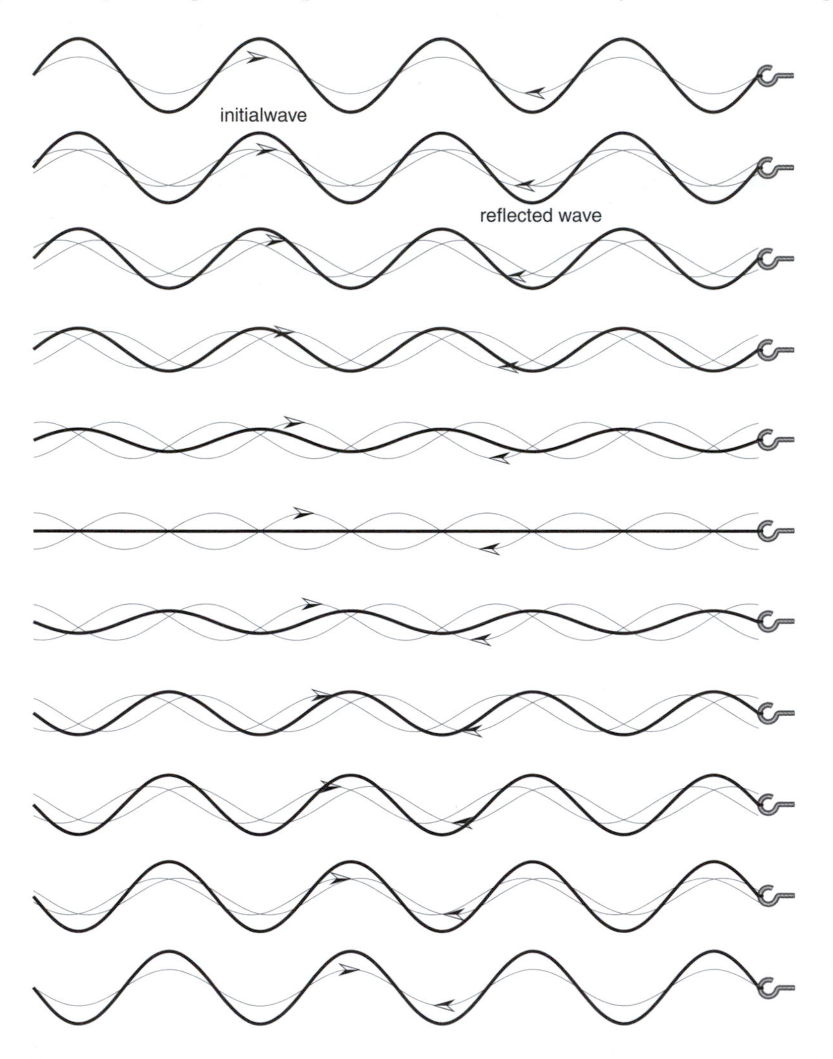

What comes out of these pictures very clearly is that the composite 'wave' no longer appears to travel either to the right or to the left. Instead it looks like yet another sine curve whose amplitude increases and decreases with time, but which otherwise stays still. It will not come as any surprise to you to learn that this is what is called a **standing wave**.

Now please be perfectly clear about why this happens. The initial and reflected waves always cancel one another right at the end-point of the rope. They were carefully 'chosen' to do that. The end-point never moves—it is a **node** (to use a term I introduced in Interlude 1). But sine curves repeat themselves in space. Whatever happens at one point, exactly the opposite happens at a point half a wavelength away. So if the end of the rope never moves, neither does the point half a wavelength before it. Nor the point half a wavelength before that. And so on. All these points are nodes too. Therefore, no matter what motions the rest of the rope undergoes, there can never be the appearance of a wave shape moving past these points. The wave has been rendered 'stationary'.

Now think of it in another way. We saw before that in ordinary wave motion, what actually travels is energy. When a periodic wave moves down a rope, energy continuously flows past any point. But when the wave is reflected, the energy starts flowing back in the opposite direction. Therefore the total effect is that the *net* rate of energy flow past any point is *zero*. But there is still energy all along the rope (you can tell that because there is sideways movement). Therefore this energy has become **trapped**. And we have met the idea of trapped energy before. The wave has become a **vibration**.

Now go one step further and think about what happens if the rope is fixed at *both* ends, like a violin string. Imagine that you get one part of the rope moving in a simple harmonic oscillation (let's not worry about how). A periodic sine wave will travel down the rope, with a wavelength given by the speed of waves in the rope divided by the frequency of the oscillation (refer to the formula on page 141).

When this wave gets to the far end it is reflected. A second wave travels back up the rope and a standing wave is set up. But this second wave eventually meets the near end of the rope, which is also fixed, and it is reflected. So there are now three waves in the rope, and soon there will be four, five, and so on. In the end there will be hundreds of waves travelling back and forth, and the whole thing will be a messy jumble of standing waves. Unless ...

There is one circumstance where something special occurs. If it so happens that, when the second wave comes to be reflected, the *near end of the rope coincides with a node of the standing wave*, then the third wave will sit right on top of the first and be indistinguishable from it.

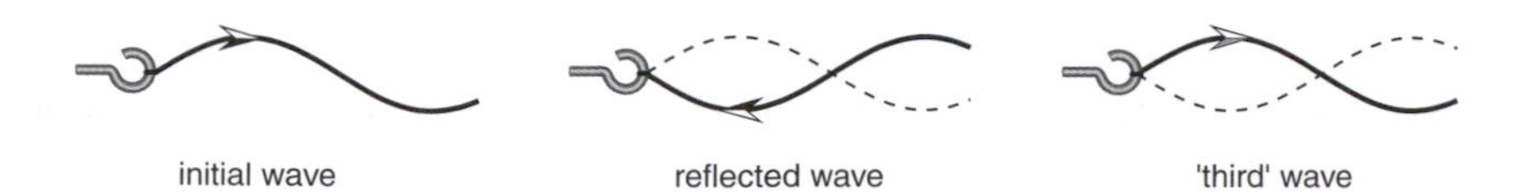

Once that has happened, the fourth wave must be indistinguishable from the second also. In the end, all the waves travelling to the right will look like just one wave, and the same will be true for all those going to the left. There will therefore be just two different travelling waves and these will combine to produce one simple standing wave.

When I drew a picture like this earlier, in Interlude 1, I called it a 'vibration'. What the 'standing wave' idea gives us now, is a way to understand exactly how such a vibration builds up. You feed energy in at some point on the string by making it oscillate. That energy moves away, but becomes trapped, so that after a while, a lot of energy accumulates at the **antinodes** and the amplitude of the vibrations there becomes quite large. It is only frictional processes tending to dissipate this energy which stop the amplitude growing indefinitely.

But even more importantly, this new way of looking at things allows us to *calculate* the frequency. The simplest such vibration—which I have called the **fundamental mode**—occurs when the two ends of the rope are separated by exactly half a wavelength;

or, to turn that sentence around, when the wavelength is equal to twice the length of the rope. Then, since we know the speed with which waves travel along such a rope, we can work out that the frequency of this fundamental mode (which I will hereafter denote by the symbol f_0) is

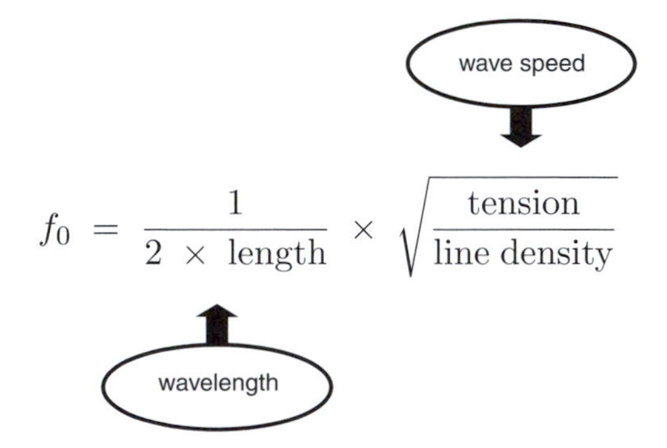

$$f_0 = \frac{1}{2 \times \text{length}} \times \sqrt{\frac{\text{tension}}{\text{line density}}}$$

You will notice that this is the same formula I wrote down earlier (on page 65).

We can even go a step further and identify all the higher modes of vibration. The key observation to make is that the wavelength is forced to take on only certain values by the requirement that both ends of the rope must be nodes.

STANDING WAVE SHAPE	WAVELENGTH	FREQUENCY
	$2 \times$ length	f_0
	$1 \times$ length	$2 \times f_0$
	$2/3 \times$ length	$3 \times f_0$
	$1/2 \times$ length	$4 \times f_0$
etc		

Let me sum up at this point. Most of the results I have just quoted were ones we knew before. What is new is that we can now understand them better. We can see exactly why the frequency of vibration of a string depends on its length—a result whose importance I have continually stressed; why there are higher modes which give rise to the overtones; and why these overtones are harmonic. The study of waves in general, and standing waves in particular, has tied up the loose ends of the study of vibrations.

In historical terms this is what happened during the 18th century. The same mathematicians who contributed to the general question of vibrations of a string—and if I had to single out one name as the most important, I would probably go for Leonhard Euler—also developed the formal understanding of waves. By the beginning of the 19th century a formidable body of theory had been amassed. The next stages of development were in the hands of the experimentalists who explored the acoustic vibrations of more complicated bodies than simple strings, and who extended this understanding into the spectacular new field of science just then beginning—electricity.

Interlude 4

Acoustics in architecture

Rooms and concert halls aren't exactly musical instruments, but they do affect the sound of any music played inside them, and that is why I want to devote an interlude to talking about them. To start at the very beginning, the most basic problem in hearing a sound is simply picking up enough acoustic energy for your ear to work with.

Any source of sound sends out waves in all directions, so that, at some distance away, the energy is spread out over the surface area of an imaginary sphere.

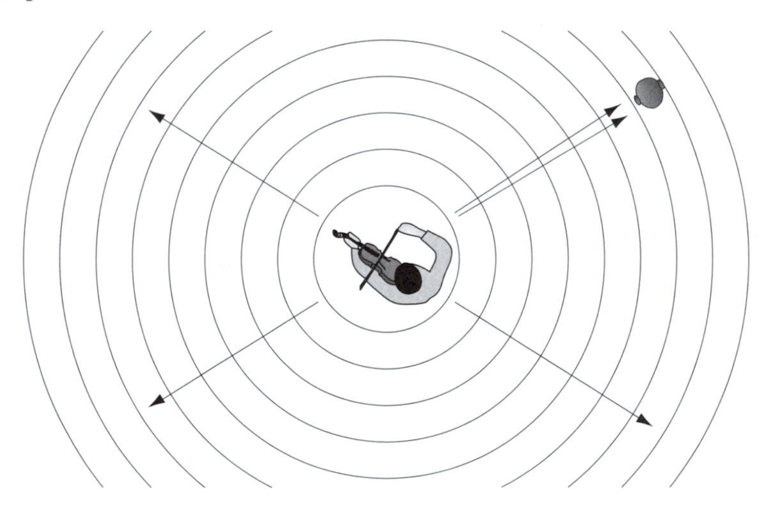

At a distance of 1 m from the source, the area of this sphere is almost 13 m^2, while the collecting area of your ear is perhaps 12 cm^2. Therefore the energy you actually receive is only about 1/10 000 of what is available—i.e. the ratio of the two areas.

If you are twice as far away, you get only a quarter as much (which is known as the **inverse square law**). This must mean that your ear is an impressively sensitive device: but it also means that, should you have difficulty hearing the sound, there is a lot of energy going to waste that you should be able to use to improve matters.

One simple thing you could do would be to put a large reflecting wall behind the player, so that all the energy that would otherwise have gone away from you is bounced back. Your ear would then pick up another ray of sound energy, as though from an imaginary player *behind* the wall. So it should sound as loud as *two* players.

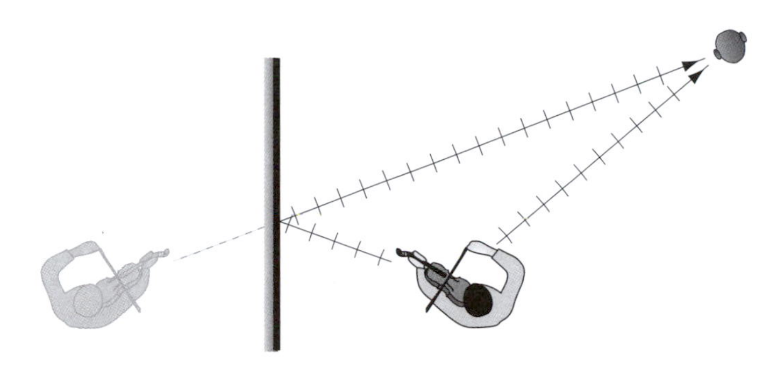

It might help here to think about the same thing in terms of light. Before the days of electricity, it was common practice to stand a candle in front of a mirror and thereby double the amount of light shining onto whatever it was you were trying to see. A white painted wall would do almost as well, because a white surface scatters back most of the light falling on it; but of course a dark wall would just absorb the light, and that wouldn't help at all.

Much the same is true for my violinist. The wall will have to be a hard, flat surface, otherwise much of the sound energy will be absorbed and it won't help you hear any better. But there is an extra crucial difference. Because sound waves *diffract* easily, the wall has to be large. Something the size of a hand-mirror—which would work quite adequately with the candle—won't do at all; the long wavelength sound waves would just diffract around the edges and very little would be reflected.

This very simple example illustrates nicely the way we should think about the acoustics of rooms and halls. Most of the effects can be understood by considering sound waves as traveling in straight lines, and bouncing off or being absorbed by surfaces, just like light rays: but every now and then we should be prepared for something different happening because of their long wavelengths.

Outdoor acoustics

It is probably true to say that acoustics first became associated with architecture when people started coming together to hear speeches, plays or music. Greek architects built splendid open-air amphitheatres for this purpose, and many survive to this day, like the beautiful examples at Delphi and Epidaurus in Asia Minor. Typically they consisted of steeply banked semicircular rows of stone benches built on the side of a hill. In front was a stone platform with massive masonry walls at the back and sides. The acoustic effect of these was to direct forward, by reflection, all the sound that would otherwise have gone into the three back quadrants. So a single speaker could often sound as loud as four.

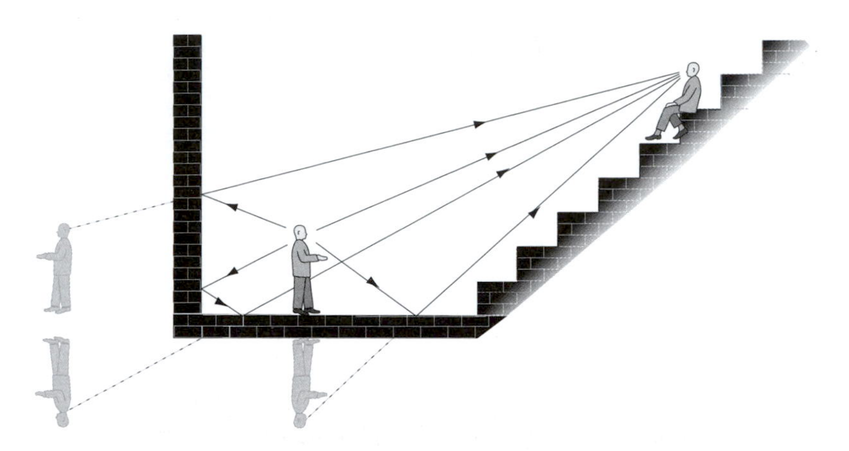

I don't think there are any of these amphitheatres which still have the stage and back wall intact, so their true acoustics can't be experienced today. Even so, they sound good to tourists, though a lot of that is because the sites are naturally very quiet anyway. Writing in the 1st century B.C., Vitruvius claimed that large urns tuned as resonators were placed underneath the seats to reinforce certain sounds. A moment's thought should convince you that this is nonsense. After all, only so much energy would arrive in the sound wave at the mouth of such an urn, and there is no way that the urn could increase that. At very best it could only re-radiate it in a different direction. If you want to increase the energy in a wave you've got to use some device which is connected to a source of power and which feeds energy in (like, for example, an electrical amplifier).

Possibly the most revolutionary bit of acoustical science that the Greeks used was to give their actors masks with in-built megaphones. But even so, the intuitive skill of the architects who designed these theatres was impressive, because some of them held audiences as large as 14 000, who reputedly had no difficulty in hearing right to the back rows.

A similar beaming and reinforcing of sound can be achieved with a single reflecting wall which is curved inwards, on the same principle as the parabolic mirror of a searchlight. The Romans exploited this in their open-air theatres. There is one in Verona which is still used today for performances of opera, and reputedly has splendid acoustics. But to use something like this properly presents a difficult design problem, because curved surfaces can actually focus the sound waves down to a small area; and that can produce occasional freak effects.

Perhaps the best known is the famous **whispering gallery** in St Paul's Cathedral in London. The dome of this building, designed in 1668 by Christopher Wren (one of the founding members of the Royal Society), is a perfect hemisphere. If you make a soft sound near one of its inside walls the spreading sound waves are all reflected inwards by the curvature and brought to a focus at the other side. At this point you can hear the original sound with amazing clarity. (I should mention that there is some controversy about how much effects other than simple focussing contribute here, but I think we can ignore that.)

Now, whether or not this effect was intentional isn't clear; but the principle behind it was certainly understood at the time, as evidenced by this drawing published in 1650 by Athanasius Kircher in his *Musurgia Universalis*.

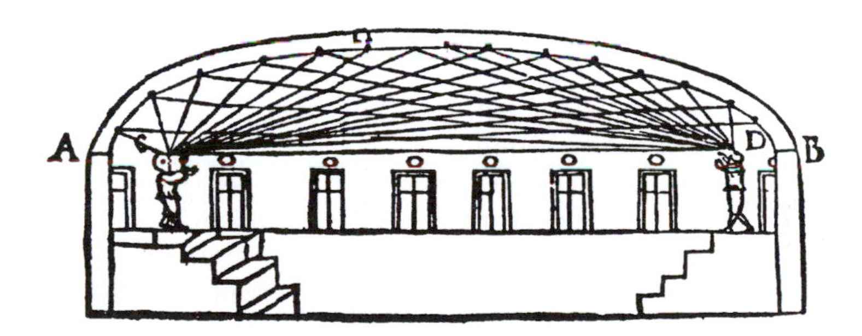

All these principles combine in the design of modern **band shells** constructed for the performance of music in the open air. They usually have reflecting walls, curved in a vaguely shell shape, in order to beam forward all the sound that would otherwise go backwards and be lost. Again the effect is equivalent to increasing the number of instruments in the band. But there are very definite limits to how much can be done in this regard. Firstly, because of diffraction, it is only the high frequencies (short wavelengths) which are efficiently beamed forward. This seriously alters the balance of high and low frequencies in different directions. Secondly, there are band shells in existence where the beaming effect is so pronounced that where the sound goes is critically dependent upon where the player is on the stage.

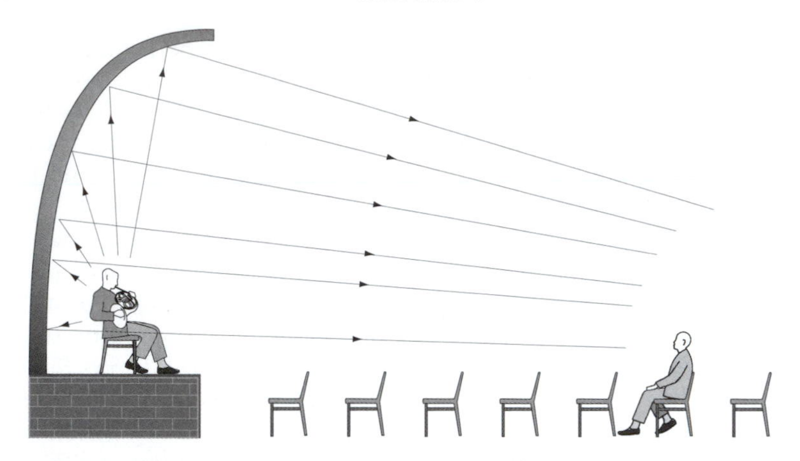

As a result of all this, when a band plays in them, the sounds from different instruments go in different directions, and there is nowhere that the audience can sit where they will hear the music as the band-leader hears it. So in the end, most band shells have some flat walls, sacrificing directivity (and loudness) for quality.

Indoor acoustics

Most musical performances, of course, don't take place out of doors, but inside rooms, theatres, churches. And when we come to consider what happens to sound in an enclosed space, a completely new phenomenon arises. All of the acoustic energy from any source will be reflected from the walls or the floor or the ceiling. You don't only hear the sound coming directly from the source, but also that which has been reflected from each wall, and doubly reflected, triply reflected, and so on: till eventually it comes at you from all directions. If the walls were perfect reflectors then, in principle, the energy could never get out and it should hang around the room forever.

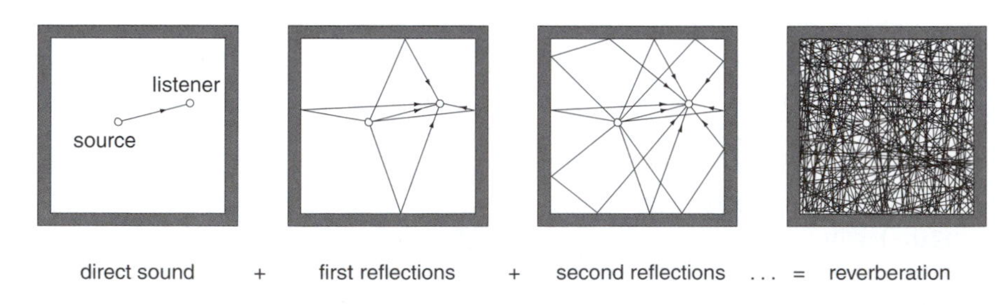

Of course, no real surface reflects absolutely all of the sound energy that hits it. Some small fraction is absorbed each time, and the acoustic energy will disappear eventually. You've probably experienced a similar effect in one of those halls of mirrors. You see endless rows of images of yourself marching off to infinity in all directions; but they do become fainter and fuzzier in the distance, because not even highly polished mirrors are absolutely perfect reflectors.

The production of multiple reflections is known as **reverberation**, and it has two important consequences. Firstly, any sound in this environment seems much louder than in the open air, simply because so much more of the acoustic energy gets to your ears. After a very short time the original acoustic energy is spread pretty well throughout the whole volume, so in a small room it sounds much louder than in a bigger room where the energy is more dilute. That's the reason why so many people enjoy singing along with the radio in their cars. For a modest expenditure of vocal energy you can produce a very satisfying volume of sound, and imagine (with a bit of poetic licence) that you're being joined by a choir of the heavenly hosts.

The other thing which affects the perceived loudness, is what the walls are made of. Soft surfaces—like chipboard, curtains, rugs—absorb a lot of energy on each reflection and lower the sound level. Open windows are the worst of the lot, they 'absorb' it all. But hard surfaces—plaster, stone, concrete—absorb very little energy and the level remains high. Again, the most popular room in the house to sing in is the bathroom, with hard tiled walls and no soft furnishings. You often hear the word 'resonant' used to describe the sound of the bathroom baritone; but it's a bit of a misnomer. Only very occasionally does real resonance occur—i.e. when standing waves are set up because the long wavelengths of the notes fit into the dimensions of the room exactly; most of the time it is simple reverberation.

In bigger rooms a second effect becomes obvious. The sound waves have to travel a lot further while making all those reflections. You can start to notice that it takes time for the sound level to build up at the start and, more importantly, to die away at the end. Here is an interesting experiment you might be able to do for yourself. Set off a loud, sharp noise (a firecracker or a starting pistol) in a large hall and record the sound on magnetic tape. Then play the tape back very slowly (at 1/8 or 1/16 of the speed if you can). What you hear can be represented schematically like this:

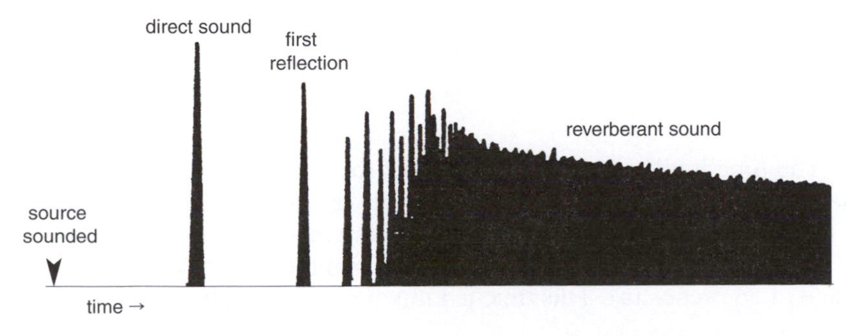

The time delays depend on exactly where in the hall you put the the microphone and the sound source. If the nearest wall were, say, 3 m away, then the reflection from that wall would be the first thing you hear after the direct sound, at a time delay of about 1/50 of a second. It's very obvious on slow playback, but even at real speed it's just on the threshold of being subconsciously noticeable.

If instead of a short sharp noise, you suddenly turn on a *steady* source of sound, then the level builds up in steps as the individual reflections arrive at the microphone.

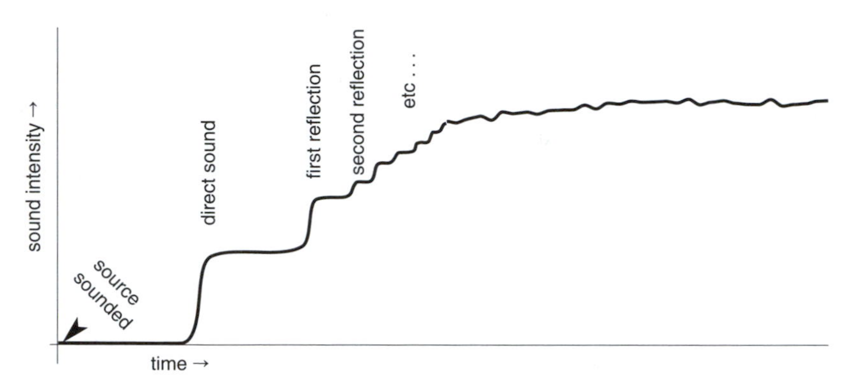

It continues increasing until it reaches a level at which the acoustic energy coming from the source exactly balances the energy being lost to all the absorbing surfaces throughout the volume.

If, after it has reached the steady state, you suddenly turn *off* the source, you will notice that the sound level doesn't immediately drop to zero. Instead it falls away slowly like this:

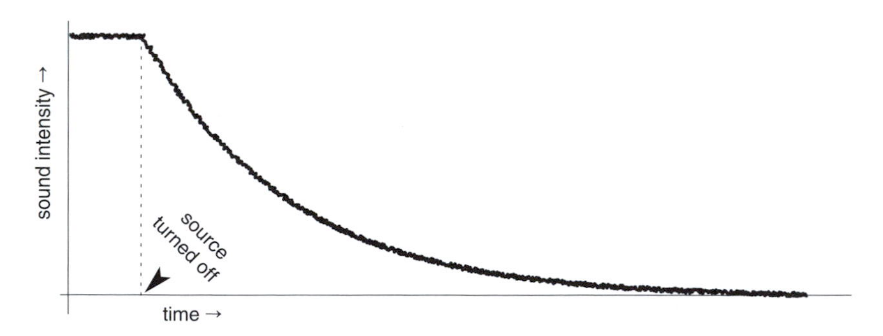

The time it takes to reach the level of the background noise in the hall (never quite zero) depends on how loud it was initially: but in the interest of making comparisons, it is usual to measure the time it would take to drop to 1/1 000 000 of its initial value (which is about the maximum dynamic range of an orchestra). This time is known as the **reverberation time**.

Now what is interesting is that, if you do the measurement for sounds of different loudness, you'll always get more or less the same value (if you don't change where you do it). But if you repeat the measurements in different halls, you'll get different values—one for each hall. For an ordinary lecture room, seating some 100 students, it is usually about 1/2 second: for the La Scala opera theatre it averages about three times that value: while for Cologne Cathedral it is reputed to be 13 seconds!

Now, of all the effects I've talked about so far, this one must be the most important for music. Every note you play, every syllable you utter, will be heard by the audience for this length of time after you have stopped making it. If this time is long, then all the separate sounds will run into one another. This may not matter for lush romantic music—something like Tchaikovsky perhaps; but it would be intolerable for speech. Therefore possibly the first and most important calculation in the science of architectural acoustics is to work out how long this time is for any particular hall.

Reverberation time

Understanding just what determines this for any specific room is actually a bit tricky, which is why it wasn't worked out till the beginning of this century. The key idea to bear in mind is that, when the sound source has been turned off the sound level would stay the same, were it not for the fact that some energy is absorbed on each reflection. Now the amount of absorption that can take place at some surface depends on how large its area is and how absorbent is the material it is made of. These two characteristics are usually combined into a single quantity called the **effective absorbing area**, by which I mean the following:

- The **effective absorbing area** of a surface is the product of the real area with a quantity called the **absorption coefficient**. This is a number between 0 and 1 specifying how readily it absorbs sound. A perfect absorber (like an open window) has a coefficient of 1; while for a good reflector (like glazed tile) it is about .01.

The absorption coefficients involved here have been measured definitively for most building materials and if you ever need to do one of these calculations you just look them up. As an example, here is a short table of representative values. It shows more or less exactly what you would expect—that the softer the material, the better it absorbs and the higher its coefficient. But what is perhaps unexpected is that this property varies markedly with frequency.

ABSORPTION COEFFICIENTS OF STANDARD MATERIALS

	FREQUENCY (Hz)					
	125	250	500	1000	2000	4000
Marble or glazed tile	.01	.01	.01	.01	.02	.02
Brick wall	.01	.01	.02	.02	.02	.03
Wood floor	.15	.11	.10	.07	.06	.07
Plywood on studs	.60	.30	.10	.09	.09	.09
Plaster on laths	.30	.15	.10	.05	.04	.05
Carpet with felt underlay	.08	.27	.39	.34	.48	.63
Heavy curtains against wall	.14	.35	.55	.72	.70	.66
Cane fibre tiles on concrete	.22	.47	.70	.77	.70	.48

By and large, those materials whose absorption coefficients increase with frequency—which are called **treble absorbers**—have rough or porous surfaces, covered with small protrusions, indentations or holes; and the size of these gives the clue to their behaviour. Waves with long wavelengths won't notice these at all, they will see the surface as being smooth and reflective. But those with short wavelengths will get inside the holes or scatter off the protrusions and lose energy more effectively. By contrast, any large surface that can vibrate like a membrane in front of a cavity—like wooden panelling—will act like a **bass absorber**. It traps energy because the cavity acts as a loose spring which resonates at very low frequencies.

Members of the audience make ideal treble absorbers. Being almost entirely composed of bits and pieces some tens of centimetres in length, and covered with hair and woven materials, they absorb most frequencies above about 500 Hz. Measurements show that the average effective absorbing area (at 500 Hz) of a person sitting down is somewhere around .5 m^2 (i.e. two people absorb as much as an open window 1m x 1m). It is customary nowadays, in major auditoriums, for architects to specify specially padded seats with an effective absorbing area equal to this, so that the acoustics shouldn't change whether the hall if full or empty (they are usually only moderately successful).

Anyhow, to sum up, if the total effective area of all the absorbing surfaces in the room is large, then the rate at which acoustic energy leaks away must be large also; and the time it takes to dissipate completely is correspondingly short. In other words, as the total effective absorbing area increases, the reverberation time decreases.

The other quantity which affects it is the rate at which sound energy hits the walls (before being reflected or absorbed). This depends on the ambient sound intensity, which, as I said before, depends on the volume of the room. If the room is big the energy is thinly spread throughout and so it can only be absorbed slowly. The reverberation time will be correspondingly long.

So when you put these two effects together you can produce a general formula which needs only a few actual measurements in real halls to get the numbers right:

$$\text{reverberation time} = \left(\frac{0.16 \times \text{volume of the room}}{\text{effective area of all absorbing surfaces}}\right)$$

(where it goes without saying that volumes are measured in m^3 and areas in m^2).

Wallace Sabine, of whom I'll have more to say presently, was the first to establish this formula in about 1898. Nowadays it is realized that it is a bit over-simplified, and overestimates the reverberation time in situations where there is a lot of absorption; but it isn't bad, and certainly good enough for our purposes.

Anyhow, you can see now how to do a rough calculation of the reverberation time of any room (at 1000Hz, say). Firstly, for an ordinary suburban living room:

Typical dimensions:	4 m × 5 m × 3 m
Volume:	60 m^3
Effective absorbing area of ceiling (plaster):	2 m^2
of floor (carpet):	8 m^2
of walls (3 plaster, 1 curtained):	10 m^2
Total effective absorbing area:	20 m^2.
Therefore the reverberation time is about	**0.5 second**.

Fabric covered sofas and chairs and other furnishings will lower this figure even further.

Now do it for a school auditorium:

Typical dimensions:	10 m × 30 m × 8 m
Volume:	2400 m^3
Effective absorbing area of ceiling (wood):	30 m^2
of floor (wood):	30 m^2
of walls (brick):	12 m^2
Total effective absorbing area:	72 m^2.
Therefore the reverberation time is about	**5 seconds**.

Of course, when 500 students are present for daily assembly, the effective area increases to 322m^2, so the time falls to 1.2 seconds. But even that's a bit long. It's not going to be easy for the principal to make a speech with the sound lingering for that length of time. It will probably be necessary to cover those brick walls with something much more absorbing, like cane fibre tiles.

This raises the question of what is the ideal value of reverberation time. As I've already suggested, that depends on what you want to use the hall for. There has to be a compromise between clarity on the one hand, and a satisfactory sound level on the other. For speech, the former is paramount, and if the absorbers necessary to keep the reverberation time down mean that the sound level is too low, then you'll just have to use electronic amplification. But for music you can put up with a longer time, so there will be a good intensity of sound even in a large hall—which is just as well because many musicians don't seem to like playing into microphones.

Tests carried out on many different halls (and audiences) suggest that the optimum reverberation times (again at a standard frequency of 500 Hz) are as represented in this diagram:

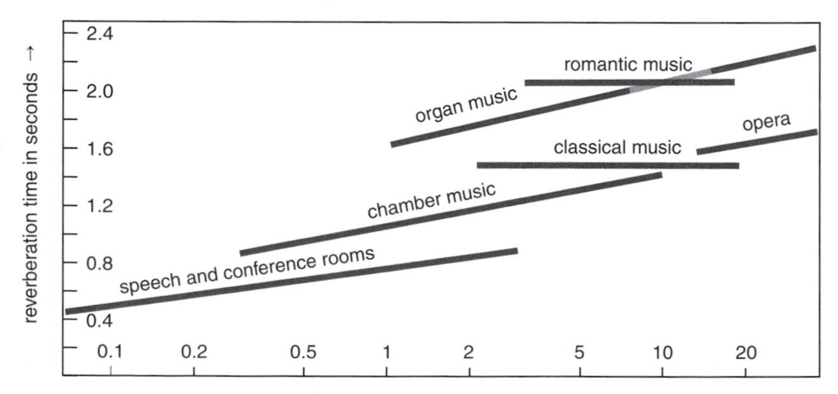

I should point out that, despite the many sources of these figures, they still represent subjective judgements and there isn't universal agreement about them. Nonetheless, acoustic experts need something to work towards when they design halls, and these are the values they usually strive for.

Of course, calculations are only a guide to what the finished hall will turn out to be like. Once construction is complete, actual measurements have to be taken, which present difficulties in their own right. It is necessary to do these measurements when the hall is full because, though it is possible to design seats which absorb as much on average as a member of the audience, you can never make a seat with *exactly* the characteristics of a person (without making it look like a person). Audiences, however, tend not to respond kindly to someone firing pistols from the stage, so the measurement usually has to be done surreptitiously. It means finding some piece of music which ends with a short sharp chord with a clean cutoff—several parts of the Mozart *Symphony No. 40* are said to be good—but in general it's not easy to ensure that all the players stop their instruments sounding at exactly the same time. So there is in existence a composition, called *Catacoustical Measures*, composed especially for the purpose in 1962 by one Daniel Pinkham, which contains a lot of short sharp discords of just the right character. It can easily be put into the programme without sending the audience running for the exits—though I'm not so sure about the critics.

Reverberation time and music

Once you understand what determines the reverberation time of different rooms, it is possible to look back over the way music has developed over the last several centuries, and see what an important part architecture played. An interesting exercise is to go through your music collection and pick out a fairly representative pieces from each period; and listen to them while thinking about what the buildings were like. I'll give you my favourite examples as I go through.

You can start with the old Gothic cathedrals. They were enormous buildings—some of them had volumes of over 100 000 m^3; and they were made of hard reflective materials—stone and marble. So they often had reverberation times of 7 or 8 seconds—sometimes as high as 13. Furthermore the masonry was often rough cut, so it tended to absorb better at high frequencies. This meant that low pitches were emphasized, giving the music a 'dark' timbre—which, admittedly, went well with the often gloomy atmosphere. Under these acoustic conditions, only slowly moving music was possible; certainly with no rapid modulations. Gregorian chant was like that of course; but even in the heyday of classical polyphony, music retained this character. Just listen, for example, to the *Missa Papa Marcelli*, written by Palestrina in 1561 (the work which is supposed to have saved music for the Catholic Church when the Council of Trent wanted to ban polyphony completely). That music stays on exactly the same harmony for many bars at a time, and then sedately moves to another chord—obviously to prevent the reverberation of the building muddying the composer's intentions.

These buildings were so vast that an ordinary congregation made little or no difference; and the acoustics only changed with a really big crowd. The gigantic Basilica of St Peter's in Rome, for example, has had a reverberation time of only 3.5 seconds measured during one of its big ceremonies. But even so, when you remember that people are treble absorbers, you realize that such a crowd will only exacerbate the acoustic imbalance towards low frequencies, and further destroy any hope of understanding the words of what is being sung.

Baroque churches, on the other hand, were different in several important respects. Not only were they smaller on average (volumes of 20 000 m^3 were typical), but they also had lots of wooden furnishings—choir lofts, pews, side altars—as well as panelling on the walls. All this tended to lower the reverberation time—when these churches were full, it was probably not much more than 2 seconds. Furthermore it also shifted the region of maximum reverberation towards higher frequencies (the panelling was responsible for a lot of this). In such an acoustic environment, performances of polyphonic works of amazing clarity were possible; but more importantly, a different kind of music flourished—one in which the different parts are supposed *not* simply to blend with one another, but to be heard as separate entities. This was the heyday of the great contrapuntal music writing, of which perhaps the most obvious form was the fugue. Consider for example the very popular Bach *Toccata and Fugue in D Minor*. That has four different voices on the organ playing individual melodic lines; and in order to be heard distinctly, it is absolutely imperative that the reverberation time should allow your ear to distinguish at least between the faster and slower moving parts.

The chamber music of the Baroque and early Classical periods was, more often than not, performed in the palaces of the nobility. The rooms tended to be much smaller than the churches—a volume of 1000 m^3 is probably a good rough figure—but their construction material was similar, lots of plaster and wood panelling. Reverberation times were as short as 1 second, again favouring higher frequencies. Hence the kind of acoustics they possessed were characterized by brightness, crispness and clarity, and so was the music written to be played in them. Again if you look at something like the Vivaldi *Four Seasons*, there can be, on occasion, as many as 16 different notes per second, in several voices, every one of which has to be heard clearly. But it wasn't just that fast music was popular at the time. The clear acoustics allowed the timbres of different instruments, particularly the all important transient parts of the sound, to be clearly distinguished. So the practice of writing music with specific instruments, or group of instruments, in mind, became much more widespread. That is undoubtedly why the concerto was such a popular musical form during that period.

In the last half of the 18th century, and all through the 19th, audiences for orchestral music increased, and so, of necessity, did the size of the concert halls it was played in. The hall in which Mozart and Beethoven heard many of their own works, the Redoutensaal in Vienna, had a volume of just over 10 000 m^3 and a reverberation time of 1.4 seconds. Any of the Beethoven symphonies offer examples of great contrasts between passages in which individual instruments play melodies with their own particular tone colour, and then the whole orchestra blends them all together in a great climax—the blending process of course relying on the hall to make the individual sounds cohere. In fact, the very name 'symphony' means blending together of sounds.

The hall which was built in the same city nearly a century later, in which Brahms and Bruckner conducted, the Musikvereinsaal, had a volume of almost 15 000 m^3 and a reverberation time of just over 2 seconds. I always think that the musical sound of this period is epitomized by the Tchaikovsky *Fifth Symphony*, which is full of great washes of sound, building to huge climaxes and ebbing away again. There can surely be no doubt that these grand orchestral effects and lush harmonies so characteristic of late romantic music, owe a lot to the sheer size of these concert halls.

The music of the 20th century has tended to retreat from this sound, back to crisper delineation of individual lines and greater clarity of presentation of often very strange harmonies. And then there is pop music, which draws enormous crowds and is irrevocably tied to electronic amplification. The requirements on modern concert halls are perhaps more diverse than they have ever been. So today, although there are several halls throughout the world with volumes of over 25 000 m^3, elaborate precautions are taken not to let their reverberation times exceed about 2 seconds.

The acoustics of opera theatres are in many ways even more complicated. By and large, the constraints on what the reverberation time should be haven't changed much over the years. For the kind of fast-moving recitatives written by Mozart or Rossini, it is better if it is short; while on the other hand, the *bel canto* style of singing favoured by Bellini or the dramatic arias of Verdi need something a little longer. Nevertheless, it is always important to be able to understand what words are being sung, so although the theatres of the 18th century were probably a bit 'dry', with reverberation times between 1.0 and 1.3 seconds, they weren't too different from later ones, which might only get as high as 1.5 seconds.

Over that period however, audiences did get bigger and the theatres had to grow to accommodate them—without increasing reverberation times. One of the main consequences of this was that orchestras had to get larger to fill these increased volumes with an appropriate sound level. In the damnation scene at the end of *Don Giovanni*, a most dramatic effect is achieved when three trombones (who have been doing nothing at all for the rest of the performance) suddenly start to play. The original effect must have been shattering because, in all probability, they sounded just as loud in the tiny (6600 m^3) theatre in Prague where it was first performed, as the whole brass section in a Wagner extravaganza in the Festspielhaus in Bayreuth (18 000 m^3).

The science of architectural acoustics

The man who was almost single-handedly responsible for this field of knowledge was an American physicist, Wallace Clement Sabine. He was born in 1863 in the mid-western state of Ohio, and had an all too typical academic career at Harvard University—well regarded in his own circle, but unknown by the outside world—until a lucky circumstance occurred. In 1895 the university opened a brand new, expensive auditorium, considered something of an architectural masterpiece. It had only one drawback: the reverberation inside was so bad that students had difficulty understanding what their lecturers were saying. Somehow or other Sabine was landed with the job of doing something about it.

He quickly discovered that the architectural design practice for such halls was a curious mixture of traditional wisdom and hit-and-miss. So he began a long series of experiments to build up a body of knowledge from scratch. He recognized the importance of reverberation time, devised a way to measure it, and set about trying to change it. One of his favourite techniques was to put cushions made from different materials on all the seats and then to test the reverberation; but since the hall had to be used for classes (despite the reverberation), and since the only time when it was quiet enough to take sound measurements was in the night-time, he had to do all his work between midnight and 5 a.m. So for over two years he and two laboratory assistants had to move 400 cushions into the hall each evening, and out again in the mornings.

Though he never made that particular hall more than barely adequate, his research was successful in that he built up a formidable body of useful knowledge. It was put to the test when he was called in as a consultant for the projected new Boston Symphony Hall. It was in the course of this design work that he worked out his formula for calculating reverberation times.

The interior of the hall was of a rectangular shape, and of moderate size (18 000 m^3). It seated over 2600 however, by having two complete wrap-around balconies. In design it was rather old-fashioned, even for the time, mainly because it attempted to stay in character with the older hall it was replacing. These drawings will give you some feeling for what it looks like.

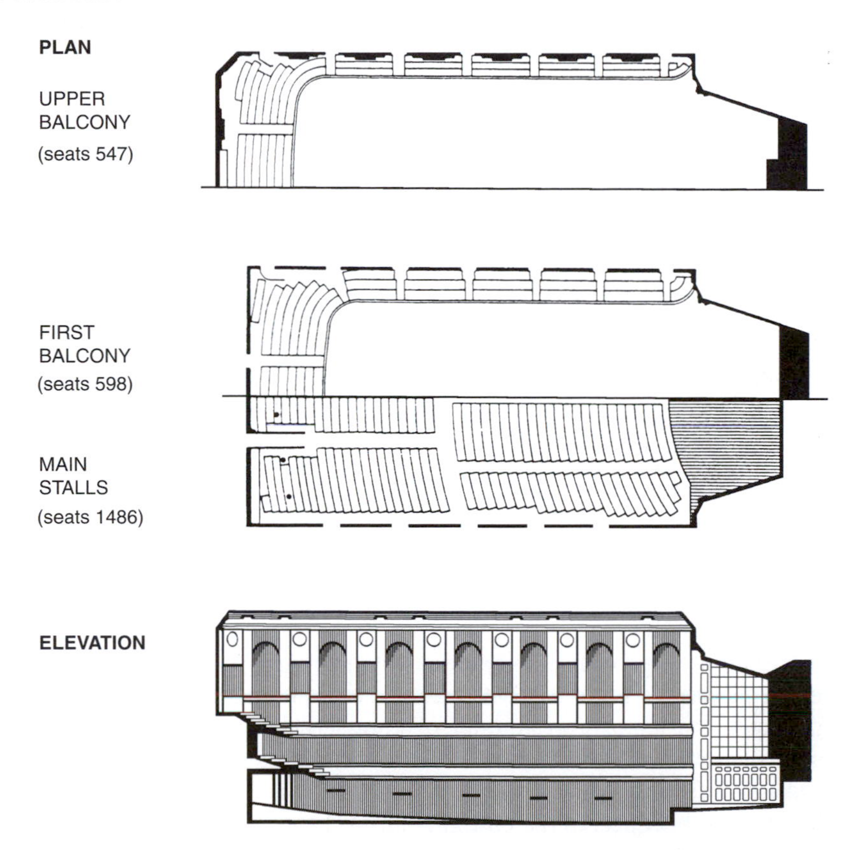

The colour scheme is cream and grey, with red carpets and balcony rails, gilt balcony fronts, black seats and an array of gilded organ pipes at the back of the stage. Of special interest is its very high upper half. The ceiling is broken up with recessed square panels, and the walls are lined with niches in front of which stood replicas of Greek and Roman statues. All these tend to scatter waves in all directions and fill the large reverberant volume with a thoroughly smoothed out field of sound.

The building was opened in 1900 and was a source of bitter disappointment to Sabine. Firstly when its reverberation time of the full hall was measured, it was found to be 1.8 seconds; whereas he had predicted it should be 2.3 sec. The reason for the discrepancy seems to have been that he had underestimated the absorption from the audience. Secondly there was its reception in musical circles. By and large the reaction was uniformly unfavourable. Typical was this comment in a leading Boston newspaper by a widely respected music critic:

> "Like many essays on musical subjects by scientists, it arrives at conclusions with which musicians find it difficult to agree. We have not yet met the musician who did not call the Symphony Hall a bad hall for music."

Sabine was clearly devastated by this reception. He withdrew back into the academic world and devoted himself to teaching for many years, emerging only to advise on the occasional architectural project. He never mentioned the Symphony Hall again in any of his later published papers. He died in 1919.

The irony was that he had been right all along and the music critics wrong. On the evidence from surveys carried out some fifty years later, most informed opinion today regards the Boston Symphony Hall, acoustically, as the best in the western hemisphere, and for its size, one of the best in the world. I suppose the lesson to be drawn from this unfortunate affair is that we are still a long way from being able to say that architectural acoustics is a cut-and-dried science.

Modern architectural acoustics

Because this whole field relies so heavily on subjective judgements, it has developed rather differently from other sciences since the early years of this century. An enormous amount of knowledge and expertise has been accumulated, but it doesn't seem to have brought us much closer to being able to say definitely what makes a hall acoustically 'good'. And this is almost certainly because, in the end, this has to be an aesthetic judgement—and one which is, to some extent, subject to the dictates of fashion.

Nevertheless there are some criteria that are used as yardsticks against which to rate the acoustic characteristics of any hall. The most important two are, of course, first the quietness, the degree to which it can be isolated from any external noise; and secondly the reverberation time, which I have said enough about already.

Then there is something called **intimacy**. This word is usually taken to describe the subjective perception of music heard in a small room; and it seems that it is related, psychologically, to the time separation between the sound you hear directly and the earliest strong reflections. The shortest time gap between two sounds the human ear can detect is somewhere about 1/30 second. (We know, for example, that our ears will hear repetitive pulses of sound as separate entities if the repetition frequency is less than about 30 Hz; anything higher than that starts to sound like a steady tone). So if the sound and its first reflection are closer together than 1/30 second, you can't hear them separately and you somehow get the feeling that you are close to the player, even though other features of the sound may be characteristic of a large space. A time delay of 1/30 seconds, you will recall, means a path difference of only 10 m, and therefore, in large halls, this requirement is sometimes difficult to meet. In newer buildings it is often achieved by suspending reflecting panels (called **clouds**) well below the main ceiling to provide early reflections—without of course, preventing the main body of the sound reverberating through the upper part of the room's volume.

Perhaps the most extreme example of 'non-intimacy' occurred in the Albert Hall in London when it was first opened. There the time delay between the direct sound and the first reflections was so large that the audience could hear a distinct echo of everything that happened on the stage. It was made worse by the fact that the high domed ceiling actually focussed the sound somewhat, and strengthened the effect. This was only finally solved by the installation of 'clouds'.

This is however, not the primary purpose of clouds (or 'donuts' as toroidal shaped ones are sometimes called). They are often suspended directly above the stage so that some of the sound will be reflected *backwards*, and the players can hear one another clearly and get an idea of the whole sound they are making. They provide what is called **ensemble**. Again there is a maximum distance at which they can be set for this purpose. If they are much more than 4 or 5 metres above the players there will be a gap in the sound of more than the critical 1/30 second, and all hopes of playing together will be completely lost.

Another word you often hear is **warmth**, which seems to refer to the proper balance between low frequencies and the rest of the sound. This can be a worry with artificial reflecting devices like clouds, which will reflect high frequencies well enough, but which lower ones tend to diffract around. The danger then is that all the important early reflections make it sound as though the bass instruments aren't playing loudly enough, and the sound is perceived as lacking 'warmth'.

There are other ill-understood effects. There is some evidence that audiences like lots and lots of reflections coming to them from all directions—in other words being bathed in a diffuse field of sound. On the other hand, there is also evidence that many listeners also like the sound to be somewhat directional, not to be exactly the same at both their ears. However, such effects need to be much more thoroughly understood before they will contribute much to the practical business of designing halls.

It must always be borne in mind that any judgements about a hall's acoustics are very subjective, and by and large audiences like listening to music in the halls they are used to. So when a new building is opened a very natural reaction is to blame the unfamiliarity of the sound on 'bad acoustics'. And then acousticians may talk about the need to 'fine-tune' the acoustics of the hall. But, apart from such devices as using moveable drapes and panels to change the amount of absorption (and hence the reverberation time), there isn't much else that can be done. In the end they often do nothing at all, and simply wait for the audience to get used to the hall.

Chapter 6

The romance of electricity

During my wanderings through the history of physics and music, I am constantly delighted when I stumble across interesting cross-connections. Right at the end of the Age of Enlightenment, in the very year of his death, Mozart wrote a charming little piece for a strange collection of instruments. It was an *Adagio* (K.617) for flute, oboe, and an instrument with a haunting, ethereal tone, which had only recently become fashionable for (alas) too short a time—a **glass harmonica**. It is played by stroking the rims of a set of glasses, just as you yourself must have done many times at the dinner table; and it had been invented some twenty years earlier by the American patriot, ambassador, political theorist, scientist and notorious womanizer, Benjamin Franklin. And he is important to our story because he was one of the earliest experimenters in the field of study just then opening up—electricity.

We are now moving into that period of time during which historians talk of there being a transition in the intellectual climate of Europe from **classical** to **romantic** ideas. It wasn't exactly a revolution, there was no sudden, clear-cut change of direction; and I must stress that, when you compare the artists and writers of the late 18th and the early 19th centuries, their similarities are much more pronounced than their differences. Nevertheless the differences are real and it makes sense to try and understand the intellectual developments of the time by using these two well worn labels. So we group together the music of Haydn, Mozart and early Beethoven and call it 'classical'; by which we mean to draw attention to the formality, the elegance and the love of order that we believe it shows. And we contrast it with the music of, say, Schubert, Berlioz, Wagner, in which we discern the 'romantic' ideals—a love of freedom, a glorification of nature, a seeking after the mystical. It goes without saying that no one composer epitomizes *all* these qualities; but as a group, on average, they all do.

The Romantic Movement affected much more than just music. It was a whole new philosophy, a reaction against the values of the Enlightenment. It's not easy to summarize the basic tenets of this world-view—they were something like the connectedness of all knowledge; the primacy of the spirit over the material, of the subjective experience over the empirical fact; and the idea that the philosopher or the artist or the scientist had feelings and perceptions and that these had to count. These ideas permeated all areas of intellectual endeavour, particularly the other arts. During the generation or so spanning the turn of the century, the cool, clean style of painting and classical inspirations of David gave way to the bold colours and exotic subjects of Delacroix and Gericault. In literature the elegance of Pope and the wit of Voltaire yielded to the flamboyance of Byron or the romance of Scott. And therefore you won't be surprised that I am led to ask the inevitable question: did a similar shift of emphasis occur in physics?

In one sense, the answer seems to be no. Without doubt the greatest advance in the first half of the 19th century was the uncovering of the law of conservation of energy. A study of how the machines of the Industrial Revolution drew their power from the furnaces of the English coal-fields soon established that heat was somehow stored in chemical compounds and could be transformed into mechanical energy—though always with less than 100% efficiency. Hence there grew up the concept of a fixed store of energy which couldn't be created or destroyed—but which could be bought and sold. And it is this connection with the squalid taint of commercialism which makes us think of 19th century physics as the very antithesis of the Romantic ideal.

Yet in another sense the answer is yes. Science as a whole was enormously influenced by Romanticism. There was a reaction against what was seen as the overly mechanistic view of the Enlightenment, which wanted only to take things apart and tinker with them. The poet John Keats (while he was still a young man, I might add) wrote these lines:

> Philosophy will clip an Angels wings,
> Conquer all mysteries by rule and line,
> Empty the haunted air, and gnomed mine-
> Unweave a rainbow ...

Spurred, perhaps, by sentiments like these, there arose an organic model of nature, in which everything was connected with and depended on everything else, in which things should be studied in whole rather than in parts. In many ways the typical Romantic scientist was a naturalist, rather than a physicist. Yet physicists are part of the times in which they live; and the study of electricity, which developed most during that period, showed unmistakable signs of being influenced by the Romantic Movement.

The history of electricity

On the fringes of science there have always existed odd pockets of knowledge, which nobody quite knows what to do with; until suddenly a use is found and they take off like rockets. Electricity was one such. The ancient Greeks knew that, when you rubbed amber (which they called *electron*), it would attract light feathers. The Romans knew that a certain kind of stone from the island of Magnesia would attract pieces of iron. They also knew that certain fishes gave you a nasty shock if you touched them, and their doctors actually prescribed it as a cure for gout. They knew all these things but they had no idea what they meant.

It was only in the 16th century that William Gilbert, physician to Queen Elizabeth I of England, who was responsible for the insight that the earth behaved as though it had a giant magnet at its core, classified all the materials capable of being charged by rubbing. Over the next two centuries, more knowledge gradually accumulated: that electric charge can travel through some materials (called **conductors**) but not through others (**insulators**); that large amounts of charge can be stored in specially shaped metal containers (called **capacitors**); and that there are two kinds of charge (given the labels **positive** and **negative**) rather like the two poles of a magnet.

This last was largely attributable to Franklin, who did a lot of his work on lightning; and the most common picture we have of him is flying a kite in the middle of a thunderstorm, trying (with incredible foolhardiness) to discover the nature of electric charge. As a result of this work he was able to show that the discharge from a cloud consists of only one type of charge, negative; and he invented a very successful lightning arrestor. It is amusing to read that, after the American War of Independence, King George III ordered the removal from English public buildings, of all of these devices based on the design that Franklin advocated, presumably because the science behind them was ideologically unsound.

But the whole subject remained an interesting diversion so long as no steady source of charge could be found, other than cumbersome rubbing machines or the occasional thunderstorm. That changed in 1791 when the Italian anatomist, Luigi Galvani, published his work on 'animal electricity'. Starting from the simple desire to investigate the effect of electric shocks on living organisms, he was the first to perform that experiment, familiar to any student of high school biology, that if you touch electrical conductors to the right spots on a frog's leg, it will kick. He showed that this happened because a tiny electric current flowed; and so sensitive were his experiments that he could predict an approaching thunderstorm by watching his frogs' legs react. He concluded that electricity was produced in all animal tissue, and was possibly responsible for life itself.

It is interesting that in the public's reaction to Galvani's work, you can find early evidence of a growing split between science and the Romantic movement in the arts. In 1815, Mary Woolstonecraft Shelley, wife of the poet Percy Bysshe Shelley and friend of Lord Byron, wrote her celebrated Gothic novel *Frankenstein.* In it she proposes that Galvanism could hold the secret of how a creature constructed from parts of human corpses could be brought to life; but at the same time she expresses profound mistrust of scientists who hold the mechanistic view that creatures are no more than the sum of their parts, and who don't think through the social consequences of their work.

Anyhow, the controversy surrounding Galvani's experiments spurred on his compatriot, Alessandro Volta, to the discovery that electricity comes, not from animal tissues as such, but from a chemical interaction between the fluids they contain and the metal probes used to investigate them. From then it was a short time until he produced the first chemical cells to yield small amounts of charge; and a pile (or **battery**) of these to provide the first sizeable source of electric current. After that the subject went ahead by leaps and bounds. The connection between electricity and chemical forces was unearthed by Humphrey Davy, the son of a Cornish miner; its connection with magnetism by Hans Christian Oersted, the son of a Dutch apothecary; and the whole lot was unified by the son of a London blacksmith, Michael Faraday.

Of all the figures in the history of science, Michael Faraday must surely be one of the most conventionally 'romantic'. He was gentle, unassuming, unswervingly honest, deeply religious, never very wealthy yet on speaking terms with royalty, a man of many friends yet always working alone. In personality he sounds very like his contemporary, Franz Schubert; but he had one gift that Schubert did not. In the things that mattered he was lucky.

Born into a large working-class family in 1791, there was never any question of his being given more than the most rudimentary education. As a young man he worked in a book-binding business, where at least he could read scientific treatises, which luckily his employer didn't mind. Then in 1812, again by luck, a customer gave him a ticket to hear Humphrey Davy's very prestigious series of public lectures at the Royal Institution. When they had finished, Michael transcribed and bound his notes and sent them to Davy, asking for a job.

Davy was not usually a generous man, but he happened to need an assistant, so he took Michael on as a bottle washer. However just then an important continental tour was being arranged, a rather complicated undertaking in pre-Waterloo Europe, and Michael was taken along as general dog's-body. It was hard work, but at least he did get to meet some influential scientists, particularly Volta. On returning to London, he virtually lived in and for his laboratory. His talents soon became obvious and he steadily gained professional recognition (much, I might say, to Davy's discomfiture). He was eventually made director of the laboratory and was elected to the Royal Society in 1824 (with Davy's being the only opposing vote).

Another happy chance brought him widespread popular recognition. After Davy's death, the public lectures had been taken over by his successor, Charles Wheatstone—rather less successfully. However, on one occasion Wheatstone was suddenly taken ill, and Faraday was asked to step in, even though he had little lecturing experience. He proved to be a natural. Very soon he'd gained a devoted audience, including members of Queen Victoria's family, and the Institution began to show a profit. You can find admiring references to his Christmas lectures in the essays of Charles Dickens, himself no mean public lecturer.

He died in 1867. During his life he remained modest to the end, refusing a knighthood and (more importantly) the presidency of the Royal Society. One of his last wishes was that he shouldn't be buried in Westminster Abbey among the other glitterati of the English establishment, but rather in his own village among his own family. But then he hardly needs a marble monument: the general public can be reminded of him, if they want, every time they switch on an electrical appliance. For, working in his dingy laboratory (which you can still visit at the Royal Institution), Faraday produced a startling succession of major breakthroughs. He used the magnetic effect of a current to make an iron shaft rotate, thus inventing the **electric motor**. He operated it in reverse and showed that turning a metal coil between the ends of a magnet would produce a current—thus inventing the **electric generator**. He studied the way in which currents break up chemical compounds in solution (the process called **electrolysis**) which is the basis of many of the present-day methods of extracting metals from their ores. More than anyone else he was responsible for the realization that electricity was a form of energy.

To my way of thinking, Michael Faraday presents a splendid example of the influence of the Romantic movement in physics. He was an intensely practical man who mistrusted mathematicians. The mathematical approach to physics was very characteristic of the Enlightenment, particularly the Newtonian point of view which treated everything as particles moving in response to forces; and Faraday always claimed it often obscured more than it enlightened. His approach to theoretical physics was quite different. He believed that the influence of an electric charge (or a magnet) resides, not so much in the body itself, but in the space surrounding it—in the **field**, to use the correct term. Furthermore he invented a way of visualizing these electric and magnetic fields as being shot through with long elastic strings (called **lines of force**,) for all the world like some fantastical harp. To the casual reader it may sound naïve, but in the hands of a genius it yields insights which are breath-taking. It is a broad brush attack, rather than one which picks out all the details. And if there truly is a 'romantic', as opposed to a 'classical', approach to physics, I think this is it.

Of course fashions always swing back. In the 1860's the Scottish mathematician, James Clerk Maxwell, re-established the mathematical approach. He showed that electric and magnetic fields should behave exactly like any other medium which was capable of carrying a wave. This idea remained theory for nearly a generation, but in 1888, Heinrich Hertz (whose name we've been commemorating all through this book) became the first man ever to generate and receive **radio waves**. With that discovery this branch of science took off in another direction, which I will come back to later.

Meanwhile, back at acoustics

By this time I am sure you are wondering why I am talking about all this in a book which is supposed to be about Physics and Music. Well, because electricity was such a rapidly expanding field in the early 19th century, it couldn't help influencing the rest of physics. It isn't easy to put ourselves into the frame of mind of the times, but it must have been extraordinarily exciting; and, even though it didn't have much effect on society for more than a generation—the telephone, for example, wasn't invented until 1876 and the light bulb till 1879—intelligent men and women must have been able to sense its immense possibilities. It must have been particularly irresistible to ambitious young science graduates looking for a field to work in. So they flocked to it, leaving other areas relatively denuded—none more so than acoustics.

Let me give you a list of very famous names from all over Europe, who started off working in acoustics and who switched fields early in their careers. Jean Baptiste Biot as a young man measured the speed of sound in metals, taking advantage of the relaying of iron sewer pipes in the city of Paris. Felix Savart wrote his first thesis on the vibration modes of a violin body. Georg Simon Ohm gave his name to an important observation on the working of the human ear, which I will discuss in Chapter 7. Charles Wheatstone did his early work on resonance in air columns and actually invented the concertina. Gustav Kirchhoff worked on the vibrations of plates, which I will also come back to in Interlude 7. All of these names will be well known to any student of modern electrical engineering: but I doubt that any of them would be aware of the contributions these people made to what was fast becoming a Cinderella subject—acoustics.

But all was not lost. In the end acoustics gained, because some of the knowledge and understanding that was developed for the glamorous new technology proved to have application in the the study of sound, and there were enough workers familiar with both fields to recognize this fact. The reason for this cross-fertilization is simply that both areas of research are involved in much the same problem—the transfer of energy through an intervening medium. As an illustration of what I am saying, think about one of the earliest technological developments in electricity—made in 1835 by Charles Wheatstone, among others—the **telegraph**.

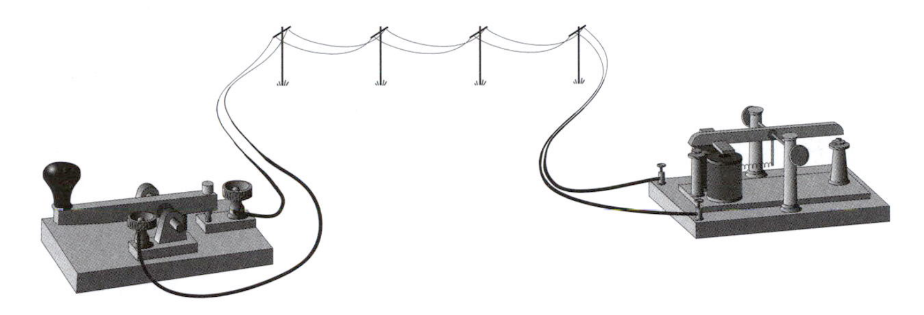

In principle a telegraph is very simple. You turn on a switch or press a key and a current starts to flow in a long wire, causing a light or a buzzer or something (Wheatstone used an electromagnet) to go on at the other end. If you establish a code of *ons* and *offs* you can send a message, and the code invented in 1837 by the American, Samuel Morse is still in use today. But difficulties arise if the wire gets too long. When it reaches the other end, the current must have enough energy to light the light or sound the buzzer; but an electric current loses energy as it travels through even the best conductor, because it tends to heat the material. It was Ohm who first realized this, and he called the property of metals which caused this energy loss, **resistance**. Technically it is measured by the ratio between the energy the battery gives to electric charges to make them move (measured as its **voltage**) and the current that it produces.

The problem is not insuperable. What was done in the early days was to put batteries at regular intervals along the length of the telegraph wire to feed in extra energy as needed. The device which did this were called a **relay**—an invention, incidentally, which is traceable to an Englishman who migrated to Australia in 1838, Edward Davy. But then in the 1860's a project was launched which posed a technological challenge of such magnitude as to engage many of the very best minds for over a decade—to establish a telegraphic link between Europe and America by laying a trans-Atlantic cable.

Now a cable consists of two parallel wires, or more usually one wire inside (but not in contact with) a thin conducting sheath. When you send a current down one of these, quite intense electric and magnetic fields are established between the two conductors. It is found that, if you set up an *oscillating* source of electrical energy of very high frequency at one end, an oscillating (or **alternating**) current will flow down the cable, with an extremely low rate of energy loss. The energy seems to travel in a wave down the *fields* rather than in the conductors themselves. Nevertheless, even though the energy loss is low, there is still a limit to the amount of current which you can get with a source of fixed voltage. The ratio of these two is known as the **impedance** of the cable; and electrical engineers treat this concept as a generalized form of resistance.

The point of all this is now, I hope, clear. The problem of sending an alternating current down a cable is exactly the same as sending an elastic wave down a string or an acoustic wave down a pipe. So when the scientists working in acoustics migrated to electricity, they were able to bring with them a lot of valuable know-how. Then as time went on and solutions were found for an ever wider range of electrical problems, the flow of knowledge was reversed. Particularly during the early part of the 20th century, a new way of looking at problems and a whole lot of technical jargon was carried back into acoustics. That is why I want to stop talking history (once again) and to explain, without necessarily calling on much electrical theory, the concept of **acoustic impedance**.

Energy coupling

A good way to approach this topic is to start with something simple, like a tuning fork. When you strike it, it makes a very soft sound. But if you hold it upright on a table top—or anything else with a large area—it is much louder. The explanation is fairly simple. To produce a loud sound, strong pressure waves have to be set up in the air. But the tines of the fork are quite small and can't get much air moving, no matter how vigourously they vibrate. However, when the fork touches the table, it causes the wood to vibrate in sympathy; and even though this vibration has a much smaller amplitude, it is in contact with a lot more air. As a result the sound is louder.

Clearly the sound-board of a piano does the same sort of thing, and also the body of a violin. They serve the purpose of **coupling** the vibrations of the strings to the air, so that acoustic energy can get out. However, if you position your ear carefully near a violin (or better still use a microphone), you will notice that only some of the sound comes from the wooden sides. A lot (particularly at some frequencies) comes out of the f-holes.

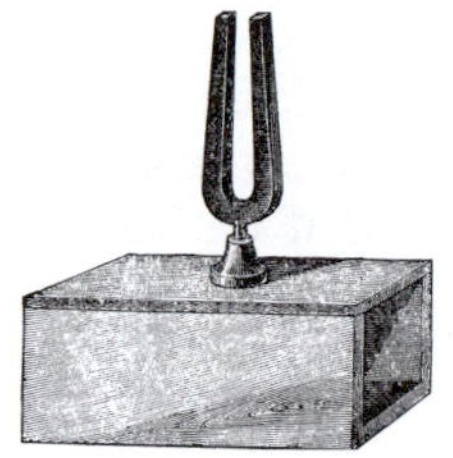

Even the humble tuning fork is often provided with its own wooden sound-box, usually open at one end only and closed on the other sides. In that, essentially all the sound comes from the open end. What both of these examples show, perhaps unexpectedly, is that it is easier for the energy to travel from air to air than from wood to air.

This is a deceptively simple observation, and to understand it we have to return to basics. I want you to think back to the Newton's cradle of Chapter 5, and to why it is that a 'wave' of energy travels through the row of steel balls.

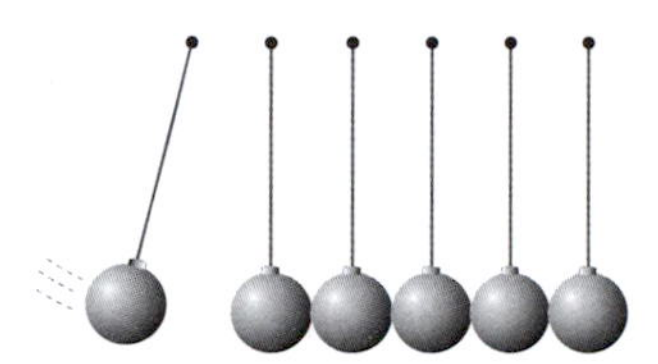

You will recall that the most relevant question is whether or not the balls are all identical. If they are, the energy travels smoothly down the row; but if any one of the balls is different from the others, it doesn't. We have already seen that if the different ball is much heavier than the rest, the 'wave' approaching it will be reflected. What is perhaps not so obvious is that, if the different ball is lighter, the approaching 'wave' will still be reflected (at least partially). To understand that, we have to consider once again a Newton's cradle with only two (different) balls.

I think there is no difficulty about intuiting how this behaves. If, for example, a heavy ball hits a light one, it is clear what happens.

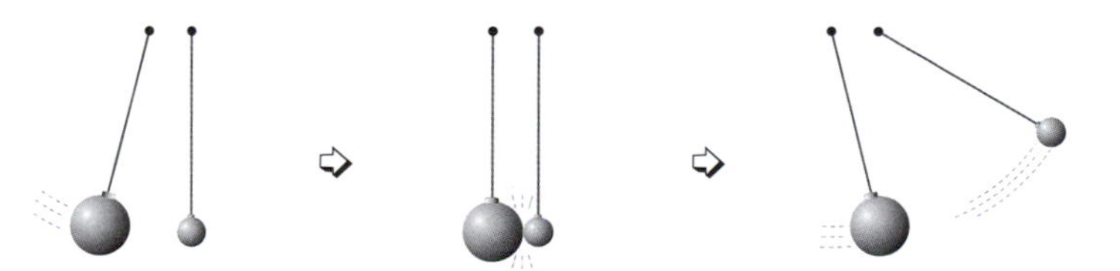

The critical observation is that, after the collision, the first ball retains some of its initial energy: not all has been transferred. Had the second ball been heavier than the first, the result would *look* different—the first ball would be moving backwards at the end—but the main conclusion would be the same. The first ball would still have some of its energy.

Now I want to apply this insight to something more complicated than two balls colliding with one another. For a longer Newton's cradle, the process by which energy is passed down the line involves each ball being slightly *compressed* rather than being moved forward bodily. If one of the balls happens to be heavier than the others, its greater inertia means that this compression occurs more slowly. The elastic forces last a longer time than is necessary to bring the first ball to a stop, and will cause it to rebound instead. But this effect could be *compensated* if the elastic properties of the second ball were weaker. A smaller force acting for the longer time might be just enough to stop the first ball without letting it rebound. All the energy would then be passed on. Hence in this case it is not just the mass of each ball which must be identical if the 'wave' is to travel cleanly, but some other quantity, one which involves the product of the mass and the elastic properties. (Because it is a *product*, an increase in one compensates for a decrease in the other).

By this stage you are used to making the next logical jump. What is true for a row of balls, will almost certainly be true, analogously, for other sorts of waves. Of most interest to us are acoustic waves in a column of air. We saw before that, in this system, the air pressure usually plays the same role as does 'elastic property'. But what plays the role of 'mass' is slightly more complicated. It obviously depends on the density of the air, but also involves the cross-sectional *area* of the column. You can appreciate this by considering how the air inside a pipe responds to a change in external pressure. If its density is large it will respond sluggishly, but this could be compensated by widening the pipe, giving the pressure a greater area to act on. What is relevant here is the density *divided* by the area.

This 'other quantity' is given the generic name **impedance**. I've been coy about defining it more carefully, because there are several kinds of impedance which are used in various technical calculations, each with a slightly different meaning. But I don't think you need worry about these subtle distinctions, so long as you grasp the main idea.

- Any medium (or part of a medium) which can carry an acoustic wave has a property, which I loosely call **acoustic impedance** which can be used to describe how energy moves through it.

For an acoustic wave in a column of air:

$$\text{acoustic impedance} = \sqrt{\frac{\text{pressure} \times \text{density}}{\text{area}}}$$

(There should actually be a another number close to 1 in this formula, but I think we can ignore that).

The most important use for this quantity is in working out what happens when an acoustic wave tries to travel from one medium to another. If the impedances of the two media are not the same, then only some of the energy will go forward and the rest will be reflected; and a knowledge of the acoustic impedance lets us calculate how much.

I will discuss this subject a bit more quantitatively in Appendix 5, because I think it is worth pursuing by anyone interested in the subject; but too much detail at this point would obscure the flow of ideas. So I will just summarize the results and point out that you can calculate what happens when a sound wave meets a sudden change in acoustic impedance. The *fraction* of acoustic energy which is reflected when the wave tries to move from medium (1) into medium (2) is given by this formula:

$$\text{fraction of energy reflected} = \left(\frac{\text{impedance}(1) - \text{impedance}(2)}{\text{impedance}(1) + \text{impedance}(2)}\right)^2$$

For us now, the most important feature of this formula is that it shows the only conditions under which no energy is reflected—so it *all* gets through—is when the two impedances are exactly equal. When this happens we say the impedances **match**.

Impedance mismatches: transmission and reflection

Telegraphy provides a very straightforward example of these ideas. The electrical impedance of a cable, which is a measure of how much voltage is necessary to get a certain current flowing, obeys these same laws. So also does the resistance of devices that use energy. Therefore when you try and put a signal into such a device you must watch out for the possibility that, if the impedance of the supplier doesn't match that of the user, then all the energy may not get through. You have probably experienced, during an overseas long-distance telephone call, hearing an echo of everything you say. The telephone operator will apologize for this by calling it a 'bad connection'.

Another good example is a television set. Its antenna picks up a tiny, high frequency signal from the 'air space'. This has to get to the set, usually via a cable. The antenna has an impedance, and so does the cable. If they are not the same, then all the signal will not be transferred; some will be reflected from the connection point. Likewise the input impedance of the set must match that of the cable, or else more of the signal will be lost at the next junction. This is a very common feature with the use of electronic equipment. Whenever you have to plug one device into another—microphones or earphones or loud speakers into amplifiers or radios or distribution boards—you should, in principle, make sure they are compatible (that their impedances match) and that you use the proper connectors; otherwise the energy won't get through properly. (I say "in principle" because, especially with loudspeakers, it is usually better to put up with energy loss at the junction in order to avoid other problems associated with the amplifiers having to work over a large range of signal levels).

The acoustic equivalent of this kind of impedance can actually be felt when you try to play any brass instrument. Put your lips on the mouthpiece and try to get a note by setting your lips vibrating (by blowing a 'raspberry'). If you happen to hit the right frequency at which the instrument produces its notes, you can feel the back pressure. You have to apply that same pressure to get the appropriate amount of energy going into the tube. Your mouth has to match the impedance of the instrument.

But on a simpler level, it should be clear by now how to describe the observations I made earlier about the violin body or the tuning fork's soundbox, in terms of acoustic impedance. The primary vibration of the strings or the tines sets into oscillation both the wooden sides of the box and the air enclosed. But when the energy tries to get out through the sides, it meets a great mismatch between the very high impedance of the wood and the very low impedance of the open air. As a result, only a small fraction gets through, even over a large area. On the other hand, the impedance of the enclosed air is much closer to that of the outside; and so that is predominantly where the sound comes from—the open end or the f-holes.

There is one particularly important feature of all this which you can appreciate by putting a few numbers into the previous formula. Only when there is a *big* mismatch will the fraction of energy transmitted be very low. Check for yourself. If the larger of the two impedances is, say, 5 times the smaller, then about half of the energy will be reflected and half will get through. The ratio therefore has to be considerably greater than 5 if you actually want the wave to be reflected.

It is important to realize that this is not just a property of sound waves or electrical systems. As an example, most of our knowledge of the interior of the earth comes from seismic (earthquake) waves. As these travel through the material of the earth, when they meet any change in structure, a small amount of the energy is reflected and the rest goes on. So a good sharp seismic shock will pretty soon degenerate into a number of smaller echoes; and by monitoring these it is possible to work out where and how large the 'changes in structure' are.

Even more impressive are similar technical advances made in ultrasonic medical diagnosis. Here a narrow beam of acoustic waves (of very high frequency and therefore very small wavelength) are sent into a human body. At every change in tissue structure or at the surface of each new organ, there is an impedance mismatch. Only part of the beam's energy is transmitted. The rest is reflected, and can be picked up back at the transmitter. By timing the arrival of all these different echoes, and by scanning the beam through different angles, it is possible to build up a map of, say, the uterus of a pregnant woman and to 'look at' the baby within. And it all stems from the fact that some energy will be reflected at any point where the acoustic impedance suddenly changes, and some will go on.

Ideal air pipes

Musically all this has greatest application to sound waves in air pipes—particularly in wind instruments. I have already pointed out (in Interlude 1) that standing waves can be set up in pipes which have either one or both ends open to the outside air. The process is exactly the same as in a stretched string—it occurs because travelling waves are being reflected from the ends. Now it is easy to see why a sound wave should reflect (or echo) from a *closed end* of the pipe; but it is a problem which worries many people—why should it reflect from an *open end*?

The most naïve explanation is that, when a pulse of moving air gets to the end of a pipe, it spills out. This leaves less air inside the pipe than should be there. Therefore more must move in from the outside, and this in turn will cause another disturbance. So another wave will travel back down the pipe. Exactly the same happens in a Newton's cradle. When a pulse of energy has passed through the row of balls, the end one will be kicked out. Left to itself, it will then fall back and cause an identical pulse of energy to travel back up the row. The 'wave' has been reflected from the 'open' end.

This explanation is simplistic. It is much more elegant to talk in terms of impedance. An acoustic wave travels down a cylindrical pipe *because* the acoustic impedance of each 'slice' of the air column is identical with the 'slice' before it. But when it comes to the end, this is no longer true. It doesn't matter whether the end is open or closed. There is a sudden (and large) change of impedance. Therefore the wave must be reflected.

It is as simple as that.

However, there is one important difference between what happens at an open end and at a closed one; and this difference is patently obvious in a Newton's cradle. When an energy pulse reflects from a very massive ball (a 'closed end'), it does so instantaneously. There is no movement at the point of reflection. If a standing wave were to be set up, this point would be a node.

But when the pulse reflects from a light ball (an 'open' end), there is maximum movement at that point. Up till then the movement of each ball was constrained by the next ball. But here there is no next ball. So the movement of the ball will be greater than anywhere else. It must correspond to an antinode. And there's another difference. There is a time delay between when the 'wave' reaches the end and when it starts to travel backwards. The reflected 'wave' is at a different part of its cycle.

The same sorts of conclusions can be drawn about other waves. You can arrange an 'open' end for a stretched string by hanging it vertically. The bottom end must have a small weight on it (to provide some tension) but not be rigidly fixed. If you very carefully start a pulse at the top end, you will see essentially the same kind of behaviour as you saw with the Newton's cradle.

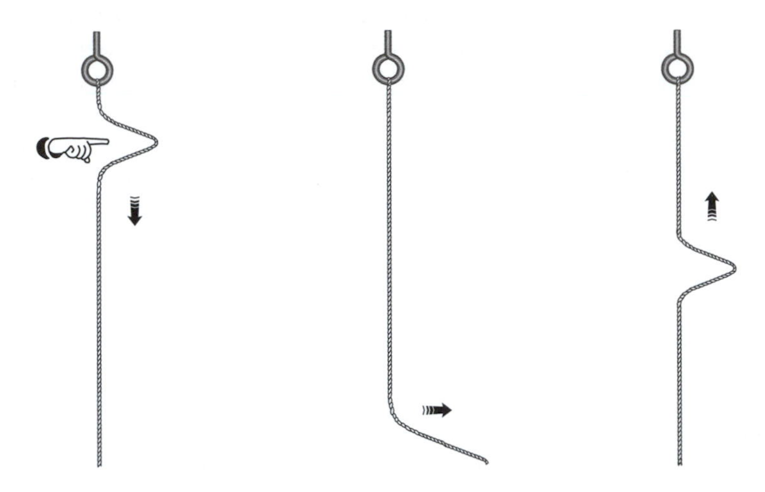

The pulse travels down till it reaches the end-point, which swings to the side. From then on the string behaves as though you had jerked the bottom sideways yourself, and an identical pulse will travel back up. The pulse has been reflected. But note carefully that it travels back up the *same side of the string* as it came down.

These diagrams could also be interpreted as describing a sound pulse reflecting from the open end of a pipe, provided that you interpret each drawing as a *graph* of the (longitudinal) movement of the air particles at different points along the length of the pipe. In the same way, you will recall that, in the last chapter, I drew pictures to illustrate the reflection of a pulse from a fixed end of a string. Those same diagrams could equally well describe a pulse of sound bouncing back from a cap over the end of the pipe.

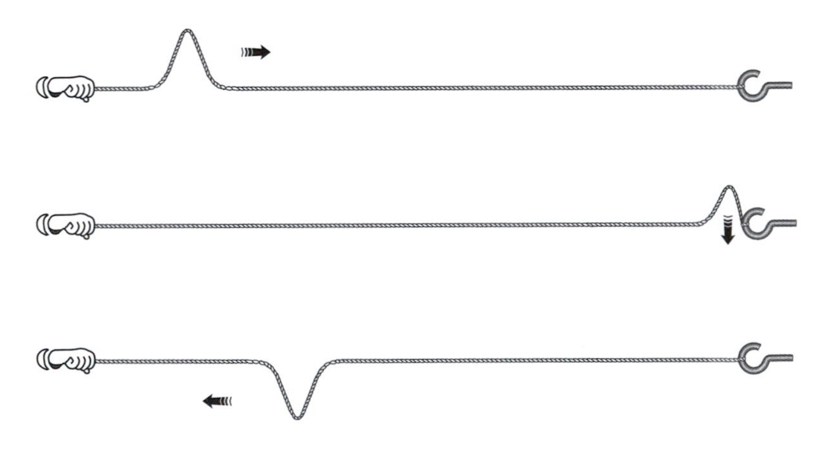

But notice the important difference between these two sets of pictures. In the second, the reflected wave is a back-to-front and upside-down version of the initial wave. That is what makes the end-point a node. But in the first, where the reflection is from an open end, the reflected wave is the initial wave back-to-front but *right-way-up*. That makes the end-point into an antinode.

Anyway, it should now be an easy step to think about *sine waves* travelling in the air pipe, and to deduce what kind of standing waves they can set up. The reasoning is the same as for a string, except that there is an ambiguity in *interpretation*. If the pictures are taken to represent how the particle movement varies along the pipe, then at an open end there must be an antinode, and at a closed end a node. Thus the fundamental standing wave for a pipe, open at one end and closed at the other, is:

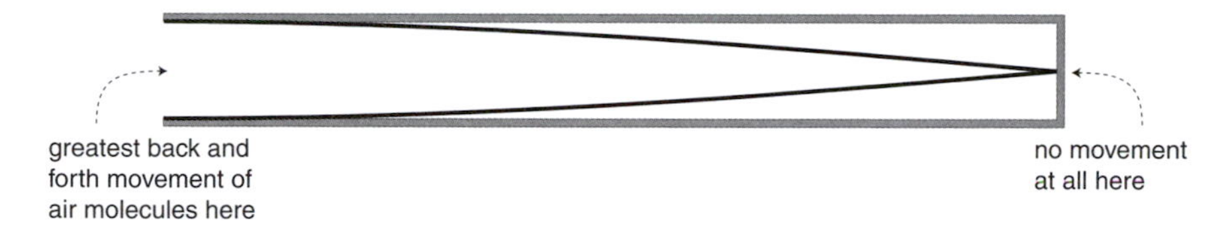

You will notice immediately that only a *quarter* of a complete wavelength fits into the length of the pipe.

There is another way to represent the same standing wave, and that is to draw a diagram to show how the *pressure* varies along the pipe. If you think about it carefully, you will realize that the air density, and hence the pressure, varies *most* at places where the particle movement is *least*; and where this movement is greatest, the pressure hardly changes at all. (Think about a stream of cars driving along a highway—they pile up to maximum density at busy intersections, toll booths or at any other places where they have to travel very slowly). Therefore the way the pressure varies along the length, for the fundamental standing wave of an open/closed pipe, is:

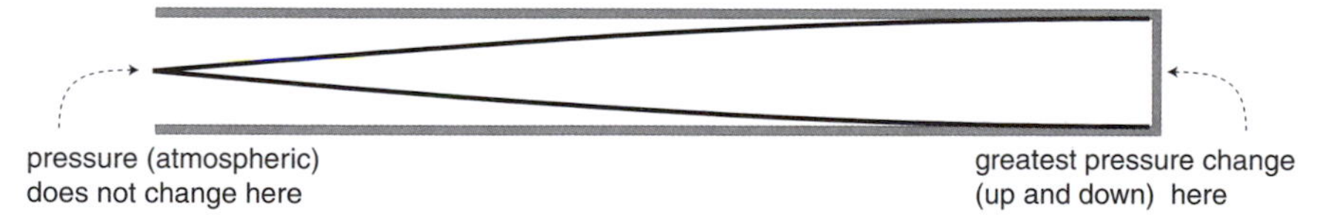

It is very important to be clear about the difference between these two diagrams. In a sense they are inside-out versions of one another, but they convey exactly the same information. Either can be used. You will find the first kind (particle movement diagrams) in most elementary textbooks; but in more advanced articles the second (pressure diagrams) are more common. I certainly prefer the latter, and I have already used that convention in Interlude 1 if you remember.

Anyhow, it is now very easy to construct a table of all the possible standing waves of a pipe open at one end and closed at the other—often referred to generically as **closed organ pipes**. In this table I am using the symbol l to stand for the length of the pipe.

PRESSURE DIAGRAM	MOVEMENT DIAGRAM	WAVE-LENGTH
	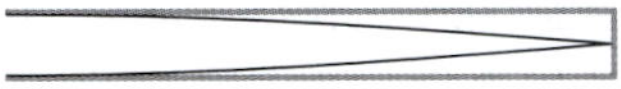	$4 \times l$
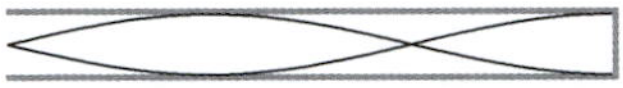	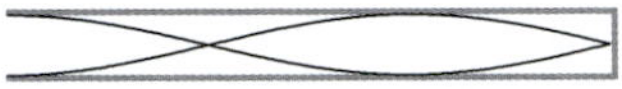	$4/3 \times l$
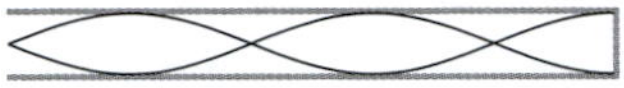	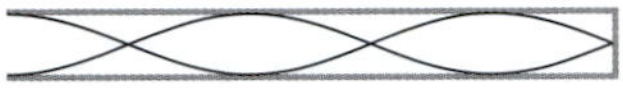	$4/5 \times l$
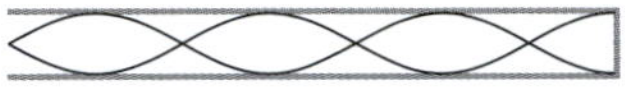	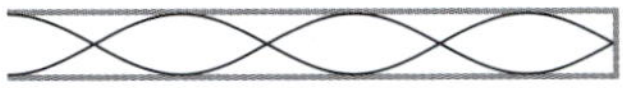	$4/7 \times l$
etc		

By contrast, an **open organ pipe** is a cylinder with both ends open. The corresponding table of standing waves looks like this:

PRESSURE DIAGRAM	MOVEMENT DIAGRAM	WAVE-LENGTH
	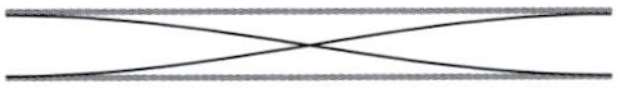	$2 \times l$
		$1 \times l$
		$2/3 \times l$
	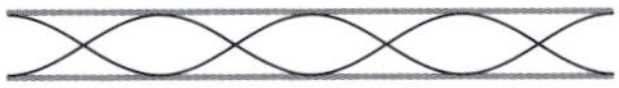	$1/2 \times l$
etc		

You will observe that the pressure diagrams have nodes at both ends and are therefore the same as the standing waves in a stretched string. The table of these (pressure) standing waves looks just like the one on page 151. The corresponding movement diagrams, of course, have antinodes at both ends instead of nodes.

Vibration frequencies of ideal pipes

The important insight to be gained from all this discussion of standing waves in columns of air, is that the length of the pipe primarily determines the *wavelength* of the sound wave. Musically however, we are more interested in its *frequency*; and that is determined at second hand as it were. To calculate it you have to use the formula on page 141, and divide the speed of the wave (i.e. the speed of sound in air) by the wavelength.

In this manner you come up with the following table of values. Note that in this table I am using the following symbols:

f_0 for the speed of sound divided by the length of the pipe,
f_{cl} for the lowest frequency in the open/closed pipe, and
f_{op} for the same thing in the open/open pipe.

	VIBRATION FREQUENCIES OF A PIPE			
MODE	CLOSED ORGAN PIPE		OPEN ORGAN PIPE	
fundamental	$f_0/4$	f_{cl}	$f_0/2$	f_{op}
1st overtone	$3f_0/4$	$3f_{cl}$	f_0	$2f_{op}$
2nd overtone	$5f_0/4$	$5f_{cl}$	$3f_0/2$	$3f_{op}$
3rd overtone	$7f_0/4$	$7f_{cl}$	$2f_0$	$4f_{op}$
...etc				

Note the important insight you can get from this table. The open organ pipe can play a complete harmonic series of notes; and presumably when sounded at its fundamental pitch, will contain many of these as overtones, depending on how it is sounded. This fact I have already used extensively in Interlude 1. The closed pipe also sounds a harmonically related series of notes, and they too will normally be present as overtones. But notice that, since its lowest note must be considered its fundamental, the overtones consist of only *odd numbered* harmonics! This will give a very distinctive timbre to any musical instrument (like a clarinet) which is basically a cylinder open at one end.

There is another important consequence of the fact that the length of the pipe determines the wavelength primarily, rather than the frequency. The speed of sound in air varies quite a lot, depending on prevailing conditions, and therefore the frequency of the vibrations will change. Wind instruments notoriously go out of tune easily. To work out how much, we need to go back to the formula for the speed of sound which I quoted on page 137 of Chapter 5.

$$\text{speed of sound in air} = \sqrt{1.4 \times \frac{\text{pressure}}{\text{density}}}$$

You will notice that it depends on both the local air pressure and its density. Now everyone knows that the air pressure can vary as local climatic conditions change; but paradoxically this is irrelevant. For as the pressure increases the air gets squeezed and its density increases, exactly in step. So the air pressure, though it contributes to the *value* of the sound speed, doesn't directly influence the way it changes. But the density certainly does; and there are two ways the density can change easily.

- As the **temperature** increases, any gas that is unconstrained will expand, and its density will decrease. This is well known. It is why hot air rises from a fire, and why cold air from an open refrigerator sinks. And the effect is quite large. I have already quoted the value of 1.20 kg/m^3 for the density of air at standard pressure and 20°C (also on page 137). If the temperature were to rise to 30°C, then the density would fall to 1.14. That is a decrease of about 5%.

  It may not seem much, but translate it into a change in frequency. If the density falls to 1.14, the speed of sound rises from 343 m/s to 348—a change of nearly 2%. So if the frequency of the pipe was initially 440 Hz, it would now be 447 Hz, which is a change of 29 cents. And to a musician trying to play in tune that's a lot—nearly a third of a semitone.

- The nature of the gases in the air determine its density, so if they change, then so will it. An extreme demonstration you might like to do for yourself (with sufficient care of course) is to fill your lungs with the gas from a helium-filled balloon. That is much, much less dense that air (which is why the balloon floats); and you'll find the apparent pitch of your voice will rise by nearly an octave. (To be absolutely accurate it is the frequency of vibration of the air inside your vocal cavity which increases, but that's OK.)

Water vapour is also less dense than air. Therefore if the **humidity** inside a pipe changes, so too will the air density, and consequently the pitch of its notes. Luckily this effect is not very large. Even if the humidity is 100%, there is still only a very small amount of water vapour actually present in the air, and the density (at 20°C) is still 1.19 kg/m^3. This would put the frequency of the pipe I talked about before at about 441 Hz. A small change, but still about 6 cents, which a good ear would start to notice.

The lesson therefore is that players of wind instruments, particularly woodwinds, have continually to monitor the tuning of their instruments; for during an evening's performance the temperature and humidity changes I used in those calculations could very easily occur. Other instruments are also affected by temperature changes, of course, but not to nearly the same extent; and in the end that is because of the *indirectness* of the dependence of frequency on the length of the pipe.

Real air pipes

Getting back to a more general discussion of standing waves in air pipes, at this stage I must confess that what I have said so far is really only an *approximation* to the truth. There are factors which I have ignored, particularly the width of the tube, which complicate the standing waves slightly. I have, if you like, only been discussing *idealized* air pipes; and I must now talk about what happens in the real world.

The point at which I made a too-sweeping statement was when I implied that, because of the change of impedance at the end of a pipe, an acoustic wave will be completely reflected there. But this only occurs if the change of impedance is very big. Now we know that the impedance of the air inside the pipe is relatively large because its cross-sectional area is quite small. That of the outside air, however, requires a different method of calculation, because there is no obvious 'area' any more. You might try imagining what is the 'effective area' of air that could be set vibrating, and this would at least convince you that the impedance of the outside air is much smaller than that inside the pipe. But it isn't zero. So there isn't a *complete* mismatch at the end of the pipe (by which I mean that the ratio of the bigger impedance to the smaller isn't infinitely large). Therefore there won't be complete reflection. Some of the acoustic energy will leave the pipe.

In one sense this is a trivial observation. Of course acoustic energy must leave the pipe, otherwise how would we hear it? But what is not so trivial is realizing that this affects the standing waves inside. You have to argue your way through this rather carefully. Because some of the wave is leaving the end of the pipe, the pressure must be fluctuating there. It is only a small oscillation, but it means that the end-point of the pipe doesn't correspond exactly to a pressure node. Instead, the standing wave will look as though it comes to a node a small distance *outside* the pipe.

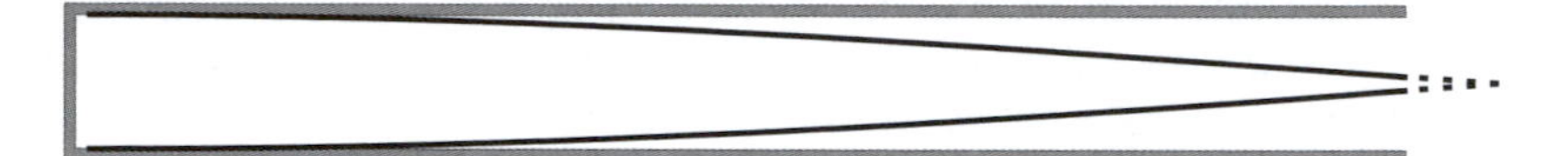

If you can't convince yourself that this is what must happen, you may find it easier to think about a stretched string tied to a peg that isn't wedged in tightly, or about a skipping rope held lightly in the hand. In these too, the end won't stay completely still and an imaginary node will be located a fraction further out, just as in my drawing. The resulting effect, so far as the wavelength of the standing wave is concerned, is that the pipe behaves as though it were a little *longer* than it really is; and therefore the pitch of its note is a little *lower* than you might expect. This extra 'effective length' is known technically as the **end correction**.

Just to show that we understand all this properly, think now about what difference the width of the pipe makes. The acoustic impedance of the air inside depends on its width, in such a way that it gets bigger as the pipe is made narrower. The mismatch at the end of the pipe therefore increases correspondingly (since the impedance of the outside doesn't change). So if the pipe is very narrow, there will be essentially complete reflection at its end, and the end correction will be very small. But if the pipe is very wide, the reflection will be only partial, there will still be a lot of pressure fluctuation at the end, and the end correction will be large. In fact a good rule of thumb is:

$$\text{end correction} = 0.6 \times \text{radius of the pipe.}$$

The concept of acoustic impedance becomes even more useful when we come to consider other complications which occur in real pipes—for example finger-holes in the side of the tube. A hole opens the pipe to the outside and changes the impedance just at that point; so its effect should be rather like changing just one ball in the middle of a Newton's cradle, or resting your fingers lightly against some point on a vibrating string. In an air pipe there is an impedance mismatch at that point, but it is only partial (and can be changed depending on how much of the hole the player uncovers), so it only reflects part of the acoustic wave. A small fraction travels past, only to be reflected at the end of the pipe (or the next hole).

The resulting standing wave will therefore have (at least) two different parts to it.

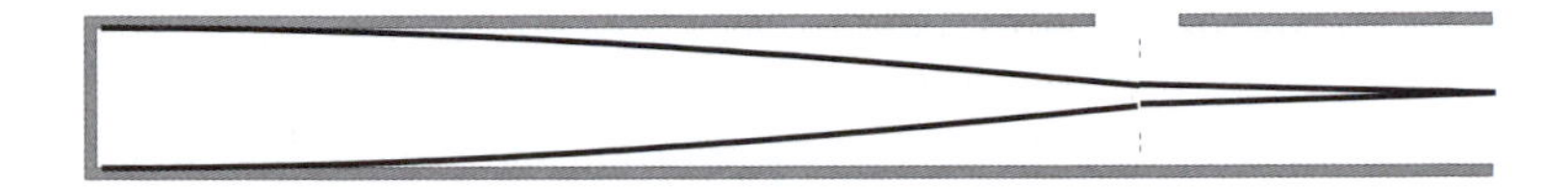

The problem of working out what these standing waves should look like, and calculating what frequencies they correspond to, is a fascinating theoretical problem, but I will defer talking about it further until the next Interlude, which is about woodwind instruments.

Impedance matching

So far I have talked about acoustic impedance in situations in which the reflection of waves was a necessary part of the function. But there are many times when you don't want reflection—when you are interested in *transmitting* energy. The most obvious example is a loud-speaker. Here the source of sound is some sort of thin disc, set vibrating by a fluctuating electric current. You want as much of this energy as possible to get to the outside, but there is a huge impedance mismatch between the air and the solid material in the disc. Is there anything you can do?

In fact the answer is yes. You can improve the transmission of energy from one medium to another by inserting between them a small piece of a third medium, whose acoustic impedance is somewhere between the two. The jargon word for it is **impedance matching**. (I have included in Appendix 5 a sample calculation showing that this conclusion is logical extension of the definition of impedance.) It is exactly what is done when optical technicians put an **anti-reflective coating** on the lenses of eyeglasses or expensive cameras, in order to cut down on the light loss that occurs every time a light ray goes from air to glass or vice versa.

If you want a simple, and really quite insightful, metaphor for what is going on, compare this to a staircase. An abrupt change in height, which may be very difficult to climb by itself, becomes easier if you put a step between the levels, so that you do it bit at a time. That's all I'm saying.

In acoustics you can play much more elaborate games with impedance matching. If you can improve transmission past an abrupt change by splitting it into two smaller changes, then you should be able to improve it even more by breaking each of these similarly (putting more steps in the staircase). Indeed, carrying the argument to its logical extreme, there seems no reason why you mightn't get *all* the energy through if you join the two pipes continuously (or replace the stairs by a ramp), thus:

Well, that's almost true; though it is important to think about how gentle the slope of the 'ramp' should be. I'll return to this point shortly.

A gradual change, from the low impedance of the wide end of a cone to the high impedance of the narrow end, is the basis of operation of those old-fashioned ear-trumpets, or indeed of the ear itself. An eardrum is very small and obviously has a much higher impedance than the outside air, so the problem of capturing an appreciable fraction of the already low level of acoustic energy is very real. It is the **outer ear** which acts as an elaborate device to get over (at least partly) this mismatch.

Exactly the same principles apply, in reverse, to the design of an ordinary loudspeaker. The vibrating disc sits at the base of an expanding cone, which acts like the 'ramp' in the previous diagram. There are still huge mismatches at the ends of the cone, but a lot more energy gets out than would without it. On very early gramophones the design of these speakers was doubly important because the original source of vibration was so weak. That's why they had such enormous horns, which were often decorated with floral designs and made into feature articles of furniture.

At this stage I'm prepared to bet that you are jumping ahead, and guessing that this is also the function of the **bell** on the end of all brass instruments. Well, yes and no. Such instruments are indeed very loud, just because they have such a large bell; but that isn't its only function. Its second function relates to the question I set aside a few minutes ago: over what distances do you need to match impedances smoothly in order to get rid of reflections?

To see what this question asks, look at these two junctions:

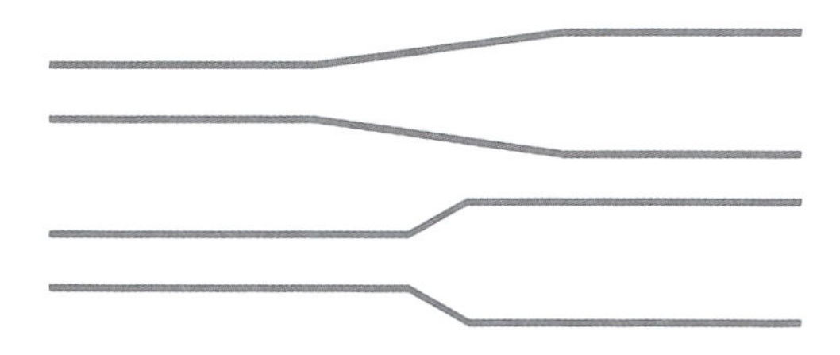

In both of these, the outside pipes are joined by an expanding cone. It may (or may not) be clear that the first will allow a wave to pass completely, but it is difficult to believe that the second one will. Somehow it goes against my intuition of how sound waves behave. For the junction to be smooth, the sound wave has to expand (or diffract) out of the narrow pipe, follow the sloping sides, and then bend in again to be channelled along the second pipe. I can't help feeling that the 'ramp' must be long enough to let the wave settle down before it comes to the second corner. This can only mean that it mustn't be too much smaller than the *wavelength*. The intuition I am using when I say this is clearly something to do with staircases again. Unless each step is wide enough for you to get your balance on, it's not going to serve its purpose of letting you climb easily. Therefore, if I push my intuition to the limit, I'd say that even the first junction would look pretty 'abrupt' to a long wavelength wave, though it would probably pass short ones efficiently.

The conclusion is that whether or not the sloping sides of an air tube will cause an acoustic wave to reflect at that point depends on the wavelength. This is a fact that the makers of wind instruments have used intuitively for as long as they have been practising their craft. They have found ways to alter the shape of the tube so as to change the frequencies of the standing waves into whatever they want. In particular the bell of a brass instrument (with some help from the mouthpiece) can change the notes it will play from a set of odd harmonics (which is what you'd expect from a tube closed at one end by the player's lips) to a complete set of even and odd harmonics. Exactly how this is achieved is a bit too detailed to go into here, so I have included it in Appendix 5.

So let me stop discussion of these matters at this point, and finish by drawing your attention to the way I have tried to discuss this subject, using insight gained in one branch of physics (electricity) to illuminate another (acoustics). It doesn't have to be done this way; I could have explained it all without ever mentioning the word 'impedance'. But to take that attitude is wrong. The new way of looking at things predicts new features which are really difficult to follow otherwise. Acoustics was irreversibly changed by electricity theory, and there's no going back.

The legacy of electromagnetism

It is interesting that acoustics wasn't the only part of physics that was changed fundamentally by electromagnetism. It all started with experiments which newly measured the speed of light at the end of the 19th century. At the same time that Wallace Sabine was doing his work on architectural acoustics in Boston, another American, Albert Michelson from Chicago University, was carrying out a series of experiments to measure accurately how fast light travels. The story of how it became for him a lifelong obsession is a fascinating tale in itself, but not what I want to talk about now. Ever since Maxwell had shown that light was an electromagnetic wave, the big question was: what was the medium that was 'waving'? And the evidence that came from Michelson's experiments was soon irrefutable: *there was no medium at all!* But, as we saw earlier in the case of sound, the properties of the medium are what determine the speed of the wave. So if light needs no medium, what determines its speed? Why should there be a special speed at which nothing moves through nothing?

The answer, when it came, involved a complete rethinking of what we mean when we measure speed, distance and even time itself. These quantities are not absolute, they must always be thought of in relation to other quantities. Hence the name given to this new theory was **Relativity**. Many workers were converging on these new ideas; but the man we usually associate with it, from the time he gave the world its first clear exposition in 1905, was Albert Einstein.

He was born and educated in Germany, but did much of his early work in Switzerland and spent his last thirty years in America. He developed early a deep love of the violin, which he played all his life—passably well, I believe. He even got to play (as one of the perks of being a world-famous scientist) with top rank professionals like Isaac Stern. And it is an often-repeated story that, during one such session, Stern remarked in exasperation: "The trouble with you, Albert, is that you can't count!".

I often think it's ironical that Einstein, who made such a mark on the public consciousness because of the wisdom and gentleness he displayed in later life, should have been the one to turn physics upside down, and whose work would eventually lead to that ultimate horror—the atom bomb. For physics was changed beyond recognition by his paper on relativity. Above all, it was the old complacency that was lost, the belief that we knew it all; and this was probably a good thing. The new mood that took its place was beautifully summed up by the English humourist, J.C. Squire, who transformed that couplet of Pope's that I quoted in Chapter 4, into a quatrain:

> Nature and Nature's laws lay hid in night,
> God said "Let Newton be!" and all was light.
> It did not last. The devil, crying "Ho!
> Let Einstein be!", restored the status quo.

Interlude 5

Woodwind instruments

This group of instruments includes: the recorder, flute, clarinet, oboe, saxophone and bassoon. They are grouped together to distinguish them from those other wind instruments, the 'brasses'. But the name is misleading. They are not all made of wood—the flute is usually silver, and the saxophone brass. Likewise, as we saw in Interlude 1, it is perfectly possible for a 'brass' instrument to be made of wood.

It is an argument that has simmered on and off, doubtless forever, between scientists and musicians whether what a wind instrument is made of affects the sound it produces. Does a flute made of gold sound better than one made of silver? Is a wooden recorder better than a plastic one? It is difficult to do properly controlled experiments on this question, because expensive instruments have a lot more care devoted to their making than cheap ones, and that accounts for most, if not all, of the difference in sound quality. So I won't give a definite answer to the question. But I will point out that during one such controversy in the late 19th century, Besson and Co made a cornet out of paper and bugles out of plaster. And Messrs Mahillon of Brussels made instruments for a whole band out of cheese! Apparently they all sounded much the same as the real instruments.

So it isn't the material they are made of which characterizes either group. You will remember that, in a brass instrument, the prime determinant of the pitch is the player's lips; and, crudely speaking, the instrument itself is just a column of air which strengthens their vibrations. In a woodwind on the other hand, the player provides a featureless stream of air—it can even be done mechanically—and it is the pipe itself which determines the note. The pitch is changed by opening holes in the side of the tube. So the features which I will need to talk about in some detail are: firstly that part of the instrument the player blows into, and later the side holes.

Edges and reeds

Everyone knows that there are two simple ways to produce a tone with a stream of air. One is to blow gently across the open neck of a bottle (the wind will do the same thing as it blows across a chimney). Another is to blow hard between two leaves held close together. This produces a high pitched squeal, and actually takes a bit of practice; but when you get good at it you can make a similar sound with thin pieces of card or slivers of wood—all of which I will hereafter refer to as **reeds**. These two simple phenomena have provided, from well before the start of history, the means of generating the acoustic vibrations in what we now call woodwind instruments. But the physical explanation of *why* they produce a tone at all was only relatively recently understood.

Let me talk about the first method first, and consider what happens when a stream of air blows against a sharp **edge**. If the stream is steady, there seems no reason why it shouldn't divide smoothly, so that some goes to one side and some to the other, like this

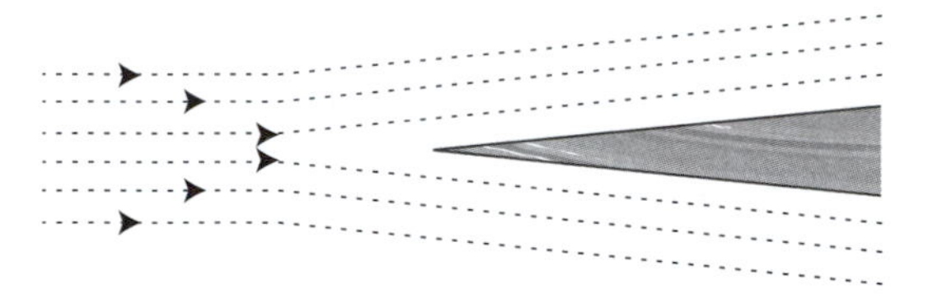

But it doesn't. The flow pattern is unsteady and turbulent. It sloshes backwards and forwards, breaking up into swirls and eddies.

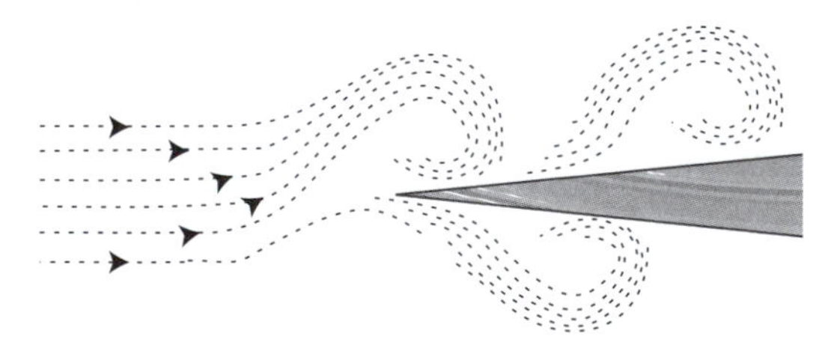

The pattern isn't exactly periodic, but there is a certain regularity to it, which depends on many factors, including the speed of the stream, the sharpness of the edge and the angle at which they meet.

You can observe the same behaviour in the way a flag flaps in the breeze. Again, you'd think that, provided the wind remained steady, the flag should hang perfectly flat and straight, edge-on to the wind. But it never does. It always flaps, more or less rhythmically; and the rhythm gets faster as the wind speed increases. The reason air behaves like this isn't terribly well understood. It has something to do with the thermal motion of the molecules that make it up: but that isn't important for us here. Once you are aware that it *does* happen, you can appreciate what its consequences are.

Imagine the sharp edge to be at one end of a cylindrical tube—as is the case at the bottom of an organ pipe. When air is blown in through a narrow slot (the **flue**) a fast jet hits the **edge**. As a result there will be turbulent motion near the **mouth** of the pipe, and a confused jumble of pressure waves will travel up the pipe. At the top end, of course, they will all reflect. When they get back to the mouth, they will make the motion there even more incoherent, *except* if there are any waves present capable of setting up a *standing wave*.

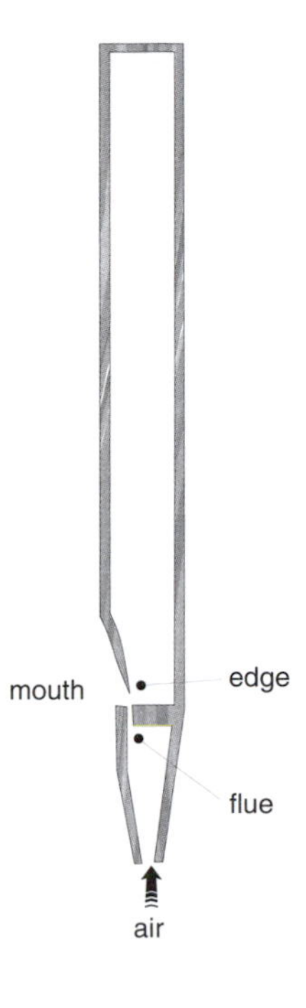

If this occurs, the fluctuations of the air near the mouth will have a strengthened component at this particular frequency; and this will, in turn, react back on the very unstable processes which gave rise to the turbulence in the first place. More waves will travel up the tube and back, strengthening this component even further. Very soon this motion will dominate, suppressing all other fluctuations, and eddies will be produced with perfect regularity, at a frequency corresponding to the fundamental standing wave of the tube. This process is a beautiful example of what engineers call **feedback**.

The same general principle underlies the action of a reed. In a stream of air, a reed will tend to flap rather like a flag. So if you blow air through a narrow gap between two reeds at one end of a pipe (as in an oboe),

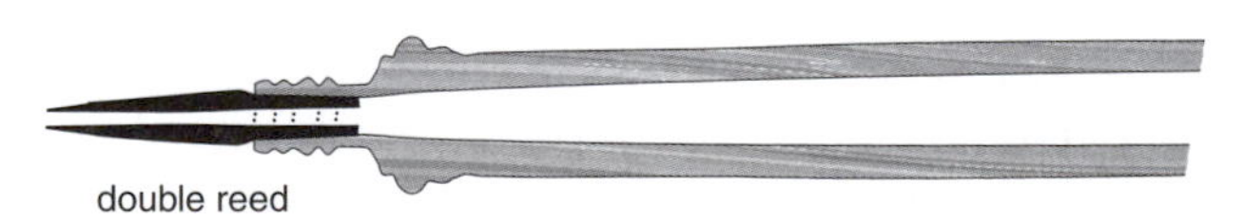

or between one reed and an unyielding surface (as in a clarinet),

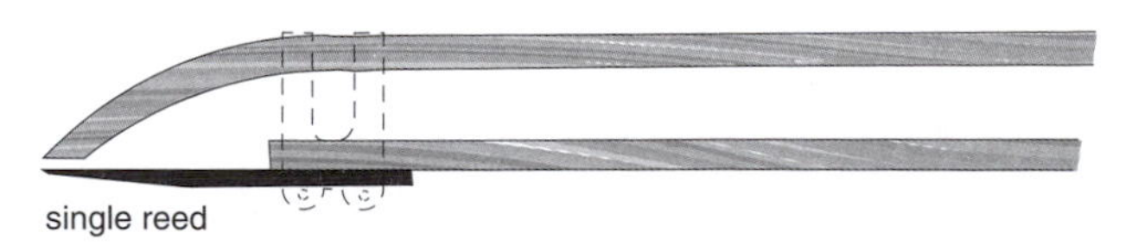

then this flapping will open and close the gap, more or less regularly. Air will enter the pipe in discrete puffs, and these will send pressure waves down the pipe. The same feedback mechanism as before will come into operation, and the reeds will soon be forced to oscillate at the fundamental frequency of the pipe.

It is worth noting here that, in principle, it is possible to mount a reed in the *open* end of a pipe. The reeds of a harmonica (or mouth organ), for example, simply vibrate like a flag when you blow past them: they are called **free reeds**. But since they don't completely cut off the air stream, as the more usual reed action does, there is no feedback mechanism to tune them to the length of a pipe. Therefore they aren't important in this interlude. But, just for the record, instruments like the mouth organ, the concertina and the accordion are known generically as **free reed instruments**.

Both edges and reeds are used as generators of the oscillations inside woodwind instruments. One of their characteristics is that, because the feedback mechanism is so critical, the player has very little control over the pitch. A small degree of tuning is possible by varying the force of the air stream, or by changing lip pressure on the reed; but that is all. There is no way, for example, that a *tune* could be played on a woodwind mouthpiece attached to a length of hose-pipe (as Dennis Brain did with a French Horn mouthpiece in the Hoffnung concert in 1956).

It is also important to recognize the intrinsic differences between edges and reeds as tone generators. All else being equal, a reed gives a *louder* tone than an edge. This is because when a reed vibrates it closes the entrance gap completely each cycle; so the air movement varies between full-on and nothing. The variation near an edge however isn't nearly so extreme. This distinction explains why woodwind instruments have traditionally always been divided into two classes—the *haut* or 'loud' instruments based on reeds, and the *bas* or 'soft' based on edges.

Similarly there is an intrinsic difference in timbre. Edge tones are produced by a side-to-side movement of the air which is symmetric (at least in the way I drew it). The pressure varies with time in a natural oscillation, a simple fundamental with few overtones. They tend to give a pure tone—though a skilled player, by changing the angle of the air-stream, can emphasize different harmonics. A reed on the other hand, vibrates very much in an on/off fashion. While the gap is closed, air cannot enter the tube and the pressure is high. When it is open, the pressure falls to the local value. Exactly how the pressure varies with time is not easy to deduce, but it must have something of this character:

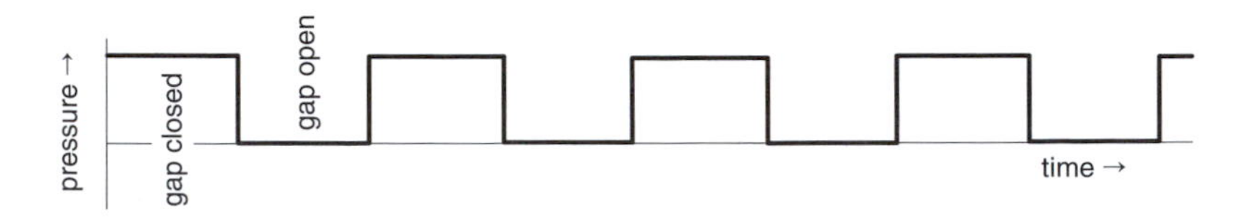

I have drawn this figure assuming that the gap is closed for precisely half the cycle, and open for the other half. It gives a wave shape we have seen before. It is the **square wave** which I showed on page 97.

Now the square wave is a shape whose harmonic spectrum is well known, and if you look very carefully at that diagram (on page 97) you will see its most important feature. *There are no even harmonics in the spectrum, only odd ones.* Its actual spectrum looks like this:

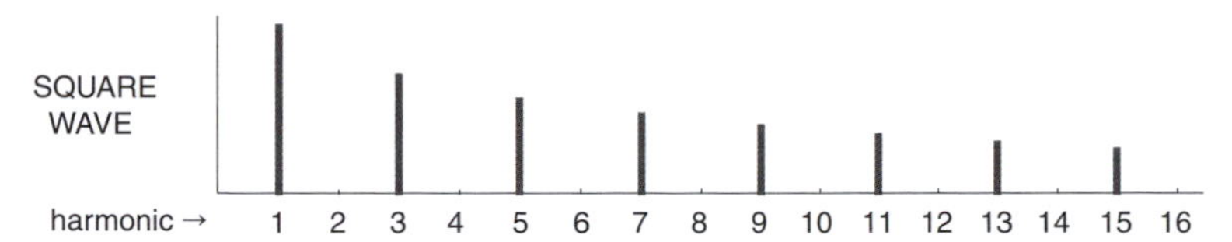

The *reason* for this is simply geometric. The square wave has the property that the second half of each cycle is the same as the first half, only upside down. Therefore if it is made up of simpler waves, they must have this same symmetry. And only the odd harmonics have that.

Now go back a bit. How reasonable was my assumption that the reed closes the gap for the same length of time each cycle as it opens it? My guess is that that's probably not too far off for a single reed, but a double reed would probably have the gap open for longer than it is closed. So let me draw the kind of pressure wave you might get if the open phase is twice as long as the closed phase—a 'rectangular' wave:

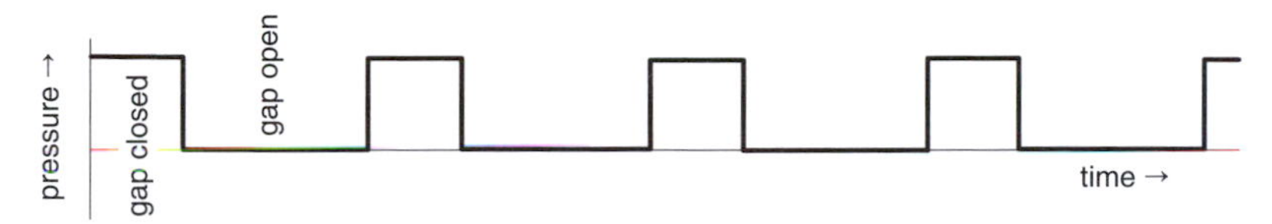

The harmonic spectrum of this shape has a similar interesting property. Every *third* harmonic is missing.

Again, if you think about it carefully, you could probably see the geometric reason for this behaviour, but I don't think we need to. I'm only guessing at the shape of the wave, so we can't take the results too seriously.

But the general trend of my argument is valid. Because the tips of a double reed are in contact for less than half the cycle, it is probable that the resulting pressure wave spectrum will have some similarity to the last graph. The third harmonic mightn't actually be missing, but you wouldn't be surprised if it were quite weak. So the sound would have as its strongest components the first, second and fourth harmonics.

And isn't that exactly what we saw when we looked at the spectrum for an oboe, on page 99?

Instruments without finger-holes

Any instrument which uses one of these generators but doesn't use finger-holes, must of necessity have a number of pipes—each one of the exact length to give a single note of the right pitch. One of the earliest recorded such instruments was the **syrinx** or **panpipes** of ancient Greece.

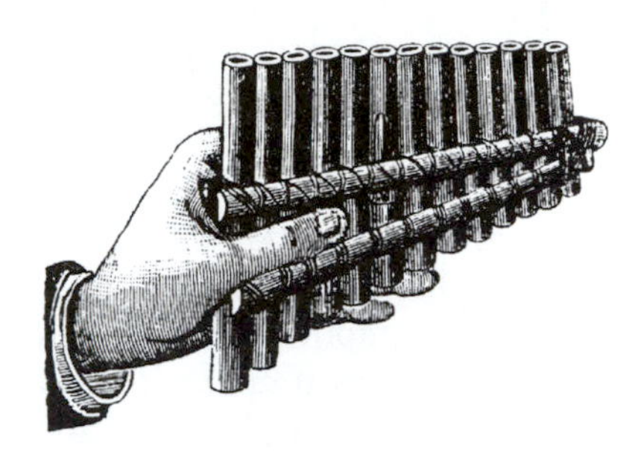

It consisted of a number of hollow lengths of cane, each plugged at one end, tied together; and it was played by blowing gently across the open end of different canes in turn. It gave a soft breathy tone, and many a poet has described its sound as hauntingly beautiful; but its limitations were obvious. Firstly it could only play one note at a time (although some later versions had neighbouring pipes tuned a third apart which might be played together).

Secondly the number of available notes was severely restricted. The shepherds of ancient Arcady could play seven notes, perhaps in a pentatonic scale (and, by careful blowing, overtones of these): musicians in modern Rumania, on instruments with twenty pipes, can span nearly three octaves, but only diatonically. So, although these instruments were widespread in Europe right up to late medieval times, their popularity fell off when polyphonic music developed to the stage of needing accidentals.

A natural development, at least from a theoretical point of view, was the **organ**. It was a much bigger, more elaborate device with a wind-box to supply air to the pipes, and a keyboard to select which pipes were to be sounded. It made available to its player a very wide range of notes, to be played singly or in any desired combination. The earliest known model was constructed by an Alexandrian engineer in about 250 BC, and was known as the **hydraulis** or **water-organ**. Again the name was a bit deceptive. It wasn't worked by steam, as early Christian writers seemed to think: the water merely served to provide a constant pressure head to regulate the air supply. In was only in the 3rd century A.D. that bellows became popular, and the hydraulis gave way to the **pneumatic organ**.

In about the 9th century, when the Church relaxed its opposition to instrumental music in worship, the organ spread throughout Christendom. In the next thousand years many impressive models were constructed to fill with music its churches, monasteries and cathedrals. There are reports of a gigantic organ at Winchester in the 10th century, which had 400 pipes and needed 70 men to keep its 26 bellows going.

During the 13th and 14th centuries, as music became more secularized, the fashion for building large organs declined; and smaller, portable models became more popular. This had the happy consequence that many of the hitherto clumsy mechanical details were worked on and became more streamlined.

By about 1500 most of the features of the modern organ which don't rely on electricity had been invented. One of the most important was the complicated **tracking action** which connects the keyboard, via long thin rods, to the lids which open or closed each of the pipes. (In modern organs this connection is usually made electrically.) So instead of the pipes having to be placed in order directly behind the corresponding keys on the keyboard, they can be arranged in any of the strikingly beautiful arrays which are such a great part of the visual appeal of large organs.

Other features which I should mention are:

- the **key action** which allowed air to be supplied to individual pipes under the control of one or more keyboards, played either by the hands ('manuals') or with the feet ('pedal boards'); and
- the design of the **wind-chest** which allowed the player to change from one set of pipes to another of different shape, and hence different timbre—(the name 'stop' is used for both the set of pipes in question and the handle which brings them into play).

Organ pipes are of two different kinds, called **flue pipes** and **reed pipes**, depending on whether the basic generator is an edge or a reed (which is why I am talking about them in this interlude). Over the centuries skilful and patient organ makers have invented well over a hundred different kinds of pipes, all producing different timbres. Some are cylinders and some are cones. Some are narrow and some are wide. Some are wooden and some are metal. Some are waisted and some have flared bells. Some have more than one mouth, and some have their edge lined with leather. Yet, because all the pipes of the same 'stop' have to produce notes of the same timbre, but with different pitch, their shapes are usually geometrically simple. So they present a fascinating field of exploration to the physicist interested in understanding the relation between the shape of a pipe and the sound it produces.

Simple **flue pipes** are the most common, and the standard type, which nearly every organ possesses, is called the **diapason stop**—a perfect cylinder which can be open or closed at the top. The word 'flue' incidentally, which normally means a 'chimney', is the name given to the narrow duct which directs the air stream onto the edge.

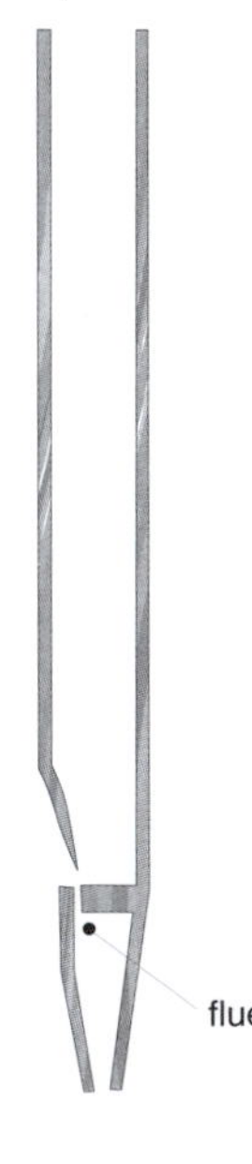

Somehow or other the tradition has grown up among organ makers of specifying the pitch of the note by quoting the (approximate) length of this sort of pipe. So the 'open diapason' pipe which produces the note C_2 is called an 8 ft (actually very close to 2.5 m for those who can no longer handle imperial measures). On the other hand the 'stopped diapason' (i.e. one whose upper end is closed) which plays the same note, is also spoken of as an 8 ft pipe, even though its actual length is only half that.

One of the interesting features of really big organs is that some of them have pipes as long as 32 ft (10 m), playing the note C_0. This is just about below the threshold of hearing, so the sound is *felt* rather than heard. Often such long pipes can't stand upright comfortably and are laid horizontally under the floor, which only adds to their effect.

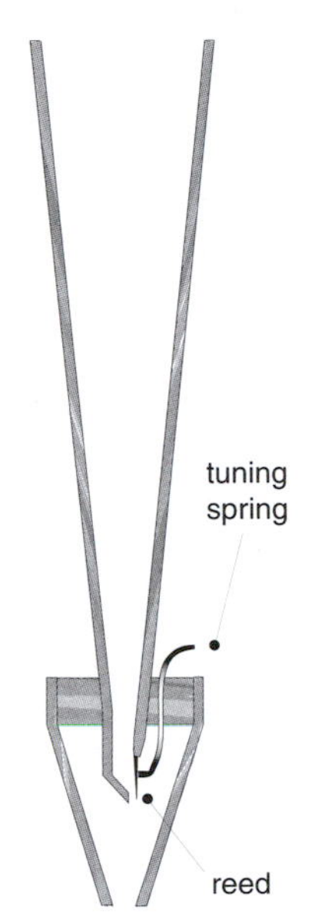

Another group of pipes are the **reed pipes**. These have the familiar vibrating reed-controlled air gap at the bottom. They are used mainly, but not exclusively, to produce timbres which approximate various orchestral instruments. It is worth noting, just as a matter of interest, that the action of a reed here is different from other reed instruments. In an oboe or a clarinet, for example, the reed is quite light and its vibration is completely controlled by the pressure variations inside the tube of the instrument, as I described earlier. In an organ pipe however, the reeds are much heavier and vibrate under their own elasticity. They therefore have to be tuned independently (which is the function of the tuning spring shown), and the pipe simply acts as a resonator which supports the note and influences its tone quality.

Other stops exist which are combinations of more than one pipe: called 'mixture stops'. The ingenuity of the organ maker seems limitless.

As I said, there is a treasure trove of physics to be explored in the construction of organs, but we must bypass it. Their versatility is gained at the expense of size, complexity and an increasing number of pipes. So if I can return to my main theme, it should be clear that, to get such versatility in a small instrument, something must be done to change the length of the pipe. And that means finger-holes.

The physics of a single finger-hole

I have already suggested that what goes on near an open hole in the side of a pipe is, physically speaking, quite complicated; but it needs to be understood before you can appreciate a lot of the features of woodwinds.

A simple experiment you could do for yourself, or imagine doing, is to put a variable hole about halfway down the side of a cylindrical tube open at both ends, and then set the fundamental mode vibrating. You'll get a note whose pitch depends on the size of the hole. If it's very small then the note is very close to what it would be if there were no hole at all. That's surely reasonable. But as the hole gets bigger the pitch gets higher until, when the hole has about the same diameter as the pipe itself, the pitch has gone up about an octave—which is what it would be if the pipe were cut off there. One easy way to think of this is that, as the size of the hole gets larger, the 'effective length' of the pipe—that is the length of an unbroken pipe that would sound the same note—gets smaller.

That much should be more or less intuitively obvious, but it is important to know what the *standing wave* inside the pipe looks like. As I said earlier, a hole behaves like a glitch in the acoustic impedance, and that means that the standing wave consists of *two* parts joined together smoothly. Both parts have the same frequency (and wavelength), but the one downstream of the hole has a smaller amplitude. Diagrams of the pressure variation along the pipe are a convenient way to help you understand the resulting sound, because it is usually not too difficult to visualize how the amplitude of each part of the standing wave is affected by the hole. Thus it should be clear that if the hole is small the two amplitudes are nearly the same, and if it is large they are very different.

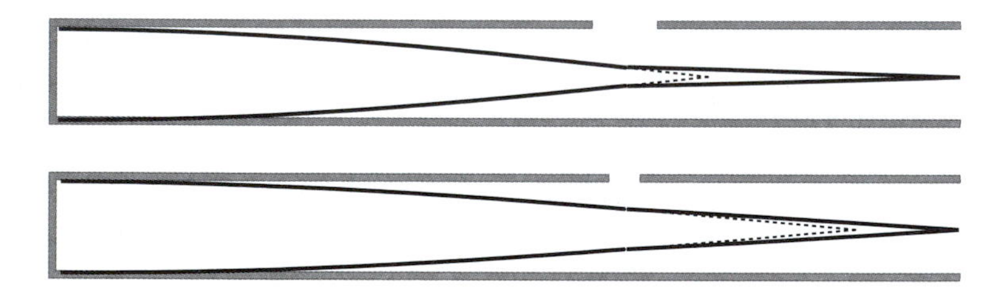

The dotted lines here are intended to show the 'missing parts' of the leftmost standing waves. I have drawn them like that so you can see clearly that the wavelength is shorter in the upper diagram and longer in the lower one. Therefore you will understand why the frequency must be greater in the upper pipe, and less in the lower one.

It is also worth noting that there are different ways to produce a note of any particular pitch—either with a small hole high up the pipe or with a large hole further down.

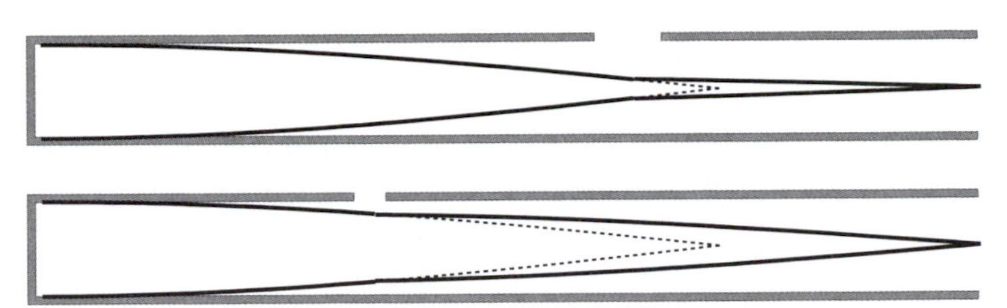

This will prove important later because it gives a valuable degree of flexibility to the makers of these instruments.

The diagrams I have been drawing have all been *pressure diagrams* because they are the easiest to intuit. What the hole does is to expose the inside air to the outside atmosphere. This constrains the degree to which the internal pressure can vary at that point. In a sense it is just like putting you finger lightly on a violin string—you damp the motion at that point, and the harder you press the more firmly you damp it. But, what happens to the air *movement* near the hole is altogether more complicated. If you remember that pressure and movement vary out of phase, i.e. as one increases the other decreases, it should be clear that there is an abrupt change in the particle motion across the hole—just as there is at the end of the pipe. Since air movement means that the molecules have kinetic energy, this observation is simply making it clear that *acoustic energy* escapes from the hole. That's where the sound comes from.

There is a second function that a hole sometimes plays. On most woodwind instruments there is a small hole which is usually located between 1/3 and 1/2 of the way down the pipe. It goes by various names, but for generality I will refer to as a **speaker hole**. Its smallness is important to its function: in a recorder, for example, it is only partially opened, by putting the thumbnail halfway over it (which is called **pinching**). Being so small, it has very little *direct* effect on the pitch of the note being played; but because of where it is situated it alters drastically the whole process by which the standing wave is set up.

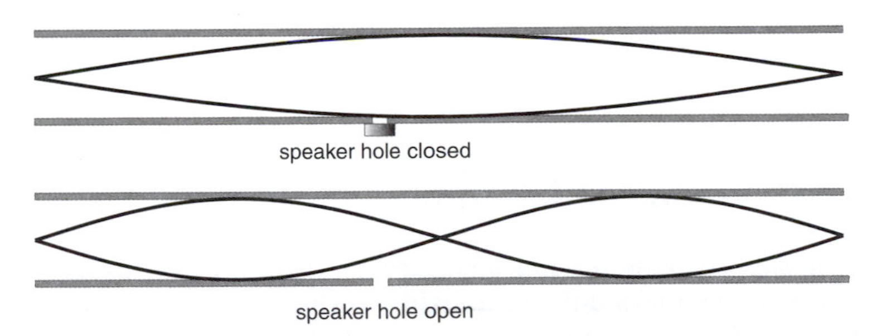

The middle of the pipe is where the pressure variation should be greatest, but a small hole just there punctures it and the whole standing wave collapses. In its place a different standing wave can be set up, whose pressure variation isn't so sensitive at that point—and that is the *first overtone*.

This doesn't normally happen quite so straightforwardly. You have to learn to use the speaker hole properly, and usually change the way you blow (called your **embouchure**) depending on whether it is open or closed. I should mention here that a flute doesn't have one of these holes; the player has to do the same job just by changing embouchure.

Anyhow the effect of the speaker hole is to allow woodwind players to do what brass players do with their lips—to change from a fundamental to its first overtone. It is called **overblowing**. Instruments which are open-open cylinders (flutes and recorders) or cones (oboes and bassoons) are said to *overblow into the octave*—i.e. the frequency doubles. In these the speaker hole (when it exists) is often called an **octave hole**. Instruments which are open-closed cylinders (clarinets) overblow into the twelfth—i.e. the frequency trebles, the pitch rises by an octave plus a fifth.

There is one last point I might mention about a single hole, and that is that a *closed* hole can also have an effect on the frequency of the standing wave. It depends on what is the closing mechanism. In simple instruments, like the recorder, holes are closed with the fingers; in the other, more professional, instruments many of the holes are covered by cork-lined metal caps, worked by keys.

In either case there will be a very slight change in the area of the pipe at that point and hence a slight difference in acoustic impedance. Therefore there will be a slight break in the standing wave and a change in frequency. In other words, the cap will act like a *very* small hole and so increase the pitch a tiny bit: a finger, especially if you press hard, might go in the opposite direction and lower it. These are only small effects, and won't affect the main thrust of what I am going to talk about, but they can be important in the fine tuning of instruments.

Finger-holes in combination

If the physics of single holes is complicated, that of a number of holes is even more so; and it is difficult to explain what goes on in a completely general way. So what I will do is to talk about the **recorder** specifically—because it is simplest in construction, easiest to play and most of its features are common to other members of the woodwind family.

The recorder is a nearly cylindrical pipe connected to a mouthpiece rather like the base of an flue organ pipe. It commonly comes in four sizes depending on the length of the pipe from edge to bottom—descant (or soprano) (30 cm), treble (or alto) (45 cm), tenor (60 cm) and bass (90 cm).

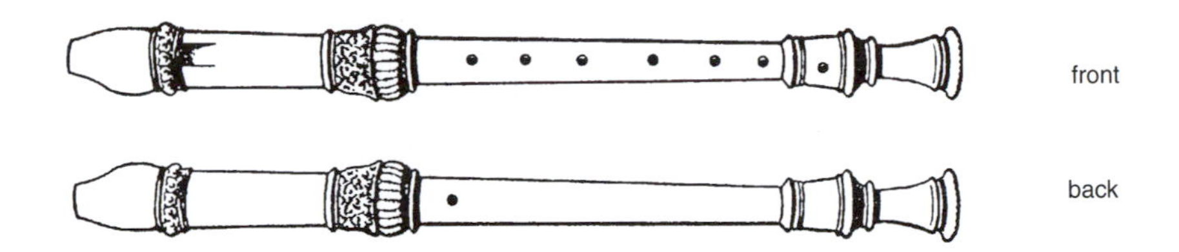

The holes in the front have to be covered by the player's fingers, and that in the back is the speaker hole, where it can be pinched conveniently by the thumb. For ease of fingering, makers have tried to allow players to sound successive notes by raising each finger in sequence. But recorder players have only seven fingers free—the right thumb supports the instrument, the left thumb works the speaker hole and the left little finger isn't much use. So the best that can be done is to try and arrange for a simple *scale* to be played in this manner.

When several holes are open in a row, each one will have some effect on the standing wave, therefore the whole thing will look very messy.

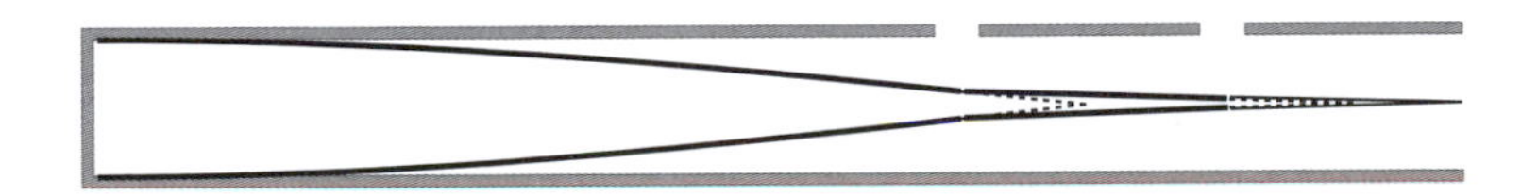

The holes furthest down the pipe will have least effect, as is clear from the diagram. And in particular, most of the sound will come from the first open hole.

Covering one (or more) of the holes further down the pipe will also affect the pitch. It may not be immediately obvious, but if you go through the same arguments I used earlier, it should be clear that this arrangement,

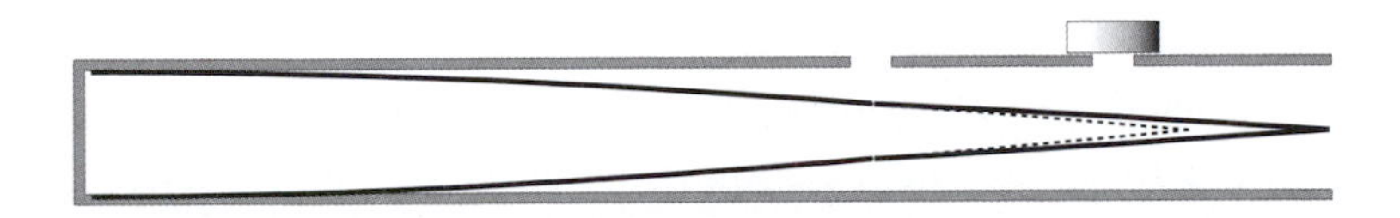

should result in a somewhat longer wavelength (and therefore lower pitch) than the previous one. Incidentally, this procedure of having open and closed holes interspersed is called **cross-fingering**.

I said that makers try to allow the playing of a diatonic scale without cross-fingering, but this isn't always possible because there are many constraints on where holes can be placed. That is where the skill of the maker comes in. The best way I can explain what these constraints are is to describe the fingering necessary to play each note.

The usual convention in books teaching you to play the recorder is to represent the condition of all the holes on a chart. The symbol ● means that the hole is closed, ○ means that it is open and ◒ signifies that it is partially closed—as with the speaker hole for example. The lowest note the instrument can play is that with all the holes closed. A descant recorder, being about 30 cm long, has a fundamental wavelength approximately twice that. This corresponds to the note C_5, one octave above middle C. Therefore the fingering for this note is represented on the charts like this,

		speaker hole	left hand fingers	right hand fingers
C_5	:	●	● ● ●	● ● ● ●

Here are the fingerings for the other notes the recorder can play (with comments where I feel the physics is interesting).

$C\sharp_5$	:	●	● ● ●	● ● ● ◒
D_5	:	●	● ● ●	● ● ● ○

Completely uncovering the bottom hole produces the second note of the scale (D_5). Partially uncovering it produces a note of somewhat lower pitch ($C\sharp_5$). The necessary degree of 'partial uncovering' isn't left up to the player, but predetermined by the maker—both of the bottom 'holes' actually consist of *two* smaller holes, to be covered by the same fingers.

$D\sharp_5$	:	●	● ● ●	● ● ◒ ○
E_5	:	●	● ● ●	● ● ○ ○
F_5	:	●	● ● ●	● ○ ● ●

Here's where you meet the first compromise that has to be made between simplicity of use and the complexity of the instrument itself. You'd think that the makers should be able to position the third hole so that F_5 would be played with the last three fingers raised; but this proves to be impractical. The reason is that the same holes have to produce the note F_6—I'll come back to this later.

$F\sharp_5$	:	●	●	●	●	○	●	●	○
G_5	:	●	●	●	●	○	○	○	○
$G\sharp_5$	:	●	●	●	○	●	●	○	○
A_5	:	●	●	●	○	○	○	○	○
$A\sharp_5$	:	●	●	○	●	●	○	○	○
B_5	:	●	●	○	○	○	○	○	○
	:	●	○	●	●	○	○	○	○

This last is an *alternative* fingering for B_5, which proves very useful in making quick transitions between it and C_6. This, more than anything else, gives you a feeling for the physical complexity of these instruments—that they can set up standing waves of quite different shapes, yet which have frequencies close enough to be musically useful.

C_6	:	●	○	●	○	○	○	○	○
$C\sharp_6$	:	●	○	○	○	○	○	○	○

Notice that the length of the closed part of the pipe is getting much shorter than it was, and the effect of cross-fingering is changing. In addition the kind of compromises that have to be made are becoming clear. The last fingering 'ought' to give C_6, but instead it gives $C\sharp_6$. There is however, a more usual fingering for that note.

$C\sharp_6$	:	○	●	●	○	○	○	○	○
D_6	:	○	●	○	○	○	○	○	○
$D\sharp_6$	:	○	○	●	●	●	●	●	○

In these notes, the thumb hole at the back is fully open and isn't acting as a speaker hole. Instead, because it is actually a little higher up the pipe, it's behaving just like an eighth hole, shortening the pipe a little more. Incidentally, the cross-fingering is getting pretty fancy.

E_6	:	◒	●	●	●	●	●	○	○
F_6	:	◒	●	●	●	●	○	●	○
$F\sharp_6$	:	◒	●	●	●	○	●	○	○
G_6	:	◒	●	●	●	○	○	○	○
$G\sharp_6$	:	◒	●	●	○	●	○	○	○
A_6	:	◒	●	●	○	○	○	○	○

With these six notes, the thumb hole *is* acting like a speaker hole. It is producing a note an octave higher than that which (almost) the same fingering gave when it was closed. The instrument can be said to be playing in its **upper register**, although that term isn't much used with the recorder. You will note however that there are subtle differences in some of the fingerings, which points up a theoretical difficulty. Given that the standing waves inside this pipe are so strange, there seems no reason, a priori, why notes in two registers *should* be exactly an octave apart. And indeed it turns out that it is only really true for geometrically perfect cylinders or cones. This point is the one which has given instrument makers much trouble over the years.

$\mathrm{A}\sharp_6$: ◒ ● ● ○ ● ● ● ○
B_6 : ◒ ● ● ○ ● ● ○ ○
C_7 : ◒ ● ○ ○ ● ● ○ ○
$\mathrm{C}\sharp_7$: ◒ ● ○ ● ● ● ○ ●
D_7 : ◒ ● ○ ● ● ○ ● ○
$\mathrm{D}\sharp_7$: ◒ ● ○ ● ○ ○ ○ ○

The cross-fingering in these six notes is so complicated that it's very difficult to interpret what is going on. The clue is to look at what the bottom holes are doing. They look like the fingering for D5–G5. The instrument is producing notes a *twelfth* up from these, so the combination of the speaker hole and the open holes in the middle are acting like a *different sort of speaker hole*. The second overtone (three times the frequency) of the standing wave is being set up. In other instruments which actually have two 'speaker' holes, the instrument would be said to be playing in its **high register**. It's probably just as well that the recorder can't get any more notes than these.

Changing fashions in woodwinds

I want now to turn to the instruments themselves. The woodwinds are an enormously varied family, because there are so many features that makers can experiment with. It seems that the simplest are the edge-tone instruments—which I will loosely call 'flutes' from now on. And the first thing to say about them is that they have a very long ancestry indeed.

In the 1980s, archaeologists at Jiahu in China, unearthed in an ancient burial site, a number of flutes, carved from the wing bones of cranes—some thirty-five in all. The six unbroken ones had as many as eight finger-holes, and you can still play tunes on them. They are dated as having been made in about 7000 B.C.

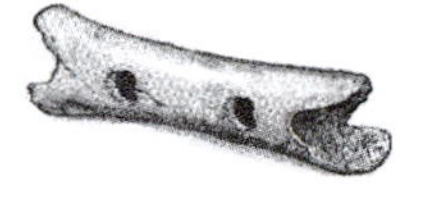

Even more ancient (and controversial) is a flute-like fragment of animal bone which was found in 1995 in a Stone-Age site in Slovenia, called Divje Babe. It is dated to be between 40 000 and 80 000 years old. It has two complete side holes, and evidence of at least two more. In the opinion of many of the workers it is part of a larger instrument which was played like a flute. If they're right it was a Neanderthal man or woman who played it.

Most anthropological evidence confirms that nearly every society, however isolated from the rest of humanity, has always started off playing flutes. Initially these instruments, like the ones I just talked about, were simple bones or bamboo pipes, with a 'mouth' cut into them and a few finger holes. They became vastly more sophisticated as time went on; but in Europe, for reasons which no one seems to understand, they remained peasant (or folk) instruments, up till comparatively recent times.

Reed instruments on the other hand seem to have been a later invention. In modern times very few isolated cultures have developed them beyond the squeaker stage, even though those same cultures might be extraordinarily inventive in their design of flutes. Primitive instruments were usually made by some such means as cutting a sliver of cane in the side of a bamboo tube in such a way that the player blew into the pipe past it. The critical advance of putting a separate disposable reed in the end of a tube to make a more substantial and permanent instrument seems to have occurred in the Middle East in biblical times and spread from there to all of Europe and Asia. These instruments all had narrow cylindrical tubes, and because the reed end had to be closed, they gave a pleasantly deep fundamental note.

The interesting thing about these ancient reed instruments was that they were almost invariably played as **double pipes**—that is, the player simultaneously blew into two independent tubes, each with its own reeds. Sometimes only one of these pipes had finger-holes, and the other simply played a drone accompaniment. Sometimes both were parallel with identical holes, so that they played the same notes. But in the more sophisticated models they were of different lengths, each with its own finger-holes, so that two different melodies could be played separately.

Though these instruments are extinct today, they monopolized wind playing for an extraordinary length of time. The old vertical flute seemed to have disappeared from about the 2nd millennium B.C., leaving very little trace in the archaeological records; and from that time the divergent double pipe was essentially the only woodwind instrument used in any kind of music. It was the 'pipe' referred to in the bible, the Greeks called it the **aulos** and the Romans knew it as the **tibia**.

Though it hasn't survived to modern times, some near relatives among folk instruments have. In many parts of Eurasia, a pair of parallel pipes were often fitted with a bell made of cow's horn, and sometimes with another cow's horn to cover and protect the reeds. They were known generically as **hornpipes**, and often gave a bleating sound as the two were slightly out of tune with one another. When the fashion for double pipes waned in the late Middle Ages, these instruments degenerated into single pipe varieties, and most survivors are of this form.

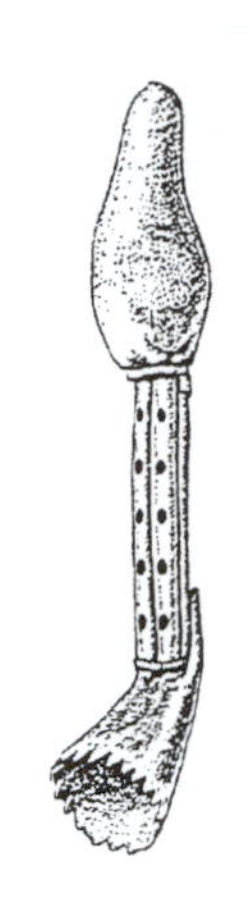

It is a complete mystery why the name "hornpipe" came to be associated with a particular dance once popular in the British Isles, now usually associated with sailors and naval traditions. This kind of hornpipe can be found in the works of Purcell, Handel and, of course, Gilbert and Sullivan.

A more important development occurred when the reeds were enclosed inside a bag kept constantly filled with air under pressure; and these **bagpipes** became immensely popular folk instruments throughout all of Europe and much of Asia. Their main characteristic is that their sound continues while the player takes a breath: which obviates all the breathing problems that players faced with the old double pipes. Modern Scottish instruments often have five distinct pipes: one to blow into, three drones and only one with finger-holes (the **chanter**).

In European countries other than Scotland, similar instruments retained some influence till at least the 1700s. A French instrument, called the **musette**, was very popular in the court of Louis XIV, and gave its name to the middle section of many an 18th century gavotte. An Italian version, the **piva**, had two chanters and two drones. You can catch the sound of its music in the "Pastoral Symphony" from Handel's *Messiah*. (He actually named the movement 'Pifa'.)

In was only after the time of the crusades that musicians in Europe changed to playing single pipes; and they adopted the fashion, current throughout the Orient, of keeping the distinction between *haut* and *bas* instruments. Music to be played outdoors — in bands and processions — or in large indoor events — royal ceremonies or dances — was usually reserved for reed. But music for small receptions, private feasts or theatres was played by members of the flute class.

The actual instruments in greatest favour at any one time fluctuated; and these changes in fashion are summed up in the following diagram. It represents schematically how different instruments of each class have risen in popularity in courtly and professional music over the past centuries, only to fade away again.

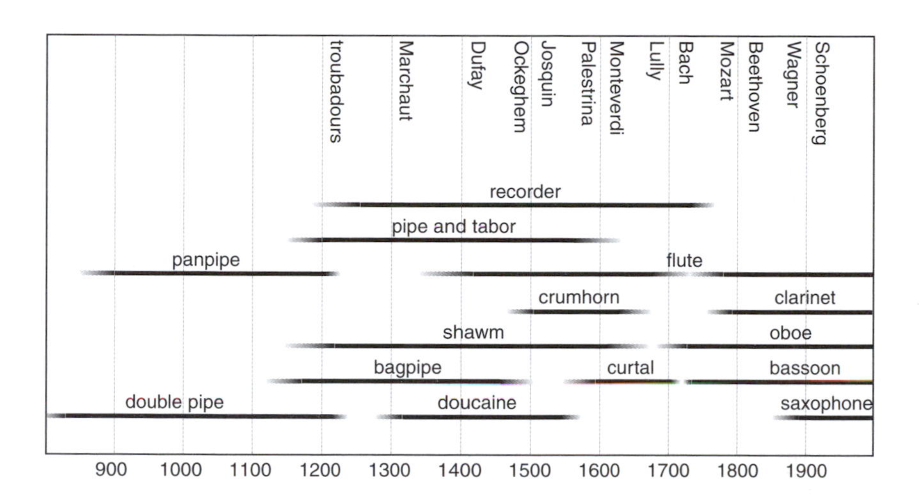

For the rest of this interlude I will describe very briefly the changes that occurred in both kinds of instruments over that time span.

Flutes

In medieval times the earliest single-pipe edge-tone instruments to attain popularity among professional musicians were all **end-blown**. They had a specially shaped mouthpiece on one end containing a flue to direct the air stream across a mouth towards an edge. This part of the instrument was called the **fipple**, and was held between the lips, with the main pipe vertically downwards. Any instrument played like this was often referred to as a 'fipple flute'. The pipe was provided with finger-holes along its length, the number and separation of these holes determined what music it could play.

The **pipe and tabor** combination was particularly popular among the early troubadours. The **pipe** was often little more than a long whistle with only three finger-holes, two at the front and one at the back, so that it could be played entirely one-handed. That left the other hand free to accompany on a small drum—the **tabor**. The interesting thing, acoustically, was that, with only three finger-holes, it had to be played completely in its upper registers, and higher ones if necessary. You will recall that there are only three notes of the scale between the second and third harmonics (the octave and the twelfth of the fundamental) so three holes are just enough to enable a diatonic scale to be played.

The **recorder** emerged very early as a variation on the same basic theme, with seven finger-holes and a thumb hole to utilize the low register properly, as I have already discussed. There's no need for further comment except to say that recorders were preeminent among 'flutes' for well over 400 years. They faded from the scene during the 18th century, but made a comeback in the 20th, owing largely (it would seem) to the efforts of one man—the instrument maker Arnold Dolmetsch. This resurgence has been extraordinary indeed. In my own country (Australia) a well organized educational programme in the primary school system has resulted in more people being able to play the recorder than any other single instrument; and I imagine the same is true in other English speaking countries.

At the same time as the recorder was in vogue, a second kind of flute was emerging—the **transverse flute**. It was held sideways towards the right side, and the player blew, through slightly narrowed lips, *across* the mouth hole onto an edge.

It first appeared in Europe in Byzantium in about the 12th or 13th centuries, made of boxwood or maple, and its rise seemed to coincide with the decline of the panpipes. It's an over-simplification to say that the one replaced the other, but they have much in common—particularly their very pure tone colour and the directness of control the player is able to get with practised use of embouchure.

One early representative, which won a lasting place in military bands from late Renaissance times, was the **fife**. Drum and fife bands were particularly popular with English armies.

But the main instrument in this class, the **flute** proper, steadily increased in popularity until it was taken up in a big way by French and German composers in the 18th century. From the time of Haydn it had almost completely replaced the recorder in orchestral usage. The most popular model, which we now call the **baroque flute**, had a slightly conical bore, with six (quite small) finger-holes and an extra one, worked by a key. This was the instrument for which, for example, all of Mozart's concertos were written. However it was notoriously difficult to play really well and in tune. So very soon other instruments began appearing, first with six keys and later with eight.

In the 19th century the flute underwent a complete redesign at the hands of the Munich flute maker, Theobald Boehm. Tradition has it that in 1831 he chanced to hear an English virtuoso playing on an instrument with greatly enlarged holes, and he realized that this was the way to increase the flute's loudness. (From all that I said in last chapter about impedance matching, I hope you can appreciate the accuracy of his insight). So he stripped the flute down to its original one key, increased the size of the other holes and supplied them with what he called 'open keys'—ones that normally rest open rather than closed. His first design was a conical flute which he released in 1832; and to this day you will hear flautists say that this event was the greatest landmark in the history of woodwind design.

But that wasn't the end, In searching for still more sonority, he realized that loudness is directly related to the volume of gas in the pipe (from energy considerations), and the volume of a cylinder is greater than that of a cone. So he went back to a cylindrical bore. His studies led him to a comprehensive prescription for the size of the holes and a parabolic shape for the end-cap (which fits on the end of the pipe upstream from the mouth). In 1847 he released the **cylindrical Boehm flute.**

There have been minor modifications since then but, in essence, it is this flute that everyone plays today. It is about 60 cm long and its fundamental note is D_4. There is also a smaller relative, the **piccolo**, half the size and playing an octave higher.

Reed instruments

An easy way to keep track of the large number of single-pipe reed instruments that appeared in medieval Europe with the demise of the double pipe era, is to concentrate on the essential acoustics. A reed pipe has to be closed at one end, therefore its low register might be more than an octave. But a player has only eight free fingers, and so can play nine possible notes without cross-fingering—an octave plus one. Reed pipes therefore posed a problem, at least until the invention of keys. For not only could keys make it easy to reach widely spaced finger-holes, and take care of more difficult cross-fingerings; they could also, by allowing a single finger to control several holes, increase the number of holes available. But this kind of mechanization didn't occur till much later.

One of the first reed instruments to appear was the **shawm**. It is thought to have been invented in the Mohammedan Empire in about 800 AD, and the early crusaders were reputedly most impressed by its strident tones in the military bands ranged against them. Within Christendom it took on a similar role, and most villages soon came to have their own shawm bands for outdoor ceremonies. It was the original *hautbois* (or 'loud wood'), gaining volume from sheer size.

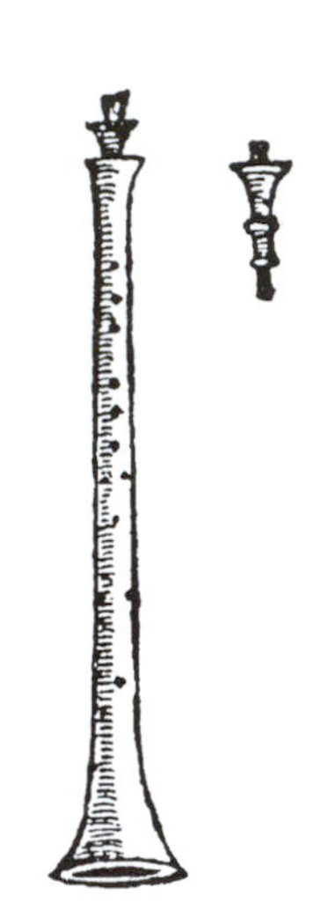

You can recognize it by the very characteristic wooden disc (called a **pirouette**), which rested against the player's lips, and out of which a wide double reed protruded. But perhaps its most important feature, musically, was that it was very nearly a perfect cone. This meant it overblew into the octave and its low register was easily spanned by eight holes, exactly as in any of the flutes.

A close relative, the **curtal**, started off as a bass shawm, but was obviously much too long for convenient handling—remember a cone that sounds, say, C_2 must be at least 250 cm long. However great advances occurred in wood-working in the 16th century and it became possible to drill two parallel bores in one piece of wood, joined by a U-bend at the bottom. So the length of this instrument was immediately halved.

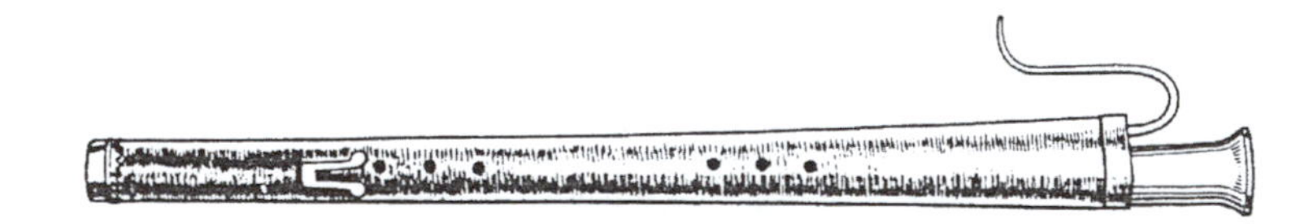

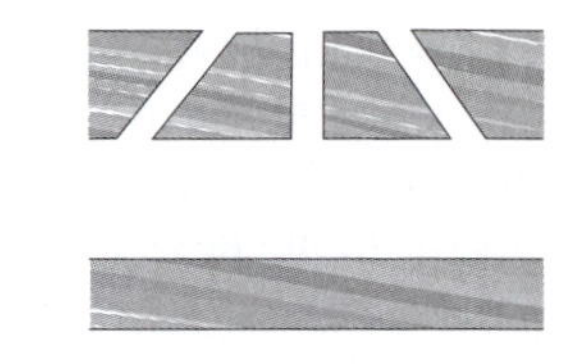

Nevertheless, their construction must have been difficult. Not only was the drilling of parallel conical bores quite a technical feat, but the finger-holes had to be drilled slantwise so that they occurred at the right spacing along the bore, while staying within finger-range on the outside.

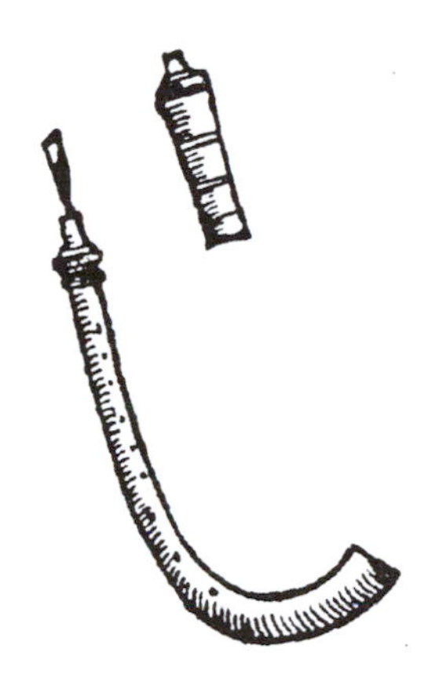

The **crumhorn** was, in a sense, a sophisticated single hornpipe—a curved, cylindrical tube made of wood. It had a mouthpiece which covered the reed—making it much easier to play, but taking away some of the directness of control. Being a cylinder, it had a range of just nine notes. (Remember, a cylinder overblows into the twelfth, and so there are eleven notes between the lowest note and the bottom of the upper register. The usual eight holes cannot cope with that number.) It had a rather raspy sound, and this, more than anything else, is what gives much medieval music its distinctive sound (at least to our ears).

The last instrument in this group is something of a mystery. The **doucaine** is mentioned in many 14th and 15th century manuscripts, especially French and Flemish. It was clearly an important instrument, but absolutely no description of it survives. Most musicologists believe it was closely related to the shawms, but (as its name suggests) it was clearly a *bas* instrument—which was an important change, in regard to the impending development of the oboe.

The first composer to write expressly for woodwinds was Jean-Baptiste Lully. Though born in Italy he moved to Paris in 1646, where there were gathered together a remarkable collection of wood turners and instrument makers. There seems to have been an agreement among them to redesign all the wind instruments. They introduced many features which are common today, particularly the practice of making their instruments in several small joints, instead of in one piece. They also typically decorated the thickened wood around the sockets; and this gave their instruments a much sought after elegance of appearance.

The first major development was done on the tenor shawm. They gave it a really good upper register and reduced its strident tone by narrowing the bore and reducing the size of the finger-holes. They added two keys, and so transformed the Renaissance *hautbois* into the classical **oboe**, which was exactly suited to the kind of music then coming into fashion. With the first performance of Lully's ballet *L'Amour Malade* in 1657, at which the new instrument was heard in public for the first time, the modern age of woodwind music might be said to have begun.

Since that time the oboe has been improved and mechanized in much the same way as the flute. In the early 19th century instrument makers added more keys and developed new fingering systems (including some followers of Boehm—although none of their models seemed to have caught on). So today, although there are a few variants around, the oboes we play are essentially the ones that had developed by about 1850. They are about 60 cm long and play a fundamental note of $B\flat_3$.

The steps by which the modern **bassoon** developed from the Renaissance curtal has been lost from documented history. It was presumably done by the same group of Parisian craftsmen, because it turned up in Purcell's London under the name 'French bassoon' and was first named in a Lully score in 1674. But no early models survive. It also underwent mechanization in the 19th century and today is an instrument about 125 cm long (though the pipe itself is twice that) playing the fundamental $B\flat_1$.

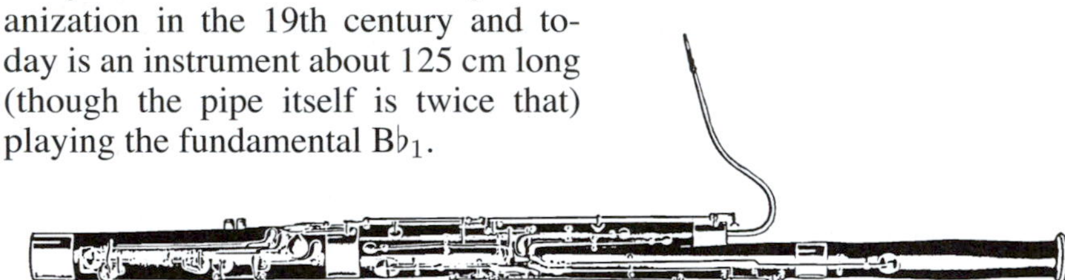

In all this time Germany's main contribution to the woodwinds was in the development of the single-reed, cylindrical bore instruments. In the beginning of the 18th century there was a kind of rustic reed pipe called the **chalumeau**, little more than a recorder with a reed in it. Nevertheless it was valued in music consorts because of its special timbre which is characteristic of an open-closed cylinder—lacking even-numbered harmonics and usually described as 'hollow' or 'woody'.

This instrument was worked on extensively in order to make use of its upper register, and entered mainstream music in late baroque times under its new name, the **clarinet** (so called because in this register it could handle well melodic lines usually given to trumpets). Keys were added and in the 19th century it was provided with a new fingering system named for the ubiquitous Herr Boehm (among others). So we have the modern clarinet (73 cm, D_3).

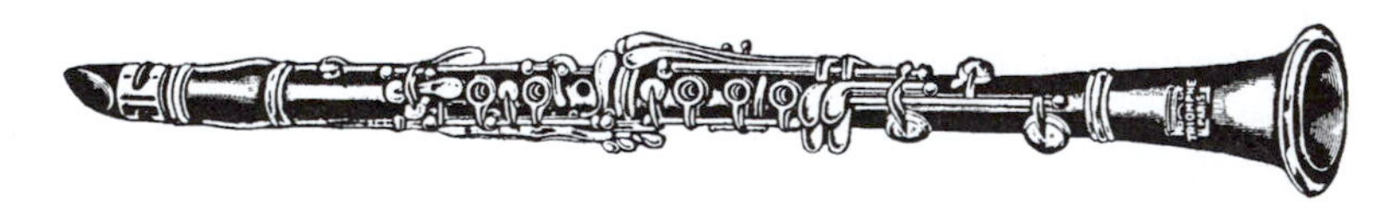

To my mind, the most interesting feature of the clarinet is that it went through its complete evolution during the late baroque, classical and early romantic periods—the music of which we are all very familiar with. So when you listen to the well known concertos and concert pieces that make up its pre-20th century repertoire, you can hear the changes taking place. Scarlatti, Handel and Molter wrote for an instrument with a complete break between the two registers. The lower one had only the usual nine notes, and the upper didn't start till the twelfth note of the scale; so two notes were simply not there. Mozart's clarinettist had an instrument on which these two notes were playable—with the aid of three extra keys—but they were still very difficult: though when he complained to the composer he got the terse reply that, if the notes were on his instrument, it was his job to play them. It was only in Brahms' time that the development was complete, and clarinet music was able to reflect the ordinary musical idioms of the day rather than the limitations of the instrument for which it was written.

The last common member of this family is the **saxophone**, a rather strange amalgam of the others. It comes in several shapes and sizes, but the one you are probably most familiar with is the tenor (88 cm, $A\flat_2$). It has a clarinet-like mouthpiece with a single reed, but has a conical bore like an oboe. This means that it overblows into the octave, yet for some strange reason it uses a fingering quite unlike the oboe, but similar to a flute.

It has a much wider bore than an oboe, therefore it is much louder; and it is made of brass. It was patented in 1846 by the Belgian instrument maker, Adolphe Sax; and was intended initially for military bands—to redress the loss of popularity of woodwinds in competition with the newly valved brasses. But it was in the jazz bands of the 1920's and 30's that its husky sound was most appreciated; and this is where it underwent most of its development. However it still isn't clear whether or not it will ever be promoted from the 'special' to the 'regular' class of woodwinds in mainstream serious orchestras.

Postscript

Evolution of the woodwinds has by no means ended. As recently as 1967 a new word entered the language of musicians—**multiphonics**. It was coined to describe the production of several tones at the same time by the same instrument. For many years bassoonists have known that it is possible to sound an approximate major chord by means of 'trick' pressures on the reed. But it seems only recently that other wind players have been experimenting with fancy embouchures and unusual fingerings; and an amazing array of new sounds have emerged.

To give an example of what I'm talking about, here is one that even the veriest amateur (like me) can get on a recorder:

◒ ● ● ○ ● ● ● ○

You have to pinch the thumb-hole nearly closed and play around with the way you blow into the mouthpiece, and suddenly you will hear two distinct notes—one somewhere round $G\sharp_5$ and the other near E_6. I think it is wrong to call it a 'chord', as do many books. What is happening simply reflects the complexity of what is going on inside the pipe. Just as it is possible for two different fingerings to sound the same note, so in this case the same fingering produces two different notes.

So far as musicians are concerned, the problem is to find out which of these new sounds are useful musically, and to develop a notation which will enable other players to get the same sounds—and indeed some composers are already calling for them in advanced compositions. But so far as the physicist is concerned, the puzzle is how to predict which fingerings will give other new sounds. Up till the present time, all the known multiphonics have been found by trial and error. If a scientific understanding of this problem could be found, it might finally convince musicians that acoustics is a useful study.

Chapter 7

Summertime in Heidelberg

Throughout history, different countries have become leaders in various intellectual fields at different times, and it is always interesting to speculate why. During the years of the late Renaissance, Italy dominated most of the arts and sciences—and presumably this can be correlated with her wealth and political power. But as Italy's material status declined, the most important science came to be done in England and France, and the centre of gravity of the musical world shifted to Germany.

I suppose it was more or less natural that, as capital of the Holy Roman Empire, Vienna should have been looked to to fill the cultural vacuum. And it was in Austria that German music began to make its mark internationally, with the emergence of the so-called Mannheim School—C.P.E. Bach (remember that Papa Bach was not well known outside of Germany proper until much later), Mozart and the Haydn brothers. But it was in the early 19th century that its dominance was truly established. Somehow many characteristics of Romantic music—the idealization of the artist as hero and the exaltation of the virtuoso performer, the technical improvements in musical instruments, the existence of large audiences willing to pay money to listen to orchestral concerts, the growing interest in chromaticism and the desire to explore new harmonic structures—all these seemed to click with the mood of the increasingly affluent and nationalistic German people. Or have I said that the wrong way round? Was it just by chance that most of the important composers of the time happened to be German, and they imposed their national stamp on Romantic music? Whichever it was, the list of these composers is impressive indeed: Beethoven, Weber, Spohr, Schumann, Schubert, Mendelssohn, Brahms, Bruckner, Wolf, Strauss, Mahler; and that very model of all that we tend to associate with the word 'Romantic'—Richard Wagner.

During most of this period, the important centres of science were elsewhere. Without a doubt, in the 18th century, the Royal Society in London and the *Académie des Sciences* in Paris formed the axis about which European science revolved. Then in the 19th, with the affluence that came from an industrial (rather than a political) revolution, England emerged as the unmistakable leader—a position it maintained for nearly a hundred years. But at the beginning of the 20th century, physics went through a violent upheaval which was led by, more than anyone else, Germans.

Why Germany left her run so late isn't clear. Traditional historians of science have argued that it was because she was 'trapped in Romanticism', and Romanticism was, as everyone knew, hostile to science. But in fact, in the first half of the 19th century two interesting things were happening. Germany started upgrading scientific education, setting up technical schools and engineering colleges; and her government and private enterprises started spending money on scientific research. And this might even have been the result of her philosophical milieu. Because Romanticism insisted on the unity of all knowledge, it was willing to adopt science into the traditional structures, at a time when universities like Oxford and Cambridge were insisting on the classics as the proper training for the elite. Whatever the reasons, if there was one person who symbolized the emerging mood in Germany, it was the scientist Hermann von Helmholtz.

Bayreuth and Heidelberg

These two figures, Wagner and Helmholtz, form an interesting contrast in styles.

Richard Wagner was the epitome of the Romantic composer. He was born into a Dresden theatrical family in 1813 (of slightly doubtful paternity), and lived his 70 years with what can only be described as a lack of moderation. He was perpetually in debt, and several times had to move to another city to escape creditors. He threw himself recklessly into political causes, and consequently was forced to live 13 years as an exile in Switzerland after the rebellions of 1848. He married young, very unhappily, and conducted two highly publicized affairs with the wives of good friends. But eventually his first wife died and he married the daughter of

Franz Liszt (who had also been notorious in his time for marital indiscreetness), and settled down to a life of comparative respectability under the patronage of the last, lonely and finally mad, king of Bavaria, Ludwig II.

Wagner developed early in life a fundamental philosophy that the function of music was dramatic expression; and he pursued this ideal all his life with unswerving determination. His only important compositions were for the theatre, and his crowning achievement was the cycle of four operas, drawn from the racial mythology of the Germanic peoples, *The Ring of the Nibelung*. He designed a special opera house for their presentation in 1876 in the provincial city of Bayreuth; and instituted an annual festival for their performance. He ended life with a god-like reputation in his own country and immense respect throughout the rest of Europe.

Today it is very difficult to appreciate the influence he held. He was the focal point of intimate gatherings of many of the most famous musicians of his time, where they planned what they called the 'music of the future'. As a result, most of his contemporaries came under his spell; and so he is one of the few composers of whom it might be said that he single-handedly changed the course of music. His writings in other fields had great influence also, particularly because of his friendship with the philosopher Frederick Nietzche—in literature and drama, as well as in political and moral issues. All in all he must be rated as one of the pivotal figures in 19th century Germany and indeed all of Europe. Even today it is claimed that, of all historical figures, he is the third most written about, after Jesus Christ and Napoleon Buonaparte.

By contrast, Helmholtz's life was much more serene and ordered. He was born in 1821 into the family of a Potsdam schoolteacher, and studied medicine in his youth. While serving as a surgeon in the Prussian army he developed a taste for scientific research. As the result of observations he had made on the way the human body worked, he was able—at the age of 26!—to propound the law of conservation of energy. Others across Europe were closing in on this most important of discoveries at the same time; and there is still controversy about who should be given the full credit. But it is probably true to say that his was the world's first clear statement of this law. It brought

him immediate fame and an academic post as Professor of Physiology, where he devoted his efforts to studying the eye and the mechanism of seeing.

In 1858 he was offered the post of Professor of Anatomy at that most beautiful of German university towns, Heidelberg, some 150 miles from Bayreuth. In the thirteen years he was there, his work on the eye led him to study the ear and the way we hear. He applied his results to the theory of music, which he published in a monumental work *On the Sensations of Tone*. In this book, for the very first time, the psychological perception of music was opened up for investigation—the field we now call **psycho-acoustics**. It was a turning point in the scientific study of our subject.

After Heidelberg he moved to Berlin as Professor of Physics, where much of his efforts were devoted to electromagnetism: and from there his influence spread, not least through the students he taught. Heinrich Hertz was one of these, and the experiments on radio waves were undertaken at Helmholtz's suggestion. Another was Max Planck who would soon be responsible for the revolutionary new Quantum Theory; but more of that later. Helmholtz died in 1894, almost 11 years after Wagner, the most respected scientist in his own country and acknowledged throughout the whole of Europe.

These two intellectual giants, so influential in different spheres of German life, existed side by side for a span of two generations. They were occasional friends. Helmholtz was in the audience for the first performance of the *Ring* at Bayreuth, at the composer's invitation, and wrote indignantly to the press about what he called the critics' 'icy non-recognition' of the work's importance. Likewise, when they were in Berlin, the Wagners were guests at the Helmholtzs' evenings-at-home, at which most of the fashionable intellectuals gathered. At one time they even found themselves on opposite sides of a public argument about vivisection. But they didn't seem to impinge on one another professionally. It is only in retrospect that we can see how much they had in common, and how similar were the effects they had on German intellectual life.

The field where they might have interacted is, of course, the theory of music; and I want now to examine what Helmholtz contributed to that subject. But before I can do that I will need to talk about a bit of physics I have avoided so far—a more complete description of resonance.

The theory of resonance

Up till now I have been rather cavalier in talking about resonance. I've discussed it only in the following terms. You have some system which can oscillate with its own natural frequency; and if you disturb it (periodically) at exactly this frequency, then energy will keep going in, and a very sizeable amplitude of vibration will build up. The simplest model, you will remember, was of pushing a child on a swing. But I've never addressed the question: what happens if you don't do it at the right frequency?

To answer that I'll have to go back to my discussion (in Chapter 3) of why systems oscillate—of how energy changes periodically between potential and kinetic forms. However, the argument I want to go through is a bit involved, and if you don't want to follow it in detail, you can skip the next five paragraphs.

The problem is one of putting energy into a system, and therefore it makes sense to discuss it in terms of impedance—though of a slightly different kind, called **mechanical impedance**. This is the property of a system which determines how great an oscillating force has to be, in order to get the system moving at a certain speed. Strictly it is the ratio of force applied to velocity produced. If you apply a big force and only produce a small movement, you say the mechanical impedance is large. But if a small force results in large movement, you say it is small. It may not look like the same sort of impedance I talked about earlier, but it is clearly a related concept.

Now for clarity, I will talk about one particular oscillating system—a mass on the end of a spring. First I want you to consider the mass in isolation, and to think about how it responds if you try to get it moving by *shaking* it at a constant frequency. Obviously, if it is heavy, it will be difficult to make it move very fast. Its mechanical impedance must be directly related to its inertia. But even a light mass won't respond if you try to shake it too rapidly — just imagine, for example, trying to shake anything at more than about ten or twenty times a second. This means that inertial impedance must depend on frequency: it gets bigger as the frequency increases.

Secondly, think of the spring by itself. If you apply an oscillating force to it, you simply stretch or compress it; and the amount of stretch or compression depends only on the amount of force you apply, not on the frequency. Therefore the velocity with which the end of the spring moves *will* depend on the frequency, since it has to move through a fixed distance in each period of the oscillating force. So the kind of impedance involved here gets *smaller* as the frequency increases. Furthermore, because the spring pushes in the opposite direction (i.e. against you) when you push on it, elastic impedance is, in a sense, the negative of inertial impedance.

Therefore when you analyze a system consisting of both a spring and a mass, you must think of it as having a total mechanical impedance equal to the *sum* of these two (actually the *difference*, since one is negative). So, if the frequency is *either* very high *or* very low, the total impedance will be large, because one of its components is large, even though the other is small. However there is one particular frequency, somewhere in between, where the two parts of the impedance have exactly the same value and *cancel one another out*. At that frequency even a tiny applied force will produce a huge response. And that is of course just what we mean by the term **resonance**.

But one further point. Even at resonance, the total impedance can never go exactly to zero. There will always be friction, or some other means by which energy can leak away; and these will contribute another kind of mechanical impedance (just like electrical resistance) which can't be compensated for. So the exact impedance at resonance, and therefore the magnitude of the system's final response, will depend on how small this 'resistance' is. The actual reason for this concerns the *time* it takes for the energy to dissipate. The vibration will settle down only when the rate at which you put energy *in* just balances the rate at which it leaks *out*. Therefore if the system has a small resistance and loses energy slowly, the amplitude at resonance will be high; whereas if its resistance is large, it will lose energy quickly, and the resonant amplitude will be low.

Let me summarize the conclusions of this argument by means of a graph. I will plot how the response of the system (measured by the velocity of its motion) varies as I change the frequency at which the driving force is applied.

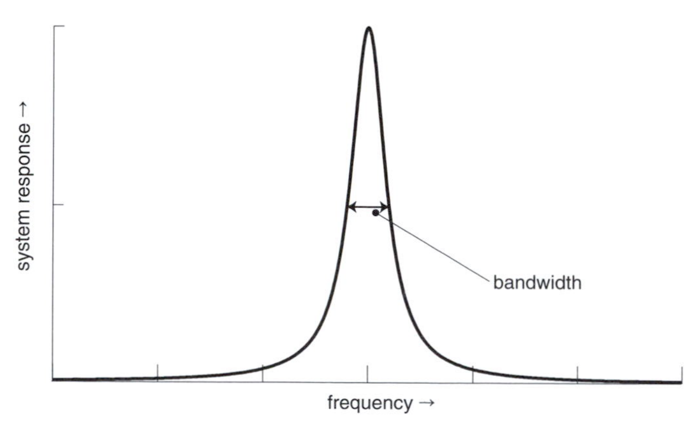

The main feature of this graph is what we already knew—that at one particular frequency, the **resonant frequency**, the system responds much more violently than at others. But what is also shown is that there is a *band* of frequencies around resonance at which the system still shows quite a large response. This range is called the **bandwidth**.

If you want to get a simple intuitive feeling for this graph, think about tuning a radio. As you turn the knob you are altering the circuit in such a way that the frequency at which it oscillates changes. You sweep this past the frequency of the radio station, and at one point they match. Then the tiny signal in the air is able to set up a big electrical oscillation inside and the audio message will come through loud and clear. But you can still hear it even if you are not quite 'on the station'. This is what I mean—this range of 'almost acceptable tuning'—when I talk about bandwidth. There are important applications of this principle in many musical instruments (especially in wind instruments, when players are able to correct for slight inaccuracies in tuning); but above all, it is vital in understanding how the ear behaves.

The ear

The ear can be considered in three distinct parts. The first, the **outer ear**, is the most obvious, consisting of the bits you can see and feel: the large shell-shaped lobe (called the **pinna**) which leads down through a narrow tube (the **auditory canal**) to the **eardrum**.

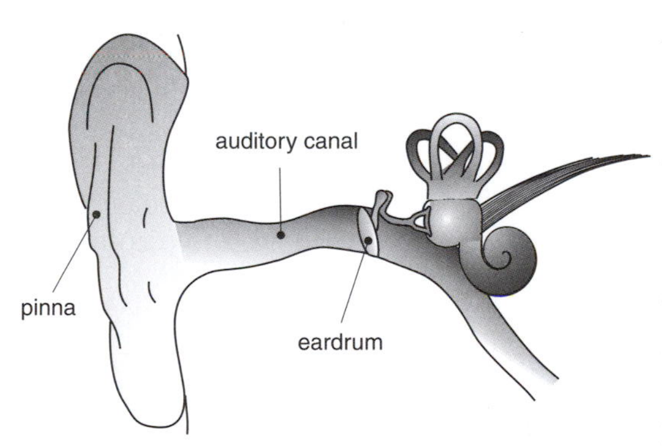

We have already noted that this is essentially the collector of sound energy; the narrowing shape provides enough of an impedance match so that a reasonable fraction of the energy falling on it ends up in a vibration of the surface of the drum. However, recent research has shown that the pinna does a bit more than this. The convolutions of its shape actually enable us to locate how high above the ground is the source of a sound. Similarly the auditory canal has its own acoustic properties. Being a tube about 3 cm long, closed at one end, it has a fundamental resonance mode at about 3 000 Hz; and that accounts for the fact that human hearing is most acute around that frequency. But, so far as *understanding* a message is concerned, we must look to the other parts of the ear.

In the **middle ear** the energy of vibration of the drum is transferred, by a mechanical lever system made up of small bones (called the **ossicles**), to where it sets into vibration a second membrane (the **oval window**). The function of this stage is primarily amplification—the lever system translates the small pressure variations on the (comparatively large) drum, to considerably larger vibrations of the (much smaller) oval window. But a secondary function is that of a buffer—if the pressure variations of the drum are too large, it doesn't amplify them nearly so much, thereby offering some protection to the more sensitive workings of the inner ear.

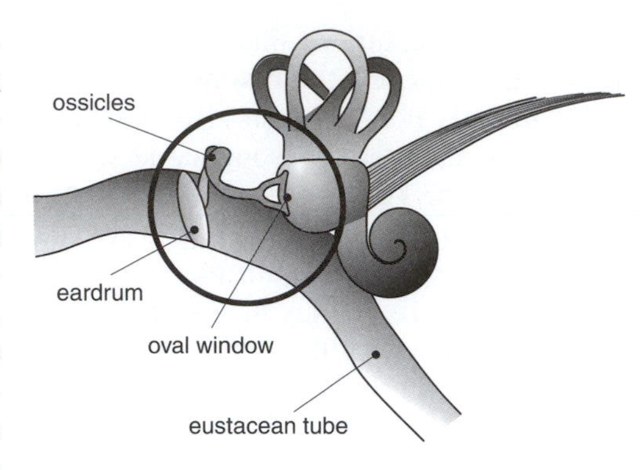

It is necessary for the middle ear to be unencumbered at all times, and so it is connected (via the **eustacean tube**) to your throat. When, for example, you go up in an aeroplane, or dive deeply under water, and a pressure difference builds up across the eardrum, you can equalize this difference by swallowing. There are several things that can go wrong with the workings of this part of the ear, which can lead to varying degrees of deafness—the drum can be punctured, the small bones can seize up, the chamber can become filled with mucus. Luckily most of these complaints are, at least in principle, treatable to some degree.

The next stage, the **inner ear**, is the most interesting from our point of view. Ignore the strange loops at the top of the diagram—the so-called **semi-circular canals**, which are concerned with the body's balancing mechanism—and concentrate on the bit that looks like a snail shell. It is a helically coiled bone cavity, filled with fluid, and is called the **cochlea**. Its internal structure is easiest to understand by imagining this coil to be 'unwound'; in which case it might look like this:

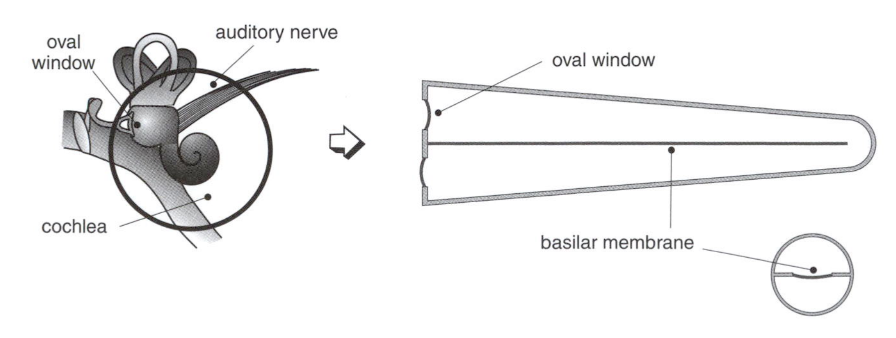

The chamber, about 3 cm long, is divided most of the way down the middle by a narrow strip of taut skin called the **basilar membrane**, which separates the fluid in the upper and lower halves. This membrane is narrowest and tautest at the front end (near the oval window) , and widest and slackest at the other. When the ossicles set the oval window vibrating, this motion is communicated, via the fluid in the cochlea, to this membrane. And this is where the sound wave is actually 'detected'.

Underneath, in the lower half of the chamber, there are millions of tiny hair-like nerve cells which respond to any movement of the membrane by firing tiny electric currents. These current surges are conducted out of the cochlea, by the **auditory nerve**, to the brain. If anything goes wrong with this part of the ear, it is obviously extremely serious; and unfortunately this is where damage resulting from excessive loudness occurs. When the nerve cells are subjected to too much stress, they are destroyed one by one. Under a microscope they look like a bomb site. And this can be disastrous because, as we will see, they are responsible for the ear's ability to distinguish pitch. Therefore musicians who play in very loud rock bands are often putting at risk their most valuable asset—their musical ear.

Just in passing, it is worth mentioning a particularly impressive example of the marriage of technology and medicine—the **bionic ear**, or if you prefer, the **cochlear transplant**. This device was developed by Australian scientists in the 1970s, and it collects sounds with a a tiny microphone sited just behind the ear. These are transmitted to a receiver buried under the skin, where they are converted into coded electrical signals by a speech processor. They are then passed on to 22 electrodes which have been surgically implanted into the cochlea at particular points along the basilar membrane. The procedure is still very expensive, and requires a lot of rehabilitation for the patient's brain to learn to interpret the unfamiliar signals it is receiving; but they do work and many, many thousands of deaf people have been fitted with them in the past decades.

Anyhow, in this discussion I've left out a lot of detail, and my description of the essential function of each part of the ear is grossly oversimplified; but it highlights an important consideration in thinking about the ear as a instrument for interpreting music. Every sound we hear is processed twice: once by the cochlea, where it is coded into an electrical signal, and then by the brain, where its message is extracted. It is currently fashionable to think of the human brain as a kind of electronic computer, and the processing of information to be carried out by some kind of computer program. In so far as this is valid, it is clear that the second stage of processing is more or less under our control—we can *learn* to change the way we think about music. But that which is done by the cochlea we are stuck with; and a lot of our response to musical sounds must be tied up with exactly what it is that the cochlea does. So we have to look at that next.

Pitch recognition

Clearly what is important is the way the basilar membrane responds to vibrational energy falling upon it. Remember that it is roughly triangular in shape; and up till now, I have said nothing about how standing waves are set up on a two-dimensional surface. But the basic principles are the same as those we have met before, and a good deal of insight can be gained by thinking about the membrane as though it were made up of a whole lot of short strings, like some long, thin, many-stringed dulcimer. It has very short, taut 'strings' at the front end, near the oval window; while those further down are longer and slacker. If the fluid around it vibrates at a pure frequency, then there will probably be one 'string' somewhere along the line which will resonate with it—near the front end if the pitch is high, further along if it is lower. This resonant vibration will, in turn, cause the nerve cell beneath it to fire, and therefore the brain will be able to recognize the frequency by noting *which* 'string' resonated.

This description is a bit simplistic, and many researchers prefer to talk about the process which gives energy to a particular part of the membrane in terms of a travelling wave, rather than a vibration (i.e. a standing wave). You see, the process has to happen quickly, so there can't be any significant 'build-up' time for the resonance. Instead they describe what happens like this. A wave of displacement travels down the basilar membrane. As it does so, it continually reaches parts of the membrane (the 'strings') where the elastic properties are different, and the speed of the wave gets slower (how much depends on the frequency). Eventually there will be a point at which the wave stops and dumps all its energy, causing the membrane to oscillate strongly at that point. (The process is exactly the same as a surf wave breaking when the depth of the water gets too shallow.)

However, I have made the point many times that there isn't much conceptual difference between a vibration and a wave, and therefore my model will let you intuit what is going on. Certainly that was how Helmholtz imagined the ear working—as a row of graded resonators: and the process of recognizing frequency as being equivalent to *locating* where on this row the resonance occurred.

It certainly explains very simply how your ear assesses the *timbre* of different notes. Since a complex periodic vibration is entirely equivalent to the summation of pure tones of harmonically related frequencies, then, depending on which overtones are present, more than one part of the basilar membrane will respond at the same time. The cochlea therefore performs the kind of harmonic analysis I described in Chapter 4, and the message it sends to the brain consists of a number of electrical signals along different fibres of the auditory nerve, one for each overtone. The brain can then be 'programmed' to identify them.

There is good evidence that this is a very useful model of the cochlea's function. It concerns a particular relationship between tones which are added together. You will recall (see page 95) that, if I add a fundamental oscillation to one of its harmonics, I generate a complex shape, like so:

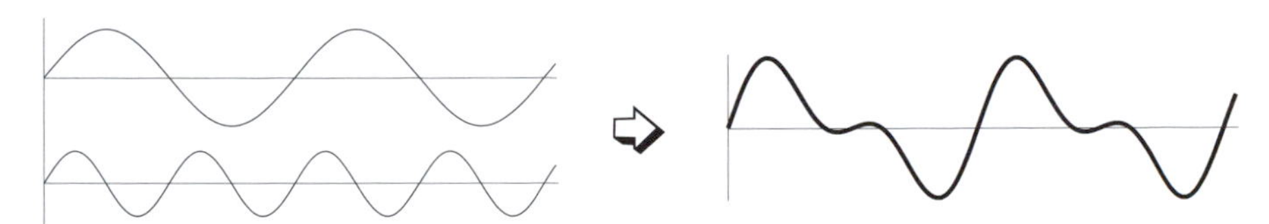

However, when I (or rather, my computer) drew these figures, I started off both oscillations exactly in step. I didn't have to do this. I could equally have started one of them at a different part of its cycle (i.e. with a different **phase**); and the result would *look* different.

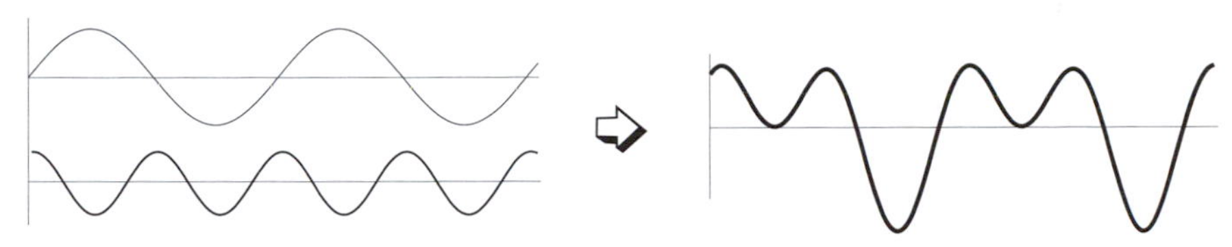

But if these shapes were pressure waves in the air (provided they were not too loud) they would *sound* the same. Your cochlea can tell that there are two tones present, and what their amplitudes are; but it can't tell anything about their relative phase.

This observation has been known for over a century, and is usually given the name **Ohm's law of acoustics** (after the same Ohm who did all that work on electrical circuit theory). I'm sure you will appreciate how strongly it supports the 'place theory' of pitch recognition. The nerve cells in the cochlea can tell that two different parts of the basilar membrane are vibrating, and how strongly; but, because they are physically separated, they have no way of telling whether or not they are going up and down in step with one another.

There is another piece of evidence which supports this same model. In Chapter 4, I talked about **difference tones**. You will recall that, if I add two high frequency oscillations, I characteristically get an oscillation whose amplitude fluctuates with time.

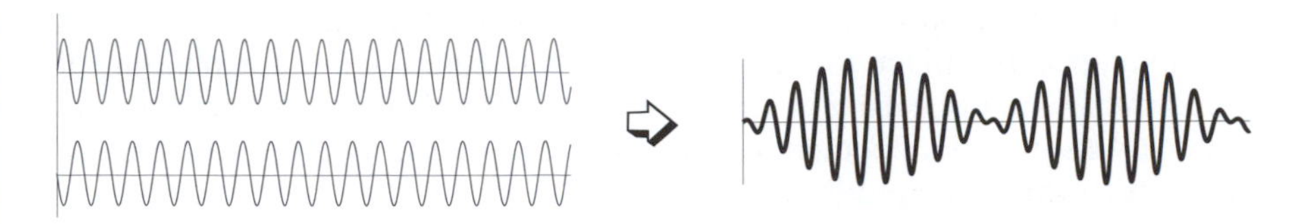

If the two frequencies are close to one another, you can hear this fluctuation as a beat; but if they are far apart you can *sometimes* hear the fluctuation as a tone. Now I want to ask the question: why only sometimes? Why can't you hear it all the time?

The answer lies in a detail which I have avoided saying much about so far. Whenever I talked about wave motion, I implicitly assumed the properties of the system which control how the wave moves are not affected by the wave's being there. I took it for granted that two waves can travel through the same medium, and the motion of each is unchanged by the other. Now when you come to think of it, this requires the medium to have some pretty specific properties. When it is stretched because of the passage of one wave, it must still be able to stretch the same amount extra as a second wave goes through. (This property is called **linearity** by mathematicians, in case you ever come across the term.) If your ear, for example, really does behave like this, then you can see that it will not hear the fluctuations we were talking about, as an independent tone. It will simply register that there are two pure tones present, because its response to each one is unaffected by the presence of the other.

But of course, very few materials are absolutely, absolutely linear. There is a limit to how far any elastic material can stretch, and if you get near that limit, then it's not much use trying to make it stretch any more. A second wave will not be able to travel through properly. And there are lots of points in your ear which behave elastically and which may not be able to respond absolutely faithfully to a large wave coming through—the eardrum, the ossicles, the oval window, the basilar membrane itself. So when your ear encounters a *loud* signal which fluctuates, the signal will be distorted. Then when your cochlea tries to harmonically analyze the signal it will still detect the two pure tones, but it will also detect a distortion which fluctuates with a frequency equal to the difference of the two pure frequencies. You will hear the difference tone. Likewise you may be able to detect a change of phase in two very loud tones, because they stretch the elastic materials in your ear differently.

Therefore I hope you can see that this model of how the cochlea works allows us to understand a lot of what we know about hearing in a simple and straightforward manner. I should point out here that this whole field of research is still changing. For example, there is evidence that the brain receives *some* direct information about how rapidly the basilar membrane vibrates—in the rate at which the nerve cells fire. As a result, some researchers have proposed a so-called 'time theory' of pitch recognition to complement the 'place theory'. Obviously a definitive understanding of the mechanism of hearing is not yet with us. Nonetheless, most of the features I want to talk about can be understood from the simple picture, even if it isn't complete. And that's all we want.

Range of pitch

Let me now turn to the question of what range of frequencies the ear is sensitive to. Everyone's ears are a little different; and over the years many experiments have been done, getting volunteers to listen to tones of different frequencies, trying to determine the lowest intensity they can detect. The average results of countless such tests are usually summarized in a graph like this:

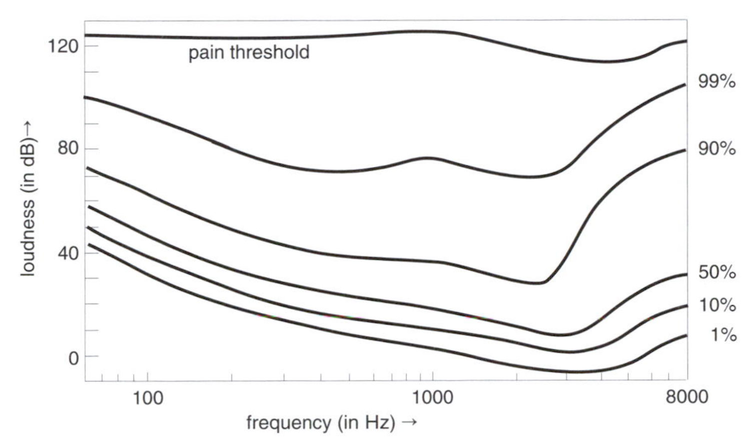

Notice that loudness is measured in its own special unit, called the **decibel** (abbreviated as **dB**). If you are interested, I have included a brief discussion of this unit in Appendix 4.

You interpret the graph as follows: 1% of people can hear any sound whose intensity is above the 1% curve; 10%, a sound above the 10% curve; and so on. The topmost curve represents the intensity at which most people start feeling pain.

The graph is not extended below 60 Hz, because of a rather strange observation. When you listen to a repetitive pressure wave of very low frequency (say around 10 Hz) you don't hear it as a tone at all. It just sounds like a series of clicks; and this is true up to about 20 Hz. After that the clicks run together, but you don't start hearing them as a tone until about 50 Hz. In between, the sound is fuzzy and not particularly pleasant.

The upper frequency limit is even less well defined. Some people can hear as high as 20 000 Hz; but for most of us, the threshold is much lower. It is one of those sad facts of life that this figure decreases as we grow older—after age 40 at the alarming rate of about 80 Hz every six months. There are a lot of reasons for this, the most straightforward being that all skin loses resiliency with age, and none of the membranes respond as well as they should.

One of the starkest examples of this is in the background squeal of a TV set. In television, the image is built up in lines, by the bright spot moving across the screen about 600 times for each frame. The illusion of motion is achieved by the picture changing 25 times a second. So in every TV signal there is a pulse telling the spot to start a fresh sweep, which occurs about 15 000 times each second. This gets into the audio system, and comes out as a tone of 15 000 Hz. Children and young adults can hear this clearly; and, I am told, find it very annoying. The rest of us, alas, have long since sunk below the level at which we can even hear it.

Between the upper and lower frequency limits, our ears respond with varying degrees of sensitivity. It is greatest around 3 000 Hz; and one factor accounting for that is, as I mentioned earlier, the physical size of the outer ear. But an equally important factor involves the way the resonators are distributed along the basilar membrane. In this diagram, I have indicated which 'strings' respond to various musical pitches:

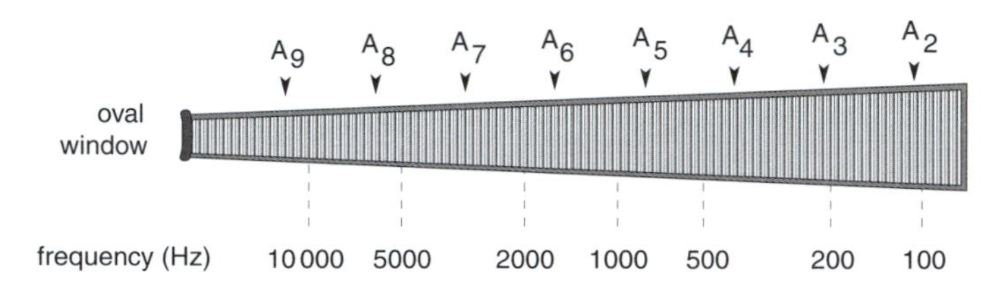

Notice firstly that the musically most important range of frequencies (from about 100 Hz to about 4 000 Hz) occupies roughly 2/3 of the length of the membrane, while the rest of the scale (up to nearly 20 000 Hz) is squeezed into the remaining 1/3. Secondly, notice that the frequencies are spaced *logarithmically* (a word which you will find defined more carefully in Appendix 2); that is, whenever the frequency is doubled (and the pitch rises an octave) the resonant point moves roughly the same distance (some 4 mm) to the left.

The importance of this last observation cannot be overemphasized. It gives a straightforward explanation of that intriguing fact I mentioned in the very first chapter of this book—that the natural way to express the 'difference' between the pitch of two notes involves forming the *ratio* of two frequencies, rather than *subtracting* those frequencies. Once you know how the ear works, it seems perfectly reasonable that the apparent interval between two notes should depend on the number of auditory resonators separating them; and that depends on the ratio of their frequencies.

Of course, this leaves unanswered the question of *why* our ears should have evolved in this way. So we haven't solved everything yet. But it is interesting that all members of the animal kingdom who employ sounds which we consider to be musical—like birds or whales or dolphins—have ears whose structure is very similar to our own. Surely there can be no doubt that a lot about music is determined by the way our ears are put together.

Pitch discrimination

The next question to be addressed is: how good is the ear at telling frequencies apart? Or, if you prefer, what is the smallest pitch interval you can have between two notes and still hear that they are different?

The most useful evidence bearing on this comes from the kind of experiment in which you sound two pure tones of different frequencies together and listen to what they sound like. The results of such experiments, again the average of many listeners, can be summarized like this:

- When the two frequencies are very close together you hear **beats**—a regular pulsation in loudness at a single pitch somewhere between the two. (I talked about this in Chapter 4). This effect persists up to a frequency difference of around 20 Hz.
- On the other hand, when the notes are widely separated, you hear them as two clearly distinct pitches, and this is true for any two tones which are more than about a minor third apart (i.e. three semitones or about 20% difference in frequency).
- In the in-between region, what you hear is a bit uncertain. The sensation is often described as 'roughness', a term which is used by a remarkably wide range of listeners.

A lot of this can be explained from what we know about the way that all resonating systems behave. Let us think first of all how we would expect the ear to respond to *one* tone. When a single pure tone is sounded, there is only one auditory resonator, one 'string' in the cochlea, which exactly matches it in frequency. This is the one which will respond most strongly. But there are other resonators nearby which *nearly* match, and these will also be set vibrating. Whether or not they do so with an appreciable amplitude, depends on the *bandwidth*. If this is large, i.e. if each resonator will respond to a wide range of frequencies, then many neighbouring resonators will respond to the tone. A large *area* of the membrane will vibrate.

It is useful to represent this conclusion diagrammatically, by imagining that the response of the nerve cells along the basilar membrane follows a resonant curve, like the one on page 226.

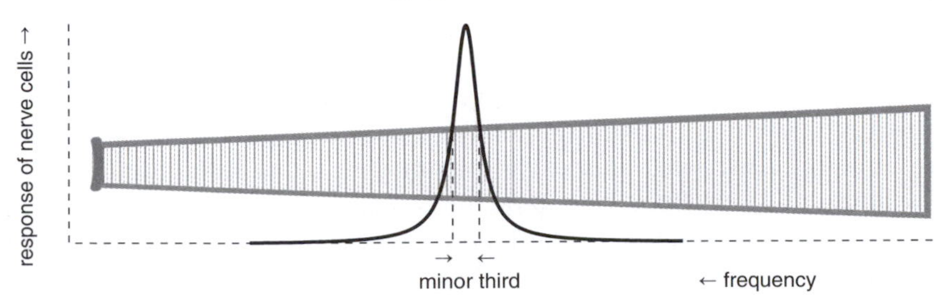

Now consider what happens when the ear hears *two* tones together.

- If they are very close together in frequency, these areas will overlap and the part of the membrane in the overlapping region will be performing two independent oscillations. It will have a big amplitude when they are in phase, and a small amplitude when they are out of phase; and will change regularly from one to the other. You will hear **beats**.

- If the two tones correspond to parts of the membrane well outside one another's bandwidth, then this effect should be entirely absent—the two vibrations should proceed independently of one another. You will hear two distinct pitches.

- But if they are not too widely separated and their bandwidths partially overlap, it is more difficult to say what will happen. Perhaps some of the signal going to the brain would say "two distinct frequencies", while another part of the same signal would say "a pulsating single frequency". It seems plausible that such a message might be described as 'rough'. It is equally plausible therefore, that we can identify the bandwidth of the auditory resonators to be the range over which this roughness is known to be detected—i.e. a minor third.

That seems a plausible explanation for the observations, but it does raise another question. If each resonator will respond to any note within three semitones of its natural frequency, how is it that we can identify pitch as accurately as we can? Most people with minimal training can pitch a note correctly at least to within a semitone. And a trained ear can do much better. There are plenty of choir conductors who expect their singers to be able to distinguish between a Pythagorean tone (a ratio of 9/8) and a just tone (10/9). The difference between these is a ratio of 81/80, or about 1/5 of a semitone.

What is clear, I think, is that the cochlea cannot, of itself, make such distinctions. Once again, it is just like tuning your radio. You can get to within the bandwidth of a particular station, just by listening to how loud the signal is. But if you want to get it 'right on', you've got to wait till you hear something you can recognize and then try to judge whether it is distorted or not. The point is that you need more information to work on. It must be the same with the auditory system. The brain needs more information than it gets from the simple observation of *which* resonators are moving. Just what this extra information is, is still the subject of investigation—it probably has something to do with the regularity of nerve firing—but I don't think it is important right now. It is enough to know that it is a secondary process, under the control of the brain. So you can learn accurate pitch recognition; but the broad, general features of relating tones to one another is built in.

I cannot leave this section without making some brief mention of the fascinating topic of **absolute** (or **perfect**) **pitch**. Most people, if they hear one tone, and then another, can tell whether the second is higher or lower than the first. Those with musical training can usually recognize the standard intervals between two tones, or can sing these intervals after having heard a reference tone. Many with well trained ears can detect a frequency shift of as little as 1% (or a sixth of a semitone), and sometimes even smaller intervals. This is called **relative pitch**, and it is, when you come to think of it, a quite extraordinary sensory ability. It is difficult to think of any evolutionary advantage which could have caused our ears to develop like this.

But even more extraordinary are those 0.01% of the population (or even fewer) who have absolute pitch—who can recognize or sing a given note, without referring to any other tone as a reference. Psychologists have been studying absolute pitch for nearly a century, but there is still no agreement about why some people have it and others don't. Some researchers claim there is evidence that it is inherited. There have been some very recent studies reported which suggest that many babies are born with this ability, but quickly lose it as they listen to the way the people in their world sing and play music without much need for absolute tuning. But there are just as many studies which suggest that it is an acquired characteristic, and can be learned (most successfully while you are young).

Possessing absolute pitch can obviously be advantageous for a professional musician—as a singer you don't need an accompaniment to sing in the correct key, or as a conductor can more easily determine what notes should be played. However it also has disadvantages. It is a reasonably common complaint among choral singers with absolute pitch that they get put off when the rest of the choir, blissfully unaware, drifts out of tune. And it is certainly not an essential prerequisite for a musician. Many composers have been reported to have had perfect pitch, including Mozart and Beethoven, but there are even more who didn't.

Loudness

The other job that your ear has to do is to recognize how loud a sound is. And here you can see what a truly remarkable instrument it is, because the range of intensities it will respond to is enormous.

The **intensity** of a sound is measured as the amount of energy which falls each second on an area of standard size (usually taken to be 1 metre square). Hence its unit of measurement is the watt/square metre, or W/m^2

(refer to Appendix 4 if you want further information about this). Just to give you a feel for the numbers involved, the intensity of radiant energy (i.e. light and heat), at a distance of 1 m from a 25 watt electric light bulb is about 2 W/m^2. A sound of the same intensity would be so loud as to be extremely painful.

On the other hand, the smallest sound you can hear (the canonical dropped pin) is less than a millionth of a millionth of this. This is an enormous range indeed, and it is difficult to think of any other measuring instrument, anywhere, which covers such a wide span.

There is such a vast difference between the top and bottom limits that it is usual to compare loudnesses on yet another logarithmic measure: the so-called **decibel scale**. Its starting point—0 dB—is taken to be the threshold of hearing (that dropped pin). The rest of the scale is divided up into equal parts (steps of 10 dB); each one corresponding to the intensity having been *multiplied* by 10. Therefore, since you can multiply the low threshold 12 times by a factor of 10 before it will reach the top, this limit is measured as 120 dB. There is a more detailed description of this in Appendix 4, but here is a short table which compares the decibel scale with the standard musical markings—*piano* (soft) and *forte* (loud).

INTENSITY	SOUND LEVEL (dB)	TYPICAL NOISE LEVEL
pain threshold	120	pneumatic drill
fff	100	underground train
f	80	noisy office
p	60	large shop
ppp	40	suburban home
hearing threshold	0	soundproof room

This arbitrary-seeming scale has a firm basis in the physiological properties of the ear. To appreciate this, just think about how loud a single singer sounds; and compare that mentally with the sound of two singers in unison. They are considerably louder. Adding a third makes it louder again—but not by so much. Keep adding one more singer, and each time the increase in loudness is less marked. By the time you've got to twenty or so, you'll hardly notice the extra voice at all. Yet each time you add a singer, the energy output (and hence the intensity) goes up by exactly the same amount. What it means is that your ear is 'measuring' loudness on a logarithmic scale. To get the same increase in loudness as you got between one singer and two, you have to double the size of your choir.

As an aside, it is an interesting fact, not really understood, that *most* of our sense organs work logarithmically. This observation is usually stated as a relation between a stimulus applied and the sensation it produces: as stimuli are increased by multiplication, sensations increase by addition. It goes by the name of **Fechner's law**, after the 19th century German physicist who tried single-handedly to invent a science he called 'psycho-physics'; and whose ideas were extensively used by Helmholtz in his, rather more successful, investigations.

In terms of what we know about the ear, this observation makes sense. A very soft tone presumably sets just one part of the membrane (one 'resonator') vibrating; and this triggers the nerve cell beneath to start firing. As the sound gets louder, the vibration becomes more violent, but the nerve cells can only respond just so much. Once they start firing at a few times their spontaneous rate, they can't do any more. They are said to 'saturate'. But that isn't the end of the ear's ability to discriminate. As the intensity increases, other resonators nearby start vibrating quite strongly, so their nerve cells will begin to fire. In other words, the sensation of loudness depends more on the *number* of nerve cells firing than on the rate at which each one fires. It may not be immediately obvious to you that the relation between this number and the intensity of the primary sound is logarithmic, but at least I hope you find it plausible.

There is one particularly important consequence of this for music. Since any real note is accompanied by overtones, you can increase its loudness by strengthening *either* the fundamental or the overtones (or both). In an organ, for example, which is pretty loud anyway, there's not much point in trying to make it even louder by bringing in more pipes to play exactly the same notes. Fechner's law is one of diminishing returns in this context. But if, instead, you bring in more pipes to reinforce the *harmonics*, you are putting your effort where it produces most effect. So listen carefully next time you hear a big 'swell' on the organ, and see if you can pick that this is what the organist is doing.

The same effect is even more dramatic in the case of a singer. Have you ever wondered how it is that an opera singer can be heard above the whole orchestra (especially the enormous ones that Wagner used), or the trained soloist can come out over the top of a huge choir? The answer is that they put a lot of their vocal energy into a particular range of harmonics where instruments and untrained singers are not strong—the so-called **vocal formant**. But that's a story I will take up in Interlude 8. For now, I want to go on and talk about what an understanding of the working of the ear contributes to the theory of harmony.

The psycho-acoustical theory of consonance

Let us return to the problem that we have kept coming back to many times throughout this book, the centuries old **puzzle of consonance** which I last mentioned when I was talking about Galileo's contributions to musical theory: why do some pairs of musical notes sound well together, while others do not? There is information bearing on this to be got from a series of experiments that were carried out in the 1960s, by a pair of Dutch scientists, Reinier Plomp and Willem Levelt (among others).

In rough outline, these experiments consist of the following. You play two pure tones of different frequencies together and, this time, ask listeners whether or not the combination sounds pleasant. (You must use untrained listeners, because you don't want them to have any preconceived notions of which intervals *ought* to sound good). Your results will probably look like this graph—which actually comes from many such experiments over the years.

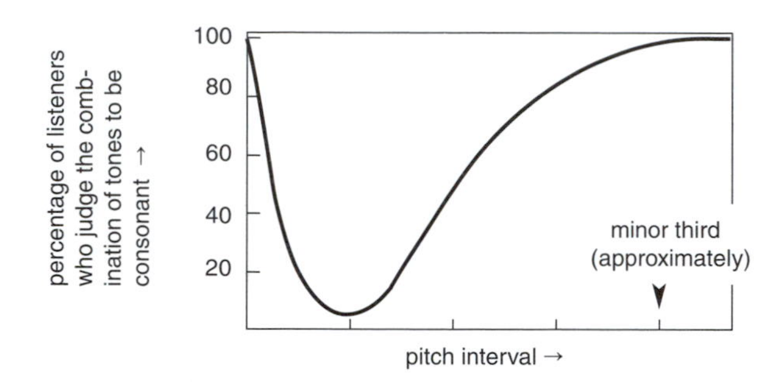

Obviously nearly everybody agrees that two tones less than about a minor third sound dissonant, and at about a semitone, extremely so. But what is interesting is that intervals greater than this are *all* judged more or less equally pleasant. There seems to be no preference at all for the musically significant intervals—the fifth, fourth, third and so on.

This observation can be reconciled with what I said before, when I tentatively identified the resonant bandwidth of the auditory nerve cells as a minor third. Let me try to represent on a diagram the nerve cells' response to two pure tones at various intervals apart.

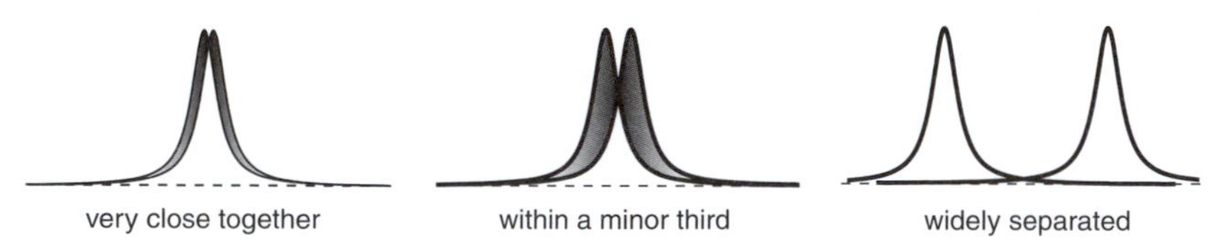

Now it seems plausible to identify on these diagrams the areas where the nerve cells respond erratically—the regions of 'roughness'—as those areas where the two curves have large amplitudes at the same frequency but are different from one another.

A conclusion we would draw from these diagrams is that, when the two tones are far apart in frequency, there is no physiological reason why we should perceive any dissonance at all. But we all know that there *is* a widespread preference for the classical musical intervals when ordinary musical notes are heard together. The only significant difference between pure tones and real notes is that the latter contain overtones; so the solution must be sought in how our ears respond to these.

If you bear in mind that, when a single pure tone is sounded, the resonators within the **critical bandwidth** all respond; then when a real note is played, resonators within many such bands will start resonating. The *total* response can then be represented by plotting the vibration amplitude of each of the resonators against their natural frequencies—in other words, by drawing the **frequency spectrum** of what the basilar membrane detects.

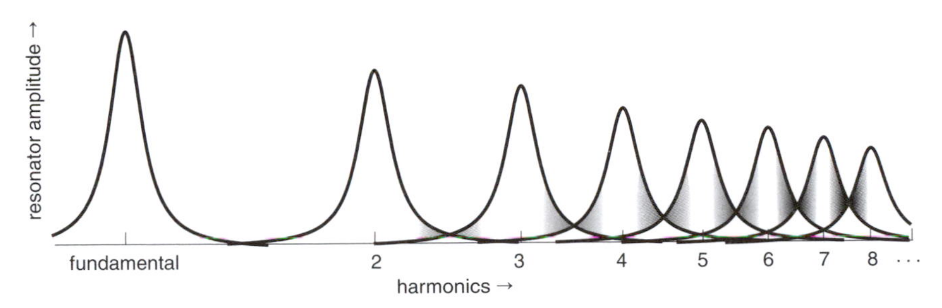

Each of these resonant peaks is centered on a harmonic of the fundamental. These should be equally spaced along the frequency axis, but I have used a logarithmic scale (because that's what the ear prefers). On that scale the higher harmonics get closer and closer together. However the critical bandwidth stays the same apparent width—it's always just under a minor third remember—so for the higher harmonics there is considerable, and increasing, overlap. And that kind of overlap implies 'roughness'.

This must mean that, in any real note, there is actually a fair bit of dissonance, especially if the high harmonics are strong. It doesn't follow that they should sound unpleasant: 'rough' doesn't necessarily mean 'nasty'. Nonetheless the effect is noticeable. We use the adjective 'brassy' for any note with very prominent upper harmonics, like those of a trumpet; while those of a flute, which has very few, are often described as 'gentle'. It is as though a little bit of dissonance is a kind of spice—too much is to be avoided, but food tastes bland without it.

But now think what happens when *two* notes with overtones are sounded together. If their frequencies are randomly chosen, even if their fundamentals are separated by more than a minor third, it's likely that there will be a great deal of overlap among the harmonics. (I can't keep using this way of drawing things, but it will serve to indicate what I mean here.)

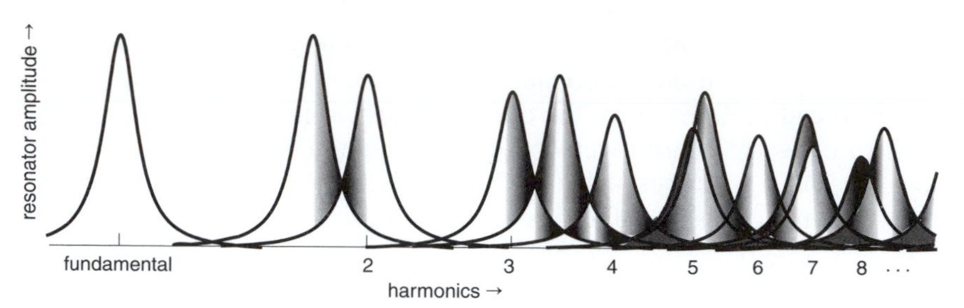

It is difficult to believe that your ear won't register this as pretty 'rough'. To continue the culinary metaphor, it is surely a bit too highly spiced. Though again, you could *learn* to like it—there is such a thing as an acquired taste.

However, there are some special intervals between the two notes for which this won't happen. The most obvious is when they are an *octave apart*. Then the resonant peaks of the higher note will exactly coincide with every second peak of the lower. So adding the former to the latter will produce *no increase in roughness at all*. That seems to me to go a long way towards explaining why two notes an octave apart are so completely harmonious that they can almost be considered the same note. In other words, the absence of any extra roughness must be what we mean by 'perfect consonance'.

A similar claim can be made if the two notes are a perfect fifth apart (with frequencies in the ratio 3/2). Then every second peak of the higher will coincide completely with every third one of the lower.

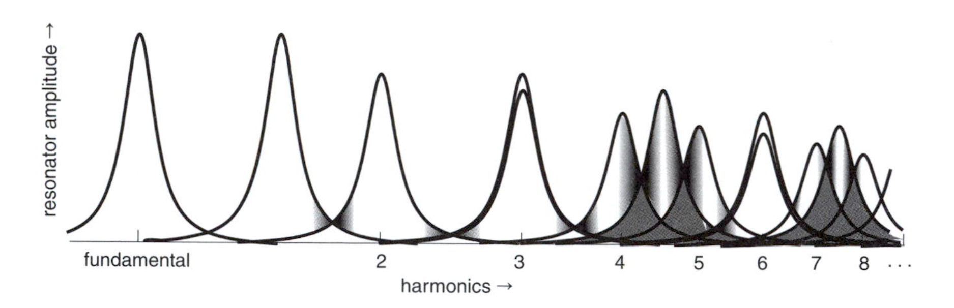

There is clearly more roughness here than for either note singly, but much less than in the preceding diagram. This is because the ratio of fundamental frequencies is just what is needed to put some of the peaks completely on top of one another, and cut down on the total amount of overlap roughness.

Much the same will be true for other pairs of notes whose frequencies are in the ratio of two small whole numbers. Therefore it is possible to calculate the degree of overlap from *any* pair of notes, and to predict how much dissonance they should generate when sounded together. The result of this calculation, as first carried out by Plomp and Levelt, is as follows. (Note that the graph is plotted so as to look like the results of the experiment at the start of this section.)

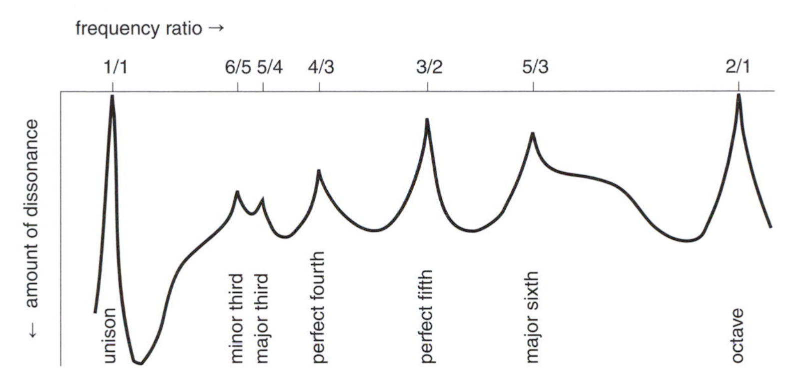

Clearly the traditional musical interval ratios stand out from others around them as being particularly free of dissonance. It would appear therefore, that we have found a truly basic explanation, in terms of the properties of the ear, for why these intervals should be pleasing to listen to. This is an important change in the theory of harmony, because it suggests that consonance is a negative feature—an absence of dissonance—rather than a positive quality in its own right. It is also important because it shows that the property of consonance is not absolutely dependent on the exact value of the frequencies involved. There is room for a little inaccuracy, and therefore the intervals will sound much the same no matter what *musical scale* (i.e. just or equal tempered) they are played in.

These insights were largely developed by Helmholtz during those years in Heidelberg—although some of the results I called upon came from more recent research. In his book *On the Sensations of Tone*, he went a great deal further than this. He devoted a lot of time to discussing combination tones, and pointed out that, when two notes sound together, there will be many *difference tones* at the frequencies separating the various harmonics. These will only be heard faintly (as we discussed earlier) but unless the fundamental frequencies are in simple ratios (again), they will be dissonant with the primary tones. Hence he was led to a theory of chords and an understanding of the role of the **fundamental bass**, exactly as Rameau had been over a century earlier.

To me, there is a paradox about Helmholtz's place in the history of the theory of harmony, as seen by musicians. If you look under his name in most of the standard musical encyclopedias, you will find only the briefest mention (if any). I suppose from the point of view of those interested in *aesthetic* questions, he didn't do much that was new. But from the viewpoint of someone like me, his contribution was immense. He supplied an answer to the great question: "Why?" Whereas Rameau had said that the rules of harmony had to be as they were because the consonances on which they were based sprang from a kind of cosmic 'rightness', it was Helmholtz who firmly showed that the answer lies—to coin a phrase—not in our stars, but in ourselves.

Envoi

Helmholtz died in 1894 and Wagner eleven years earlier, in 1883. With their deaths, a chapter of the history of both music and physics seemed to close. Almost immediately both went through a period of such great change as can only be described as a revolution. I will talk about the new music later, but now let me concentrate on what happened to physics.

In the 1880s, James Clerk Maxwell had announced, with typical Victorian complacency, that essentially all of physics had been solved. To use his metaphor, the scientific sky was perfectly clear, except for one or two small clouds on the horizon. These 'clouds' were a couple of obscure observations about the way that light reacted with electricity, and the newly discovered phenomenon of radioactivity. They were to prove precursors of a cyclone.

Both in chemistry and physics, the really exciting area of research in the second half of the 19th century, was into the structure of matter. Experiments had finally confirmed that all substances were made up of atoms; and that electricity was also carried by small particles (called **electrons**). On a fine enough scale, all of nature seemed to be 'grainy'. It seemed reasonable that these electrons were a part of the atoms, and therefore electricity was a fundamental property of all matter. Because light was also intimately connected with electrical effects, the source of all light waves must be electrons oscillating inside atoms.

But just as matter and electricity was 'grainy', experiments seemed to be pointing to the conclusion that energy was also. The first to realize this was one of Helmholtz's ex-students, Max Planck, in 1900. He proposed calling these 'grains' of energy, **quanta**; from which the whole subject came to be known as the **quantum theory**. Within five years, (in 1905, the same year in which he published his work on relativity) Einstein showed that Planck's hypothesis would be perfectly understandable if it was assumed that light were made up of particles (**photons**), just as Newton had said. But this really created a paradox, because Young's results were still valid, and they showed that light was a wave.

The next step came from something chemists had known for some time—the fact that most chemical elements, when heated to very high temperatures, emit light of only certain, very pure colours. (I'm sure you've all noticed those sodium street lights which give out a harsh yellow light, and the mercury ones which are a much softer blue.) The question which bothered everyone was, of course: why these colours and not others?

It all sounds familiar, doesn't it? The light from those chemical atoms seems to be arranged in a particular set of frequencies, rather like a *musical scale*! In about 1910 the Professor of Physics at the University of Munich, Arnold Sommerfeld, was the first to comment that the problem was very similar to that which had faced the Pythagoreans; and he developed such a reputation for apparently way-out ideas that his students used to refer to his lectures on atomic structure as 'atomysticism'.

Amazingly, the solution, when it came, sounded just like the old music of the spheres. In 1911, Niels Bohr proposed that the electrons in an atom were orbiting a central nucleus, for all the world like a miniature solar system; and the frequencies of the radiation they gave out was controlled by the fixed size of their orbits, very much as Kepler had imagined. It was a brilliant piece of work, but it still left the same question: why these orbits and not others?

At this stage the Great War intervened, and it wasn't till 1921 that Louis de Broglie suggested that, since light was a particle as well as a wave, it would simplify everything if you thought of an electron as a wave as well as a particle. This idea was seized upon by Erwin Schrödinger in 1926, who proposed that these electron 'waves' could only be trapped inside an atom if they formed themselves into *standing waves*. So the reason *why* Bohr's orbits existed was essentially the same as the reason the different modes of oscillation exist inside a musical instrument. The radiation of light from an atom must be very like the sounding of a musical note.

In the 60 years that have elapsed since then, this theory has proved almost unbelievably successful, and chemists the world over base their understanding of how substances react with one another on what is called the **Schrödinger wave equation**. But it still doesn't clear up the paradoxes. One step to resolve these was taken by Werner Heisenberg, in 1927, who pointed out that a lot of the confusion was in our own minds. In the same way that a musical note should never be considered in isolation from the ear that hears it, which involves a fundamental uncertainty because of the ear's critical bandwidth; so also an electron should always be thought of in relation to the experiments which measure it, and this will involve a fundamental indeterminacy. This is the famous **uncertainty principle**.

Let me not go any further with this. It's getting pretty deep, and there are philosophical difficulties which are still being argued about today. Instead, let me pause to comment on the *nationalities* of the scientists who made these contributions. I haven't said anything about those who did the experimental work, and a lot of them were English-speaking; but of the theoreticians I mentioned, Bohr was Danish and de Broglie was French. The others were German. I can't help wondering if that is significant.

Obviously, from the way I have explained the subject, I see a great similarity between quantum theory and musical acoustics; and I can't help feeling that the ubiquity of music in the German cultural and intellectual scene contributed in some way. As Helmholtz's student, Planck was particularly immersed in music. Several of his early research papers were about musical theory and the construction of scales. Heisenberg too was a keen pianist and, as Sommerfeld's student, swayed by the philosophical attitudes of the ancient Greeks. In an essay he later wrote on how he came to be interested in atomic theory, he tells a story which I find particularly evocative. It describes a school outing in which a heated political argument had broken out.

> "The talk was still going on when, quite suddenly, a young violinist appeared on a balcony above the courtyard. There was a hush as, high above us, he struck up the first great D minor chords of a Bach *Chaconne*. All at once, and with utter certainty, I had found my link with the central truth. The moonlit valley below would have been reason enough for a romantic transfiguration; but that was not it. The clear phrases of the *Chaconne* touched me like a cool wind, breaking through the mist and revealing the towering structures beyond. There has always been a path to the central order in the language of music, today no less than in Plato's day and in Bach's. That I now knew from my own experience."

Interlude 6

Percussion instruments

This last group of instruments in the orchestra is perhaps the most varied of the lot. They come in a bewildering array of shapes and sizes, making an equally bewildering array of noises. Bangs, thuds, cracks, chimes, rings, rattles, jingles, tinkles, crashes—all of which are purely transient sounds. They build up quickly and immediately start to die away. At no stage do they constitute a sustained vibration. And this is not just because most of these instruments make their sound by being struck (whence the name 'percussion'); it is inherent in the very way they vibrate.

When I talked about the Fourier theorem in Chapter 4, I made the point that, in order for a tone of definite pitch to be sustained, its overtones had to be harmonics (that is, their frequencies had to be integer multiples of the fundamental). What characterizes the present group of instruments, from the point of view of a physicist, is that they all give out *an*harmonic overtones: their frequencies are not integer multiples of the fundamental. This means they cannot give out a sustained tone, even if you wanted them to—without changing their sound completely.

It will help if I divide this enormous group of instruments into two subgroups, according to the two quite different functions they perform in an orchestra or a band. Firstly there are those which are used to define rhythm. Their main job is to provide pulses of sound, in a steadily recurring pattern or at strategic times, against which all the lines of melody can measure themselves. For this purpose their *pitch* is often irrelevant. In a sense, they just make noise—or at least sounds of indeterminate pitch.

There is little of a general nature I can say about this subgroup. They're all very different from one another. There is probably a lot of psychoacoustical understanding to be found in why we respond as we do to rhythmic sounds; and in such questions as (say) why the sound of a side drum is easy to march to, whereas bongo drums are much better for dancing. But so far as I'm aware those investigations have yet to be done.

The second function that percussion instruments perform is to highlight particular melodic lines. I'm thinking of something like Tchaikovsky's use of the celesta (in *The Nutcracker*) or Saint-Saëns' of the xylophone (*Danse Macabre*). Because of their unusual timbres, the sounds of these instruments don't blend with the orchestral tutti, and therefore what they play stands out in high relief. And here I can say something useful of a general nature, because in this context the essential distinguishing feature of the percussion instruments is that they all vibrate in a more complicated fashion than the simple one dimensional vibrators we have studied so far.

Standing waves on surfaces

The fact that *two-dimensional* surfaces vibrate with anharmonic overtones is so important that I should try to give a careful reason for it. But be warned. There is no simple, intuitive explanation. It depends on quite subtle geometric arguments. So if you want to, you can skip over most of this section and you won't have missed very much.

What is critically different between a wave travelling along a one-dimensional string or column of air, and across a two-dimensional surface (or even through a three-dimensional volume) is that in the former each particle of the medium passes on all its energy to the next one; but in the latter the energy *spreads out*.

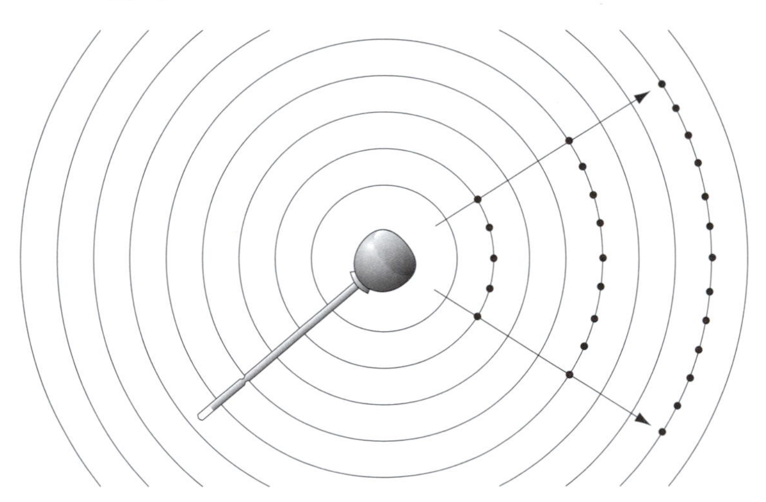

Any single particle collects only *some* of the energy from the one before it, and passes it on to several others down the track. Therefore the amount of energy that each particle receives and passes on gets smaller, the further it is from the source of the wave. The amplitude decreases as the wave goes out.

The *shape* of the wave therefore can no longer be a simple sine curve, even if the source is undergoing a simple harmonic oscillation. Instead it will look like this:

The motion of the particles along the path of the wave does *not* repeat itself at regular intervals. (This shape is called a **Bessel function**, if ever you should come across the term; and mathematicians think of it as the two-dimensional form of a sine wave.) For this kind of wave, not even the positions at which the particles have zero displacement are evenly spaced. In the vicinity of the source these are somewhat closer together than they are at large distances. So the very concept of **wavelength** loses its meaning. And that is a pity, because it was wavelength which made the pictorial representation of standing waves easy to interpret.

On a circular drumhead for example, if you strike it in the centre, a wave of this nature spreads outwards from that point. It reflects when it hits the circular boundary, and another wave then travels in a completely reverse manner, focused towards the centre. Therefore, for exactly the same reasons as I talked about in Chapter 5, a *standing wave* will be set up; and if the frequency is right, this will result in a mode of vibration of the whole surface, looking like one of these (or higher modes):

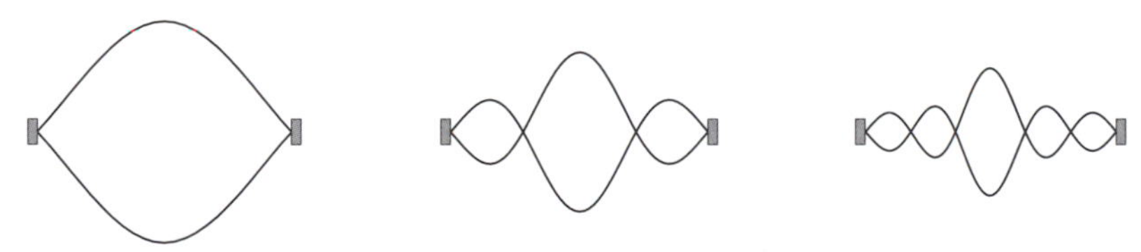

In interpreting these diagrams it is important to realize that the **nodes** are not just *points* as they appear here. The surface is doing the same thing in all radial directions, so the positions of no displacement occur in *circles* around the centre of the drumhead. We call these **nodal lines**. Seen from above, the nodal lines of these three modes look like this:

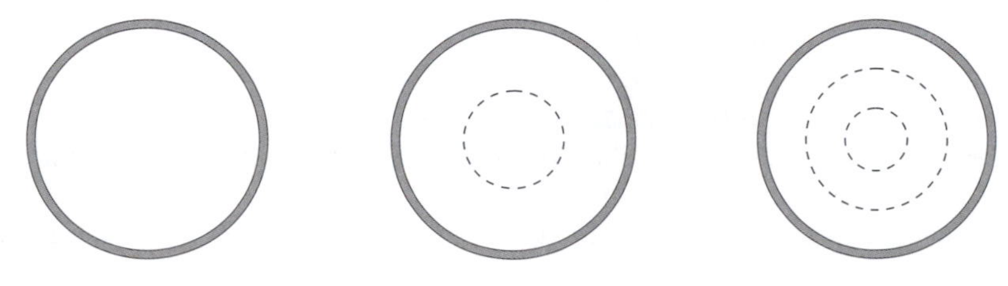

Other sorts of vibrations are possible, involving the surface bending in the middle, and the nodal lines of these are often straight lines rather than circles.

Previously when I was talking about strings and air columns, I was able to work out that the frequencies of the modes were harmonics *because* a whole number of equal wavelengths fitted exactly into the vibrating length. But now, since the distance between nodes changes, *there is no reason why these frequencies should have any simple numerical relation with one another*. In fact the frequencies of the three modes I drew are in the ratios 1.0 : 2.3 : 3.6. Therefore these vibration modes are not harmonics of the fundamental.

While I am talking in general theoretical terms, it is worthwhile pointing out that matters are simpler for waves travelling in *three* dimensions. Waves travel radially outwards through a volume in such a way that the points where the pressure is zero are *exactly evenly spaced*. Why this should be is a bit tricky. Basically it comes about because the decrease in particle speed (which occurs because the energy falls off as the wave goes out) exactly compensates for the decrease in the extra distance through which each part of the wave front has to spread. The consequence of this is that it is possible to set up standing waves inside a spherical vibrating medium, which are simple sine curves whose amplitude falls off smoothly, and these modes of oscillation are all related to one another numerically. They *are* harmonics of the fundamental.

Now this may not seem a very useful observation to make. After all, it's difficult to imagine a musical instrument made in the shape of a perfect sphere. That is true, but remember that a **cone** is part of a sphere. Therefore the normal modes of oscillation inside a conical pipe must be the same as those inside a sphere. This was a result I have been using all the way through this book, every time I said that a cone would play a complete harmonic series. I have read quite a few books on acoustics written by musicians who complain that this point is always obscure. Well now you can see why.

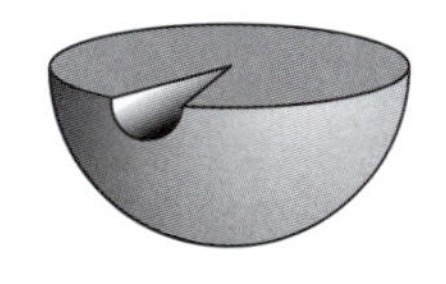

The difficulty in conceptualizing the solution to this problem is reflected in the length of time it took mathematicians to solve it. You will remember that the modes of vibration of a string were completely understood by the end of the 18th century: those of a two-dimensional surface took another 50 years. One of the earliest to interest himself in this problem was the German scientist, Ernst Chladni, who in 1808 visited Paris and caused a sensation in the *Académie des Sciences* with a beautiful demonstration he had invented to make visible the vibration patterns of thin metal and glass plates.

What he did was to sprinkle powder on top of the plates (nowadays sand is most often used) and set them vibrating by bowing along one edge. At regions of the surface where the vibration is greatest, the powder is agitated and gradually migrates away, to collect along the nodal lines where the vibration is least. The strikingly beautiful patterns which result impress audiences as much today as they did then. Here are some of those patterns as drawn in Chladni's own writings:

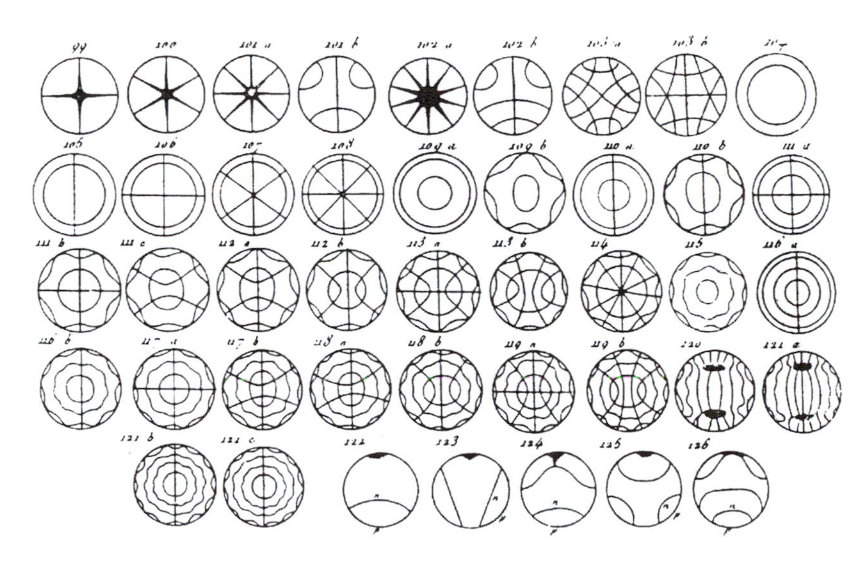

These demonstrations obviously intrigued the Emperor Napoleon enormously, and he offered a prize of 3 000 francs for a mathematical theory which would explain them. It took seven years; and in 1815, the year of Waterloo, the prize was won by a female mathematician, Sophie Germain. Today we realize her solution was mathematically flawed, but it was physically remarkably insightful. Even so, it was not until 1850, that a complete solution was found, by the German physicist, Gustav Kirchhoff (who was Helmholtz' predecessor at the University of Berlin).

It is important to realize that this work was relevant, not only to acoustics, but also to the study of how all elastic materials behave under various states of loading; and therefore to the theory of structural engineering. So it is not unreasonable to say that one of the earliest and most important contributions to this field was made by Sophie Germain—which makes her story all the sadder.

I said before that her prize-winning work was flawed. She obviously wasn't aware of the very latest developments in calculus; and this was because she was entirely self-taught. Being a woman, she wasn't allowed to enrol in the *École Polytechnique* (the establishment Fourier had helped Napoleon to set up) but had to depend on lecture notes taken by a male friend. Though she did eventually become acquainted with members of the mathematical establishment, especially after winning the prize, she was always treated as an outsider, a freak. She was never invited to participate as an equal. She died of cancer, bitter and lonely, in 1831, at the age of 54.

Even today there is to be found, in Paris, a lasting reminder of the discrimination she had to contend with because of her sex. When the Eiffel Tower was erected in 1889, on the girders round the bottom were engraved the names of all those scientists, mathematicians and engineers who had contributed to the branch of science which made it possible. If you walk round it next time you are in Paris, you would recognize many of those names from among those I have mentioned throughout these chapters. Conspicuous by its absence however, is that of Sophie Germain.

Vibrating membranes

The physical principles I have just been talking about find most obvious application in the vibrating surface (or 'head') of a drum. A piece of some elastic material, like calf-skin, is stretched across a circular supporting frame, and sounded by beating it with a stick. The first six modes of vibration of this surface, in order of increasing frequency, are these:

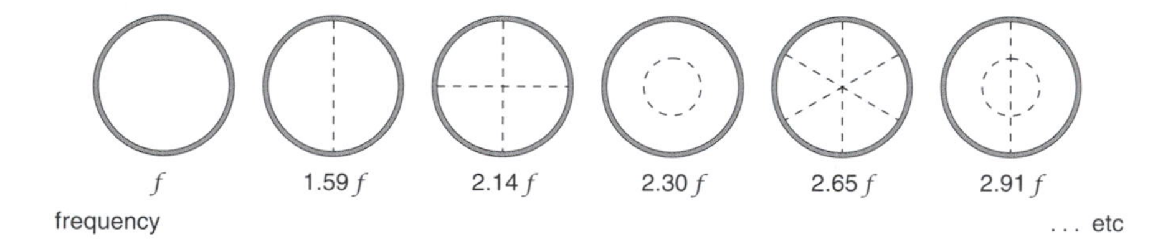

The first and fourth are the ones I described in detail earlier, and are produced when the skin is struck right in the middle. The others involve asymmetric motion of the head and are generated when it is hit off-centre.

Despite the differences between a one-dimensional line and a two-dimensional surface, waves propagate in both for exactly the same *reason*. One part of the surface moves sideways, and in doing so pulls the next part with it. In the process, it itself is brought to a halt. So its energy gets passed on. Therefore, when a standing wave is set up, many of the basic laws of vibrations still apply—in particular Mersenne's laws. The fundamental frequency (and therefore that of all the modes) increases as the skin is stretched tighter; and it decreases as either the diameter of the head is increased or the material of which it is made becomes thicker. As I pointed out in Chapter 2, you expect this kind of behaviour of all oscillators. In practice, it means that drums can be tuned, just like strings, by tightening adjusting screws around the rim.

A real drum however is more than just a stretched membrane—its behaviour is also affected by the nature and shape of the cavity underneath the skin. One common group of drums are the **timpani** or **kettledrums**, in which the skin is stretched over a bowl shaped container (which explains to some degree the name). The one pictured here is from the early 16th century.

In such an instrument, any movement of the skin will, of necessity, involve some movement of the enclosed air. The first mode can only occur by alternately compressing and expanding this air, which has a similar effect to increasing the stiffness of the head. The second mode, on the other hand, doesn't change the volume of the air at all, but instead forces it to move from side to side—adding inertia and thereby effectively increasing the mass of the head. In the end then, you would expect the enclosed air to raise the frequency of the fundamental slightly, and to lower that of the first overtone (and the others to a lesser extent).

In practice, the first mode is not very useful musically. Most timpani have a small hole somewhere, to relieve compression of the enclosed air; and this, together with the fact that the large area of the head moving as a whole is an efficient radiator of energy, means that this mode is heavily damped. It contributes little more to the sound than a dull thud when it is struck. So the lowest frequency you hear, the 'apparent fundamental', is actually the second mode: and this is the one whose frequency is lowered most by the air inside. Here is a short table of frequency ratios, showing how much this changes things:

MODE	FREQUENCY RATIO with first mode	FREQUENCY RATIO with second mode	
	(membrane only)	(membrane only)	(membrane + air)
1	1.00		
2	1.59	1.00	1.00
3	2.14	1.35	1.50
4	2.30	1.45	1.83
5	2.65	1.67	1.92
6	2.91	1.84	2.33

What is interesting about the list of frequency ratios in the final column is that, although they are clearly not harmonically related, they are not too far off. So the sound of the kettledrum is perceived by your ear as having an 'almost definite' pitch. It does blend with the rest of the orchestra and is noticeable when it isn't tuned properly. However, its overtones are not harmonics of the 'apparent fundamental', but, if anything, of a tone an octave lower. That is the pitch of the predominant *difference tone* your ear will pick out of what it hears.

This probably explains why, whenever a piano transcription is made of orchestral music, it is customary to transcribe the timpani part an octave lower than it appears on the orchestral score.

Drums in general

The kind of construction typical of kettledrums, a single skin stretched over a closed volume, is obviously as old as history itself. Yet their introduction into Western music seems to come from Africa, in three distinct periods: during the times of the Crusades, when kettledrums and their near relatives were imported; in the 18th century, when a craze for 'Turkish music' introduced a variety of noisy instruments into European armies; and in the 20th century when Afro-American raised the profile of drummers in dance bands.

The most common kinds of drums in medieval Europe were the **nakers**, often used as a pair, probably tuned differently, and played with two sticks. But during the same period there were also drums of a quite different form, like the **tabor**. It had two skins stretched over the top and the bottom of a wide cylindrical body—another construction with a very long history. In such an instrument, any vibration of the upper head is communicated (via the enclosed air) to the lower.

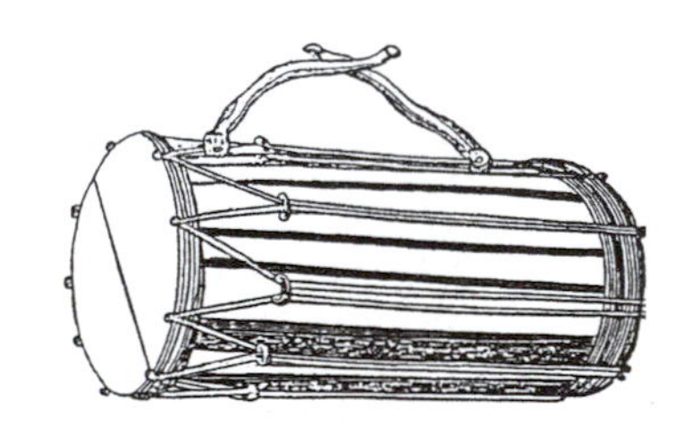

However it is extremely unlikely that the two heads should be exactly in tune with one another (indeed they are usually deliberately mistuned), so the instrument won't have any clear-cut natural frequencies of vibration. Its sound will have very little sense of definite pitch. Such instruments are therefore 'noise' makers and used almost exclusively for rhythm. As if to underline this function, the tabor was usually provided with a **snare**—a wire stretched across one of the heads, which vibrated with each striking and added a buzzing noise to the basic sound.

The 16th century saw important changes to the art of drumming. The nakers gave way to the splendour of the large Hungarian kettle-drums; and the tabor, though surviving as a folk instrument, was less prominent than the new **side drum** (which was played with two sticks). These were all very popular for ceremonial and military use, and most of the now standard playing techniques were developed during this period—the rolls, flams (double striking), drags (triple striking) and so on.

It is interesting that these military instruments didn't gain a permanent place in the symphony orchestra until nearly the end of the 18th century, though they had been used since Monteverdi's time for occasional special effects. Since that time they have been worked on extensively. The modern **timpani** is equipped with a tuning pedal which changes the tension on the head so quickly and so accurately that a melody may be played on it. It usually comes in two sizes: one 58 cm in diameter (playing from $B\flat_2$ to F_3); and the other 75 cm (F_2 to C_3). The largest member of the double headed class is the **bass drum**, which is most commonly about 75 cm in diameter, although instruments up to 3 m have been made.

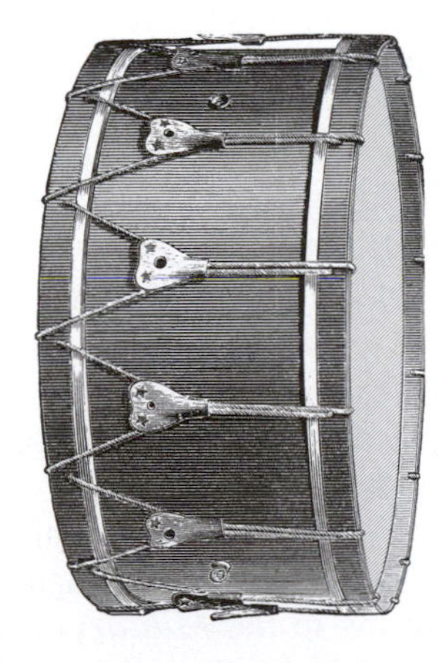

Since the 18th century the use of these drums increased to such an extent that by the present day they are one of the most prominent sections of the orchestra and even have their own solo music. There are quite a few concertos for timpani and orchestra in the standard repertoire; as well as numerous compositions for percussion alone (the names of Benjamin Britten and Karlheinz Stockhausen spring to my mind).

However it is probably in popular music, with its post World War I legacy of Afro-American styles of rhythms, in which the drummer is more visible than almost anywhere else. Although what is played in jazz or rock bands is basically simple in structure, it demands good technique and a very impressive array of high quality equipment. A modern **drum kit** will include a couple of side drums, bass drum, two or more tom-toms and a similar array of bongos, as well as a few cymbals—to be played with sticks or brushes as the music demands. With all this, the drummer maintains the beat by operating the bass drum by pedal, while embellishing it with numerous counter rhythms. It is often eye-catching and exciting; and very, very loud.

Vibrating bars

The class of percussion instruments contains, besides the drums, a whole range of instruments in which the notes are made by the vibration of long thin **bars**, usually of metal. These are essentially one dimensional vibrators, and therefore their modes of vibration look very similar to those of a stretched string, but, as we will see, their overtones are quite different.

In the percussion section of an orchestra these instruments are collected together in a group which the Italians used to call (in translation) the 'Turkish band'. This name is an echo from the period during which they first came into prominence—the second half of the 18th century, when 'Turkish' music was all the rage. It is said that, when the Austrian armies finally pushed the Turks out of Hungary in the early 1700s, they captured many of the musical instruments whose magnificent display and ferocious sound had made such an impression—of which the **Turkish crescent**, with its array of jangling bells, is a splendid example.

From that time the military bands of the Austro-Hungarian empire were transformed, and a taste for 'exotic' sounds came into fashionable music. Thus the bass drum, the cymbals and the triangle entered the works of important contemporary composers. See for example Haydn's use of drums in his *Military Symphony*, Mozart's evocation of a Turkish atmosphere in *The Abduction from the Seraglio*, and Beethoven's chorus of dervishes in the *Ruins of Athens*.

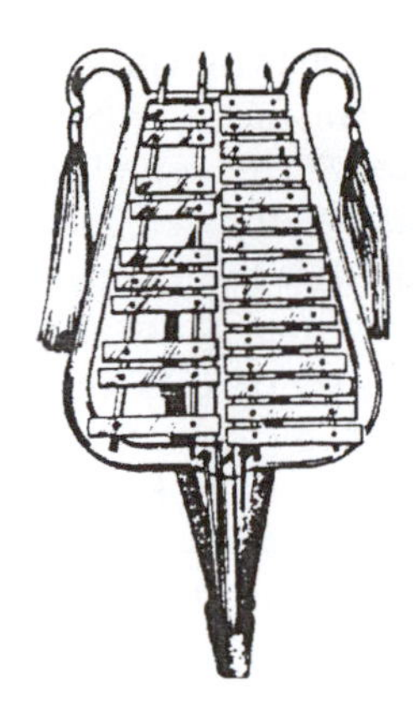

A very early development was the **glockenspiel**, a set of tuned steel bars, supported by cords on a free-standing frame and arranged like a chromatic keyboard. They were played by small hammers or sticks and gave out a bright metallic sound—just the right thing for Papageno's music in *The Magic Flute*. It lapsed into obscurity for a while but was revived enthusiastically by Wagner in *The Ring*—listen, for example, to the fire music from *The Valkyrie* or the 'woodland murmurs' from *Siegfried*.

During the 1870s a portable glockenspiel was made, called the **bell lyra**, which is the one illustrated here and which clearly shows its relationship with its Turkish ancestors.

Glockenspiels were also made with keyboards but they were never very widely used. However another similar keyboard instrument was patented in 1886. This was the **celesta**—with steel bars over individual wooden resonators, struck by felted hammers. It had a light and graceful sound, and it made a great impression with the public when Tchaikovsky used it for the dance of the Sugar Plum Fairy in *The Nutcracker* in 1892 (after quite extraordinary precautions before the opening night to keep his intentions secret). It has been used infrequently since, most notably in Bartók's *Music for Strings, Percussion and Celesta*.

The **xylophone** is also very similar to a glockenspiel, only with wooden bars rather than steel, often with a tubular resonator suspended beneath each one.

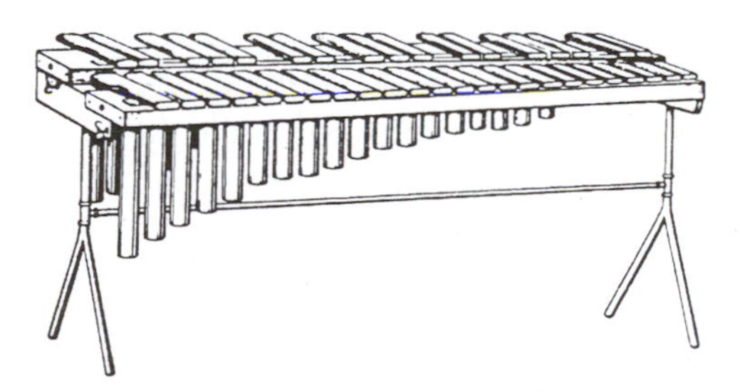

A kind of super-xylophone, the **marimba**, an octave deeper in compass and large enough for four players, sitting side by side, to perform on together, entered Western music from Africa, via South America, in the first decade of the 20th century.

Wood is an organic material and when it vibrates it loses energy much more rapidly than metal, which is homogeneous and crystalline. Hence the sound of the xylophone is much less ringing and 'hollower' than the glockenspiel. This perhaps explains why it seems to have an unfortunate association with bones (also organic material)—the picture is from a set of 16th century etchings by Holbein, called *The Dance of Death.* They are actually very old instruments, and are known to have existed in Indonesia in the 9th century. They found their way into our symphony orchestras in the first half of the 19th, most notably in the music of Saint-Saëns (still associated with bones).

The **chimes** or **tubular bells** is yet another member of the same class. It has a set of metal tubes suspended in a moveable frame and is struck with wooden mallets. It was patented in 1867 in England, apparently in response to the desire of romantic composers to incorporate 'natural' musical sounds into their music—particularly church bells—and for a long time this was the only way it was used. You can hear a really beautiful example of this in the third act of Puccini's *Tosca.* Only in this century has it been accepted as an instrument in its own right.

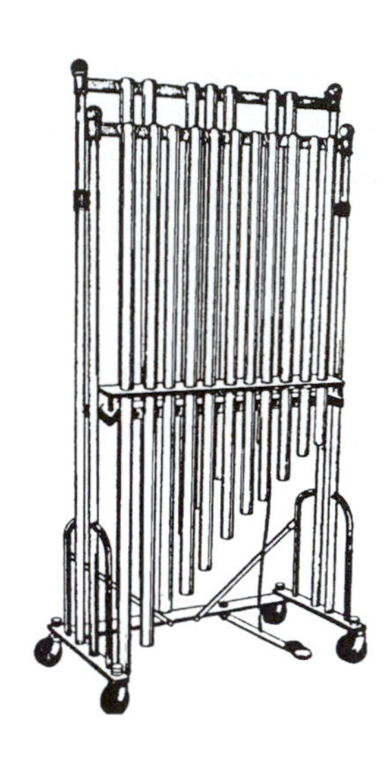

It is diverting to note that, although they were invented to reproduce the sound of church bells in an orchestra, in the 1920s they started being used in real churches, particularly in America, to call the faithful to prayer.

In all these instruments the main vibrators are long and (relatively) thin, and they vibrate by bending from side to side. In one sense they behave like stretched strings: their vibration modes are one-dimensional standing waves. For example, the fundamental mode of a glockenspiel bar looks like this, (greatly exaggerated of course),

But there is a critical difference between a bar and a string. It lies in the *force* which opposes the bending, and which is therefore responsible for the oscillation. In a string this force arises from an externally applied tension; but in a bar it comes from the **stiffness** of the material, and is critically dependent on how thick the bar is.

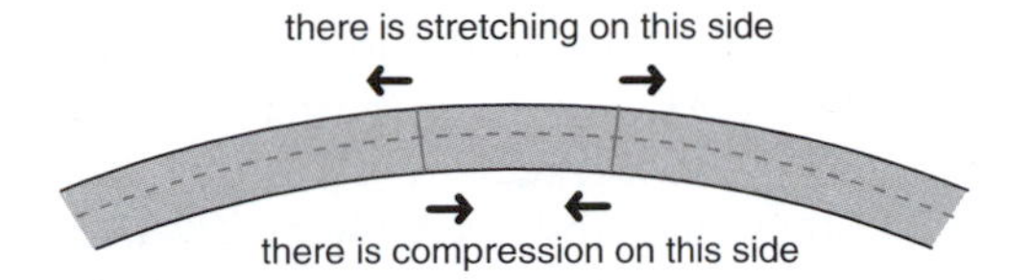

When the bar bends, it involves both a stretching and a compression, either of which is resisted by the elastic internal forces. Therefore it should be clear that a thick bar must resist bending more strongly than a thin one: or, turning that observation around, for a bar of given thickness, the restoring force becomes very much greater as the bending becomes sharper. On the other hand, when a string is pulled sideways, the tension hardly changes at all. This comparison predicts an important difference between the frequencies of the vibration modes of bars and strings.

For the usual well-worn reasons, a bar which vibrates must have overtones as well as a fundamental: but since in the higher overtones the nodes are closer together, these overtones must involve increasingly sharper bending (for the same amplitude). Therefore the energy involved in these modes is correspondingly much greater than the similar modes in a string. The speed of vibratory motion must also be correspondingly faster, and the frequency higher. (This was the argument you may remember I went through on page 83, when I was describing the overtones of a piano wire). In the end, the first few natural frequencies of a uniform metal bar are in these ratios,

MODE	1	2	3	4	5
FREQUENCY	1.00	2.76	5.40	8.93	13.34

Clearly, as with drums, the modes are not harmonics. But what is interesting is that their frequencies are more *widely spaced* than those of drumheads. In terms of how your ear responds to this sound, it means that there is little chance of these frequencies 'overlapping'. Each overtone will be detected separately and clearly. There will be little dissonance, in Helmholtz' sense of the word. Your ear will *not* hear this sound as simply noise. It will perceive it as musical, but it will have great difficulty in recognizing what the pitch is. Your ear is so used to judging musical notes on the basis of fundamental plus overtones, that it will find the pitch of this sound indeterminate.

This feature of the sound is most obvious in the **triangle**, which is no more than a steel bar (perhaps 50 cm long) bent into three sides, suspended by a piece of fine gut and struck by another steel rod. Whether played as single notes or trills, its sound seems to have a clear pitch; but nevertheless it has an uncanny ability to blend in with any harmonies that the rest of the orchestra happens to be playing.

The other instruments however, try to add 'definiteness' to their pitch in various ways. In a glockenspiel, for example, the bars are supported on relatively soft cord. Friction tends to damp any vibrations of the bar, particularly very fast vibrations. This means that the overtones die away quickly, and you only really hear the 'fundamental' pitch.

In an instrument like the chimes you will realize that, although it is the ringing of the metal tube you want to hear, the column of air inside the tube also has its own natural frequency. The tubes are constructed so that these two frequencies match, and the column of air acts as a resonator to strengthen the fundamental of each note. Xylophones, marimbas and celestas have similar metal tubes suspended below each of the bars, which act as external resonators, again to strengthen the fundamental tone.

More importantly, those same instruments—xylophones, marimbas and celestas—have bars which have been specially shaped to be thinner at the middle than at the ends. The consequence of this is that the frequencies of the overtones are altered slightly to be more nearly harmonic. (It should be reasonably easy to intuit that, if the bar is made thinner at a *node*, it won't have much effect; if at an *antinode*, it effectively decreases the stiffness, which lowers the pitch a little; while if somewhere else it lessens of mass and raises the pitch).

All in all, the tones given out by these instruments have most unusual spectrums, and vary in absolutely characteristic ways while they are sounding. It is little wonder that any melody played on them is registered by our ears in much the same way as bright points of light against the blended mass of colour in an impressionist painting are registered by our eyes.

Vibrating plates

The last group of percussion instruments I want to talk about are those in which the main vibrator is a metal plate—the kind of thing that Chladni and Germain investigated. As you might expect, they show all the complexities of both membranes and bars. They have a large number of overtones, with intrinsically anharmonic ratios to one another (determined by the shape of the plate); but each one is well separated from the others, and capable of ringing for a long time (characteristic of a metal).

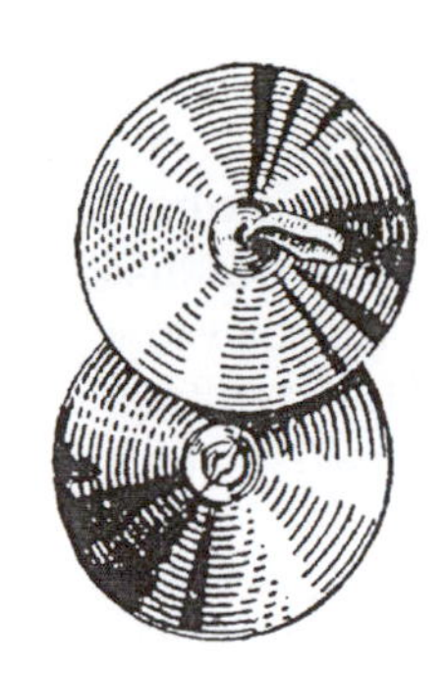

The **cymbals** are a pair of large circular brass plates, slightly convex, so that their edges touch when they are struck together. There exists a lot of folklore about how they should be cast, usually in circular striations, all of which contributes to their characteristic sound—a bright cluster of overtones with many high components, often seemingly fluctuating in intensity. You notice cymbals most when they are used as exclamation marks in grand orchestral climaxes, but they can be used in other ways. Wagner, for example, specifies the shimmering sound of a single cymbal stroked with a drumstick in *The Ring*, to suggest the glitter of gold.

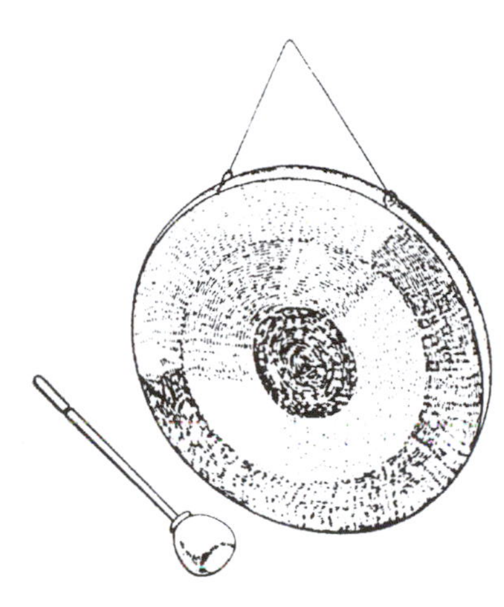

Gongs are made as broad circular disks of metal, with the edge slightly turned up like a shallow plate. They are usually hung vertically and struck in the middle with a soft headed hammer. In Western orchestras the most common is the **tam-tam**, an instrument of Chinese descent with completely indefinite pitch. The name is a fair onomatopoeic representation of its sound, which can be exceedingly loud. There is one piece by Olivier Messiaen, called *Et Expecto Resurrectionem Mortuorum*, in which three tam-tams are struck in succession throughout the whole piece, reaching such a loudness level halfway through, that essentially nothing else can be heard.

In the music of Indonesia, tuned gongs—often referred to as **gong chimes**—are much more common. They are usually mounted horizontally in a frame, to be played by one to four musicians, and form the main melody makers of the **gamelan** (which simply means 'orchestra').

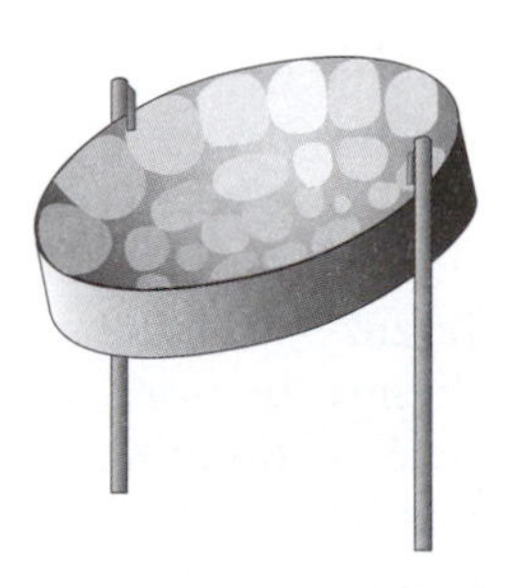

The last member of this class I want to talk about is one which involves a completely new principle, and which was invented only in the 20th century. During British colonial rule of the island of Trinidad, gang warfare was endemic, and the government outlawed the noise makers which were used to call the gangs together—firstly hand drums and later bamboo beaters. Deprived of all their traditional rhythmic instruments which had become such a feature of their annual festivals, the locals took to using any objects they could find, including garbage can lids, old car parts and empty oil barrels (from the Navy bases on the island). They formed what they called **iron bands**.

It was discovered in the 1930s that a dented section of barrel head could produce a tone, and when the annual celebrations resumed after World war II, the erstwhile gang warfare was replaced by competitions which showcased the technical ability of bands performing on the new **steel drums**.

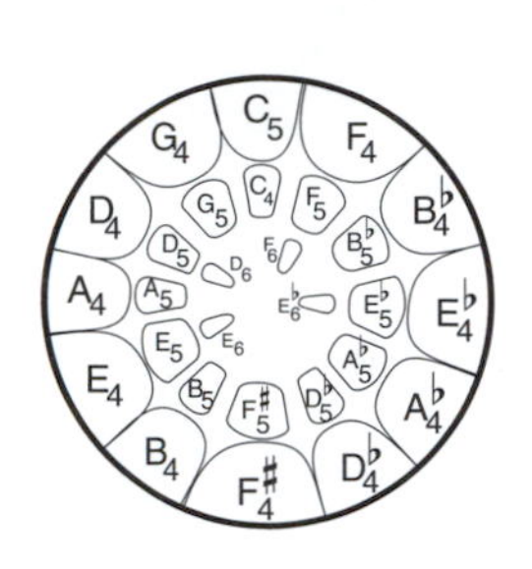

The modern steel drum, or **pan** to give it its preferred name, is usually made from part of a 55-gallon drum, the flat end carefully hammered to the shape of a shallow basin covered with flattened areas. Each area, when struck with a wooden stick, gives out its own tone, and in the largest (**soprano**) pan the possible tones comprise a full chromatic scale. Different makers arrange the playing areas in different patterns, but it seems to work better if neighbouring areas are harmonically closely related to one another. A common arrangement is shown here, with a chromatic scale arranged around the rim in a circle of fifths, each a fifth away from the note next to it. (There is more information about the circle of fifths in Appendix 6 if you want it.) The notes going in to the centre are an octave higher.

The question of how the frequencies of the notes can be predicted is still ill understood. Clearly the vibration of any one area is mechanically coupled to its neighbours, so that the overtones are strengthened by vibrations of the areas nearby. But we don't really know exactly how or why this occurs. There's a lot of interesting physics still to be done there.

Bells

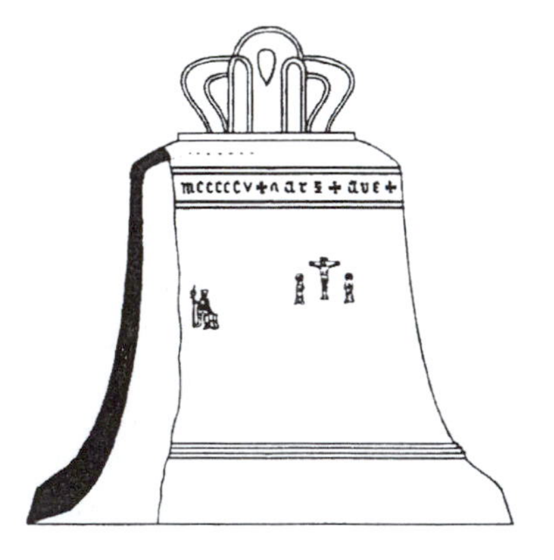

A bell should, strictly speaking, be classified as a vibrating plate instrument. It may be thought of simply as a two-dimensional metal surface bent around its middle; but its geometry is so far removed from a simple circle, and the thickness of the walls varies so much, that its vibratory motion is entirely different. Around the rim it vibrates in well defined sections (often four), with alternate sections moving inwards and outwards; and one meridian of maximum displacement corresponds, of course, with the position of the clapper. From a theoretical point of view, you can think of this as a standing wave around the circumference. Within each of these sections, the movement varies from the centre (top) of the bell to the edge.

If you were to look at the bell wall in cross-section, and imagine that it could be 'straightened out' so that it looked like a metal bar, then the five lowest frequency modes of oscillation could be represented something like this:

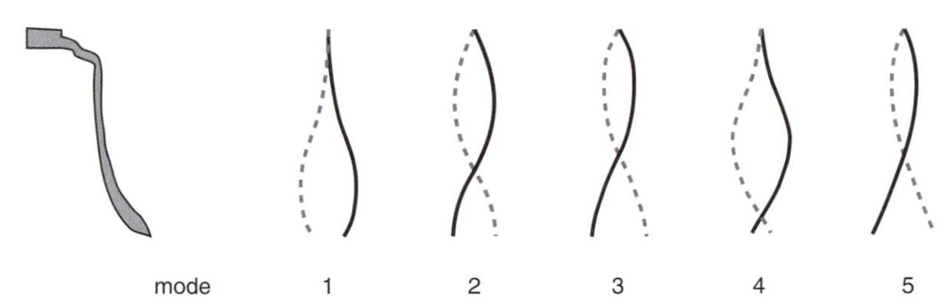

Notice that the lowest mode has no nodes (other than at the centre of the bell). The next four all have *one* node, though at different points along the 'bar'. The main difference between these four is in the number of nodal lines around the rim.

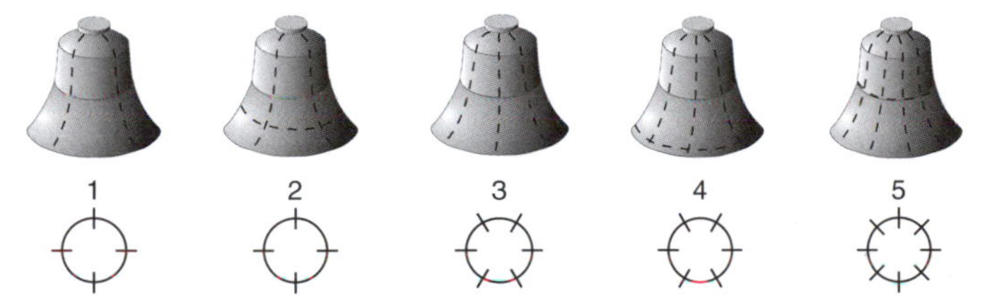

Therefore these frequencies will all be quite different, because there is energy involved in the around-the-rim vibration, as well as in the bar-like movement: but other vibratory characteristics of modes 2–5 might be expected to be somewhat similar to one another.

When the bell is first struck the sound is rather confused. You hear a 'clanging' consisting of a cluster of high overtones, as well as a tone known as the **strike tone**, which is actually a *difference tone* between two of the above modes. These however disappear quite rapidly and the most noticeable tone that remains, which your ear perceives as the principal pitch, is called the **prime tone**. This is in fact the *second* mode pictured above. Above this the next three sound quite strongly—something like a minor third, a fifth and an octave up. (I say "something like" because it varies enormously, especially in older bells.) The four all last more or less the same length of time; and when they have decayed a lower tone persists, roughly an octave below the prime—called the **hum tone**. This is the first mode in the diagram.

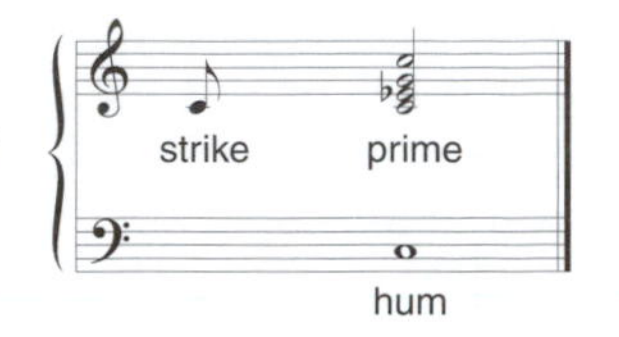

The sound of a bell therefore, even though its overtones might be quite a way off being harmonic, often sounds like a full, rich chord; and if you play this chord on a piano (whose high overtones are slightly anharmonic, remember), you can often mimic that sound quite well.

Scientific investigation into the acoustics of bells goes right back to Mersenne, who believed that the ringing of church bells could disperse storm clouds—though whether because their vibratory motion ruptured the clouds, or because they were sacred objects, isn't clear from his writings. Nevertheless his published tables of the sizes and pitches of all the great bells of Europe of his day are the source of much of our knowledge of bell casting in that period.

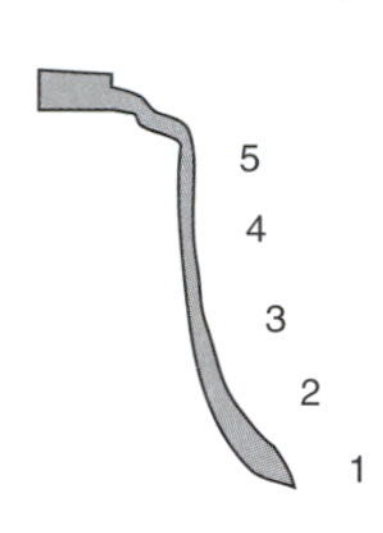

In the early 19th century Chladni took this work up again; and in the 1890s an English churchman, Canon A.B. Simpson, who had taken a lively interest in church bells, having become dissatisfied with their state in England, was able to put on a firm scientific footing what bell casters had done for centuries. He worked out exactly how each of the five principal tones could be tuned separately, by varying the thickness of the wall in just the regions where the different modes have their antinodes.

In a sense, this capability of tuning overtones individually is not terribly important for single church bells. Part of their charm is that their overtones are all slightly out of tune, and each of the great bells of the world has its own absolutely distinctive sound. But a **carillon** consists of a whole row of bells suspended in a fixed position with their clappers activated by a mechanism worked from a keyboard. For them it is most important that each bell has a clearly recognizable pitch.

Even so, when you hear a carillon playing many of the melodies of ordinary Western music, they are often quite difficult to recognize. All the overtones hang around and run into one another, and it becomes harder and harder to place various pitches into their right place in the melody. The music of Indonesia, on the other hand, which makes great use of chimes and gongs, has obviously found the proper style. It comes from a real feeling for the acoustical properties of bells and related percussion instruments—an interesting proof (if proof is needed) that the very nature of music is intimately related to the instruments on which it is played.

Chapter 8

O brave new world

> "I do not wish to quarrel with honest efforts to discover tentative laws of art. Our noblest impulse, the impulse to know and understand, makes it our duty to reflect over and over again upon the mysterious origins of the powers of art. But no one should claim that the wretched results are to be regarded as eternal laws, as something similar to natural laws. The laws of nature admit no exceptions, whereas theories of art consist mainly of exceptions."

Those words were from the preface to one of the most influential books about music in this century—Arnold Schoenberg's *Theory of Harmony*; and they sparked off a revolution in music, every bit as vast as the upheaval in physics I talked about in the last chapter. Schoenberg's book was written in 1911, the same year as Bohr's paper on the structure of the atom. Just as Bohr was a watershed in physics—bringing together the earlier experimentalists, and preparing the way for the later theorists: so also Schoenberg should be seen as a turning point between those who came before and after him.

In the period after Wagner, the accepted conventions of harmony had come under attack. The French Impressionist school—Debussy, Ravel and so on—wrote music which was becoming more and more chromatic. That is, its melodies and harmonies used increasingly many notes from outside the ordinary diatonic scale. But they never actually abandoned tonality—that organizing principle which said that all the harmonies of a piece of music should be related to one fundamental home key, which you'll remember had been around since Josquin's time. But nevertheless they seemed to be raising the question: if any succession of chords is allowable, why should any one tone be considered more important than any other?

Then came Schoenberg, who started writing in what he called the 'method of composition with twelve tones'. In his later music all the notes of the equal tempered scale were employed, with no one note more or less important than any other. It was **atonal**. He even went so far as to invent a completely new organizing principle—the twelve notes should be used in a particular order. So rather than a *scale* in the old sense of the word, he used a *series* of notes, or a **tone row**.

Now **serialism**, as this musical style came to be called, was too arbitrary to be accepted as the way of organizing *all* music; and it isn't surprising that it didn't last much longer than his own group of students, most notably Webern and Berg. But what it did show was that composers had complete freedom to experiment. And experiment they did.

As the century progressed, and World Wars came and went, music became more and more abstract—a trend which clearly paralleled the visual arts. Many different schools emerged, each propounding a new view of music. There were, among others,

- **musique concrète** which explored the art of pure sound (one very early example was Pierre Schaeffer's *Étude aux Chemins de Fer*, composed in 1940 entirely from the sounds of railway trains);

- **aleatory music** which was composed, and sometimes performed, by procedures of pure chance (like John Cage's 1951 piano work *Music of Changes*, which was determined by the results of tossing coins); and

- **minimalism** which sought to reduce the range of compositional material used in any one piece (Philip Glass's 1976 opera, *Einstein on the Beach* consisted largely of 90 minutes of a single phrase endlessly repeated with only changes of stress; and the notorious *4′ 33″* by John Cage, 1952, was nothing but 4 minutes 33 seconds of silence).

Unfortunately, in all this experimentation, audiences found it hard to keep up and often got left behind.

If the changes in music can be summarized as a turning away from tonality, they can also be seen as a turning away from the kind of 'science of music' that Rameau and Helmholtz had believed to be the very basis of the musical experience. The new music still had its theory of course, and the new theorists were still interested in mathematics, but it was an abstract kind of mathematics, without the strong experimental base it used to have.

To show you what I mean, let me return to Schoenberg for a minute and try to give you a feeling for what serial music is like. Here are a few bars of the first tenor aria, "O son of my father", from Schoenberg's only opera, *Moses and Aaron*, first performed in 1954.

You will notice that the first line consists of the twelve notes, C♯ D G♯ F♯ G F B A A♯ C D♯ E in that order, except for a few immediately repeated notes which are 'allowed' under the scheme. The second line consists of another twelve notes, starting on E, separated by exactly the same intervals, only upside down. It's not easy to *hear* this relationship. You really have to write down the notes and the intervals and *see* it. As a result, it can be very demanding to compose this kind of music. Schoenberg himself designed a kind of spreadsheet which you fill in with numbers. Only when you've got the pattern of numbers right do you turn those numbers into notes.

You might also notice that the intervals in the above tone row do not in any way favour the 'normal' intervals which we consider consonant. They consist of semitones, tones, a minor third and two tritones. (A **tritone** is an interval equal to six semitones, or three whole tones, which has always been considered very difficult to sing. It used to be known as the *diabolus in musica*.) Indeed this whole kind of music is notoriously hard to sing, because the tuning must be exactly equal tempered, and singers cannot rely on their sense of natural consonance to pitch the intervals correctly.

Anyhow, you can see why I said that, in Schoenberg's time, the theory of music lost the strong connection it once had with experimental science, and retained only an abstract mathematical underlay. But the *practice* of music also changed, and that was greatly influenced by experimental science, through the technology that came from physics, particularly the technology of communications.

The history of our present communications industry is the story of technological innovation, the work of clever inventors (mostly American), rather than insightful researchers. Often the underlying theory had been known as much as a generation before it was put into use. This is not unusual: physicists who understand fundamental principles are seldom good at applying their knowledge, while engineers, who are, tend to be conservative about adopting new ideas. So let me tell the story by singling out three pieces of scientific knowledge that were critical at three important turning points. These were:

- the realization that electricity was a form of energy,
- the discovery that electricity was carried by electrons, and
- an understanding of the nature of semiconductors.

The beginnings of mass communication

It started, I suppose, around the time of Morse and the electric telegraph; and, as I said before, the only scientific understanding that was needed for that, the first critical piece of knowledge, was that electricity was a form of energy which flowed through wires. The job of getting a current to carry a message was done mechanically. At one end the operator's hand-key sent out a current in short and long bursts ('dots' and 'dashes'), which the operator at the other end had to write down and decode later, (if there wasn't some fancy automatic writing machine to do it).

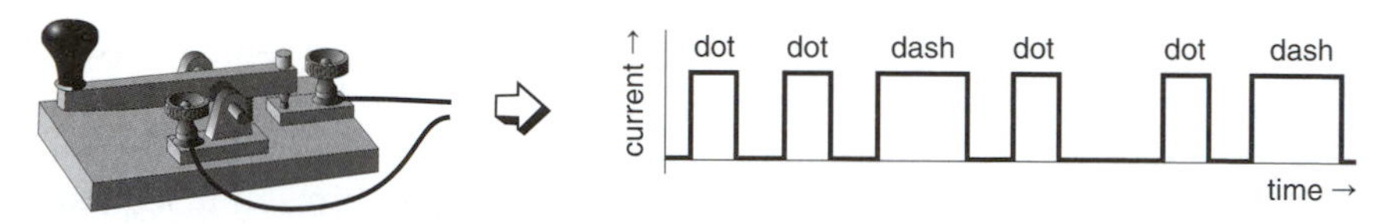

When amplification was necessary, at the relay stations, that too was done mechanically. The incoming current turned an electromagnet *on* and *off*, which closed and opened another switch for the stronger outgoing current.

The next step was the invention of the **telephone** in 1876. Alexander Graham Bell was an expert in acoustics, who worked with the deaf (his wife had lost her hearing at an early age). His knowledge of vibrating membranes, especially in the ear, enabled him to succeed where other electrical experts had failed. By putting a thin metal disk, which vibrated in response to some sound, near a strong magnet, he was able to generate a tiny electric current which fluctuated exactly as did the original acoustic wave. In technical jargon, he had found a way to **modulate** the intensity of the electric current.

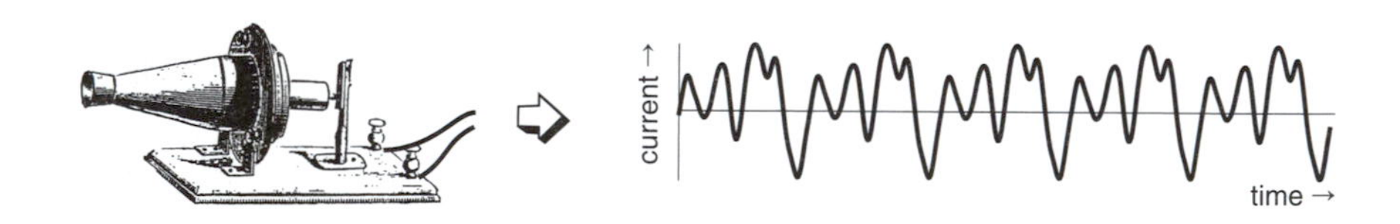

Bell was not far ahead of his competitors, and in the following year, Thomas Alva Edison also used the acoustic vibrations of a membrane to make a sharp needle scratch a wavering track in tin-foil covering a rotating cylinder. The important point was, of course, that the process was reversible. The wavering track could move another needle attached to a second membrane. So the tin-foil was actually storing the sound mechanically. Edison called his device—need I say?—the **phonograph**.

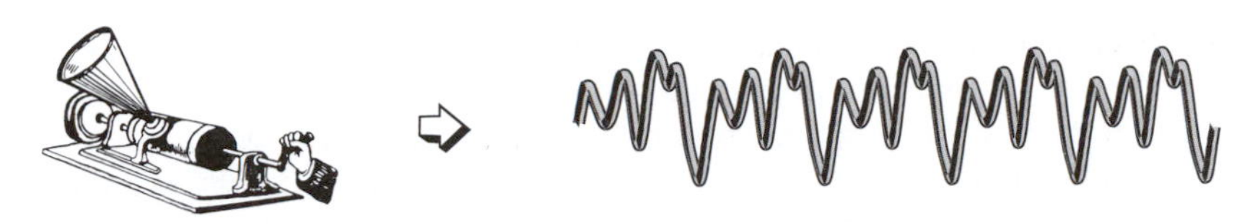

Incidentally, this story illustrates that scientists are not very good at forecasting the future. In his patent application, Edison suggested ten possible uses for his invention. The first three were: secretarial dictation, 'books' for the blind and the teaching of elocution. Only in fourth place did he rank the recording of music.

The next advance—Maxwell's prediction in 1865, and Herz's successful demonstration in 1888, that electromagnetic waves could travel through space—was put to use remarkably quickly. In 1894, Guglielmo Marconi used these waves to carry a message in the manner of a telegraph. His source of radiation was a strong electric spark, which gave out the kind of 'static' that you can hear on your radio during a thunderstorm. Then it was just a question of turning it on and off (mechanically) according to Morse's code; and the message could, in principle, be picked up by anyone, simply by allowing the wave to set up a small current in a length of wire—an **aerial**—without the expense of laying cables beforehand. Thus was the **wire-less telegraph** born.

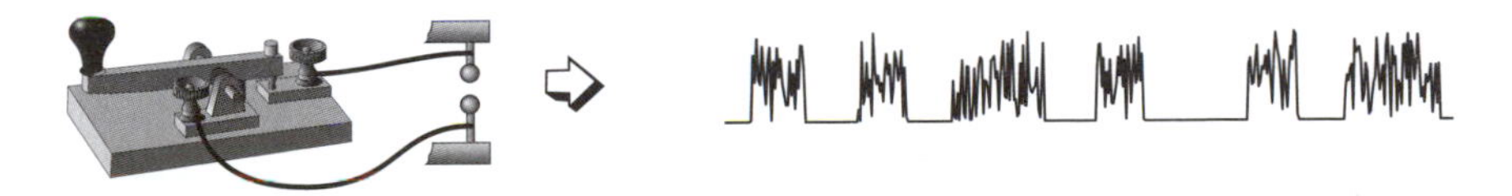

In practice however, reception of these signals was not easy, because they were usually very weak unless you happened to be close to the transmitter, and very irregular anyway. The solution lay in the idea of transmitting, not a burst of incoherent 'static', but an oscillating wave of pure frequency. When this was picked up by an aerial, it could be made to resonate with a specially designed circuit, making detection simpler and more reliable. But for this, it was necessary to know how to make an electric circuit which oscillated. And because this is something that I will keep coming back to, it is worthwhile giving you a brief description of how it works right now. You can, of course, skip over it if you want to.

I have said many times that all oscillations are, in principle, similar to one another; and if you understand why (say) a pendulum oscillates by passing energy between kinetic and potential forms, you can understand them all. An electric circuit is a beautiful example. We know that an electric current can set up a magnetic field (as in the coil of an electromagnet), and a magnetic field contains energy. This only happens so long as the current flows: when it stops, the field vanishes. So in a way, magnetic energy is *kinetic* in nature. On the other hand, stationary charges generate electric fields, which also contain energy. If you connect a current to a pair of metal plates (or sheets of metal foil) which are near, but not in contact with, one another, then a charge will accumulate on them. Such a device stores electrical energy, and is called a **capacitor**. In this form, the energy is *potential*.

Therefore if you put a coil *and* a capacitor in the same circuit, it will behave just like a pendulum. If you disturb it by putting a charge onto the metal plates (positive on one side and negative on the other), you have given it electrical potential energy, rather like pulling the pendulum to one side.

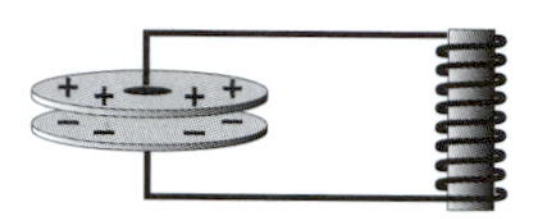

The stored charge will not stay there (any more than the pendulum will stay where *it* is), but will start to flow through the circuit, and the electrical energy will decrease. As the current builds up, it generates a magnetic field and hence accumulates magnetic energy (even as the pendulum builds up kinetic energy).

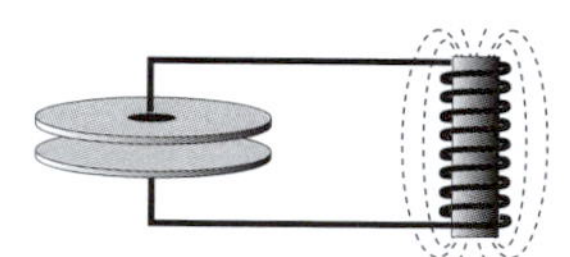

But that same current will empty back into the other side of the capacitor, charging it up till it can hold no more. Its electrical energy increases again. At the same time the current stops flowing and the magnetic energy disappears. Again the same sort of thing happens to the pendulum.

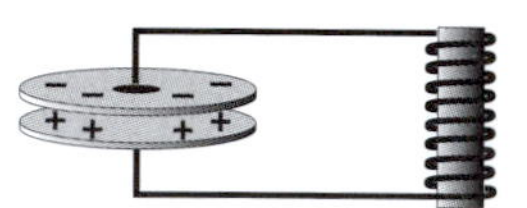

So it will go on. The current will oscillate. And what is more, the same sort of formula as we uncovered in Chapter 4 will hold. The coil offers 'inertia' and the capacitor 'restoring pressure'. So the frequency will be determined by these two, and can be *changed* very easily (by moving the capacitor plates closer together, for example). Therefore not only can an electric circuit oscillate, but also its natural frequency can trivially easily be set to any value you want.

I'm not sure who first constructed these circuits, but in 1906 Reginald Aubrey Fessenden used them as transmitter and receiver in a communication system for ships in the North Atlantic. He chose a frequency which was much too high to be heard—the actual value he settled on 200 000 Hz—and since then values between 1 and 10 times this number have become standard.

Apart from that however, the basic principle of the **radio telegraph** was unchanged. The operator still broke the (oscillating) current up into short and long bursts; and so, when the Titanic went down in 1912, the S.O.S. signal it sent out might have been represented schematically thus:

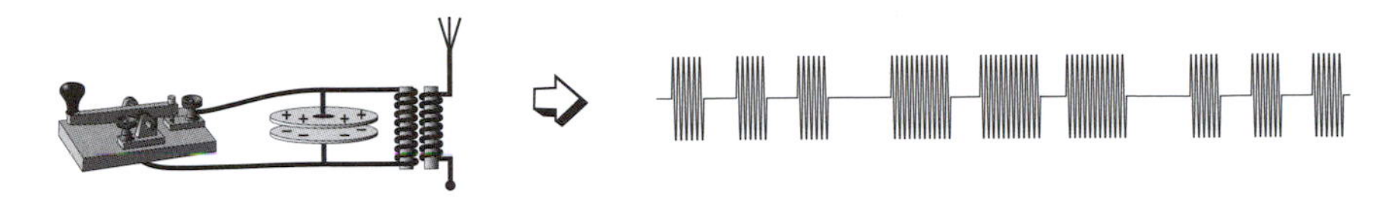

The age of electronics

A turning point came with the next fundamental advance in understanding. In 1884 Edison had noticed, during experiments to perfect his incandescent lamp, that if he put a small metal plate inside the glass bulb close to the hot filament, then a current would flow between them, apparently through the vacuum. When it was finally proved in 1897 that electricity was carried by atomic particles called **electrons**—the second critical piece of knowledge I mentioned earlier—it became clear that in Edison's bulb, the electrons were 'boiling' off the filament and being collected by the plate. It didn't take creative minds long to realize that, when a current was flowing in this fashion, it was 'naked' and very susceptible to external manipulation. That was the idea for the **radio valve** (or, as Americans will have it, the **vacuum tube**).

In 1904, an Englishman, Ambrose Fleming, used the most obvious property of this device—that current will flow through it in only one direction, with electrons going from filament to plate but not vice versa—to convert **alternating** current into **direct** current. That is where the name 'valve' comes from. Then in 1906, the American, Lee de Forest, placed a small wire mesh (technically known as a **grid**) between the filament and the plate; and by putting different electric charges on this grid, he was able to alter the current flowing, nakedly, between them. At long last, a way had been found to modulate the intensity of a current, without using mechanical means.

The first application, which de Forest was interested in, was to marry the principles of the telephone and the radio telegraph. He fed a high frequency oscillating current (which we now call a **carrier wave** between the filament and the plate, put the output from a microphone onto the grid, and he produced a radio wave whose amplitude was modulated in the manner of the original acoustic wave.

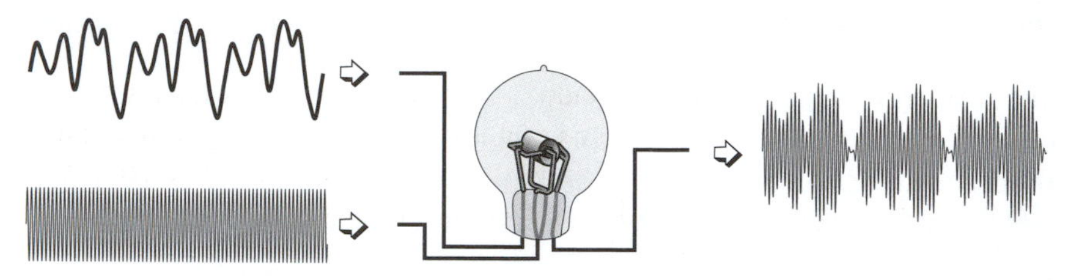

As a result, in 1910 amateur operators in New York were able for the first time to pick up his signal on their aerials, feed it backwards through the same kind of circuit, and listen to the unmistakable sound of Caruso singing directly from the Metropolitan Opera House. This was the basis of the **radio telephone**. Incidentally, scientists aren't the only ones who are bad at predicting the future. In 1913, de Forest was brought to trial for using the U.S. mails to sell stock by making the 'obviously' fraudulent claim that, before many years, it would be possible to transmit the human voice across the Atlantic.

Like many scientific inventions, the valve immediately found uses in many more fields than its author foresaw. Its ability to manipulate electrical signals without switches or moving parts made it incredibly versatile. The most obvious use was **amplification**. Up till then, for example, the ordinary telephone had been rather limited because no electrical signal can travel far through a wire without losing its energy. But now, whenever this happened, it was possible to feed the weak, voice-modulated direct current onto the grid of a valve and produce the same modulation on a much stronger current.

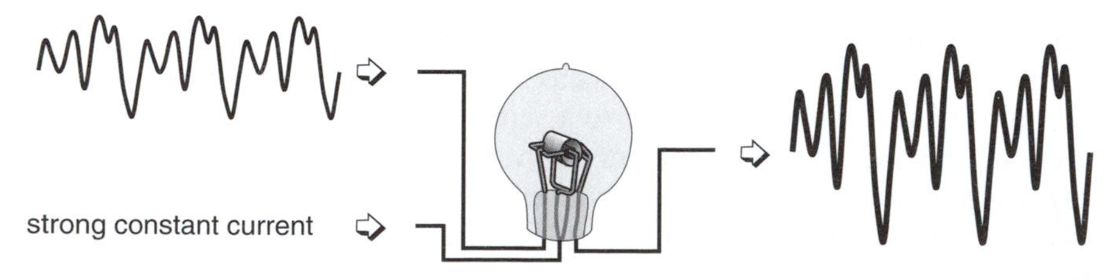

It isn't hard to appreciate the importance of this. Much of our present communications industry deals with what is called **signal processing**.

Change some kind of information (sound, light, mechanical movement, whatever) into an electrical signal, and then, provided only that you can keep this signal from getting too weak to use (that is that you can amplify it when necessary), the kind of things you can do with it are limited only by human imagination. Let me try and justify this by talking about the form of communication of particular significance to music.

Sound recording

The difference that electronics made is nowhere more obvious than here. In 1900 the process of cutting disks and cylinders, and replaying them, was still completely mechanical; and Edison's company made a couple of hundred recordings each year. By 1925, microphones collected the sound, electronics processed the signals, electric motors cut the disks; and over 100 million records were pressed annually. When reliable pick-up devices were developed to replace the old vibrating needles (somewhat earlier, in 1920), the world's first commercial radio station, KDKA Pittsburgh, was formed. From the start, popular recorded music has been the staple of **radio broadcasting**.

Once microphones had become common, other ways were found to store sound. First it was done photographically. Fluctuating currents from a microphone were used to change the intensity of a light beam and leave a trace of varying density along the side of a strip of moving film. So in 1929 the 'movies' became the 'talkies'; and in *The Jazz Singer*, Al Jolson proclaimed the slogan of the age: "You ain't heard nothin' yet!"

Next it was done magnetically. Microphone currents set up fluctuating magnetic fields which left a trace of magnetized grains embedded, first in a moving wire, and then in a plastic tape. The introduction of **tape recorders** into widespread use in the recording industry at the end of the 40s was a very important milestone. Suddenly it became easy to edit recordings, just by cutting and splicing tape. Up till then, when a mistake occurred, they had to start again from the beginning.

Nowadays information can be stored in many different ways. Chemical storage involves electric currents changing the crystal structure of certain sensitive substances; and that gives us **liquid crystals**. A new magnetic method, called **bubble storage**, involves creating little 'bubbles' of magnetization in thin crystals of certain semi-magnetic materials. But perhaps the most important method these days, at least as far as sound recording is concerned, is **optical storage**. This involves leaving spots of varying reflectivity on the surface of a metal, to be read when illuminated by a laser. It is capable of storing vast amounts of information in a very small area and is the basis of the **compact disk** or **CD**. By the time this book is a few years old, I'm sure there will be other methods.

From the point of view of acoustical science, probably the most important invention of all was the **cathode ray oscilloscope**. The ability to make a sound visible so that it can be studied directly has completely transformed acoustics, and school students can now see immediately, things that the great minds of the past puzzled over for years. That's why I gave you a description of how it works in Chapter 4. The first CRO was made in 1897 by Karl Ferdinand Braun, but it wasn't until 1936 that John Logie Baird combined it with the principle of radio broadcasting to produce that most pervasive of all forms of mass communication, **television**.

But, though television may be influential in communications generally, it is sound recording which has remained of greatest importance to music. Just after World War II, coming perhaps from the experience of electronic communications gained during that war, several important worldwide changes occurred in the recording industry. About 1948, disk cutting techniques were improved, and 'long playing' or 'microgroove' recordings increased by many times the length of unbroken pieces of music that could be recorded. The use of tape recorders developed the techniques called **multi-tracking**—that is, recording music on several tapes to be 'mixed' together later on. And most importantly, the whole industry upgraded the performance of their systems to increase the range of frequencies they could cope with. Records from that period carry the label 'ffrr' (full frequency range recording); and I'd like to talk about that in more detail.

Frequency response

In what I have said so far, I made the unspoken assumption that any electronic equipment can, in practice, respond equally well to any sound at all. But there are fundamental reasons why this might not be so. Consider first the mechanical parts—an old-fashioned needle cutting a disk, for example. No matter how light or well balanced it was, it had to have some mass, and that meant inertia. It could follow slow vibrations readily, but could do so less easily as the frequency got higher. If you like, its mechanical impedance increased with frequency. Mostly this didn't matter because the *replay* system could be adjusted to play small amplitude, high frequency vibrations loudly, (though this made it very sensitive to dust and slight scratches). But eventually there came a frequency at which the needle wouldn't respond at all. Then nothing could compensate for that.

Other parts of the system have other shortcomings—speakers and microphones, for example. You will remember that, when acoustic energy is trying to get in or out of a speaker cone, the problem of impedance matching is worst at long wavelengths. So these items will respond poorly at low frequencies. For essentially the same kinds of reasons, all the electronic components inside the amplifiers behave slightly differently at either high or low frequencies. When you put all these effects together it means that the complete recording process has a limited range of frequencies it can handle equally well. Outside this range its performance falls off.

Up till about World War II, this accessible range covered more or less the piano keyboard: roughly 100–5 000 Hz. But immediately after the war, the introduction of 'ffrr' increased this to match as closely as possible the human ear, roughly 50–20 000 Hz. I would like to give a schematic representation of this change by *plotting* the effective response of the systems against frequency.

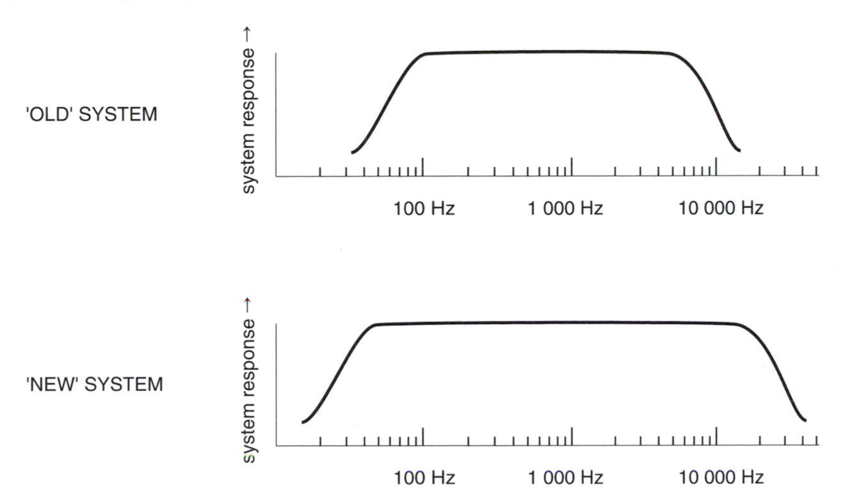

In future, I will refer to such diagrams as **frequency response curves**.

I hope you can appreciate what this change meant in practice. Under the 'old' system, you couldn't hear very low bass notes well at all. Music tended to lack 'volume': it sounded a bit 'tinny'. But don't forget, this problem lay mainly with the speakers. Even under the 'new' system, hi-fi sets commonly used one loud-speaker for low frequencies and another for high—whimsically known in the trade as **woofers** and **tweeters** respectively.

In many ways the high frequency response is more critical. The proper perception of a musical note depends on hearing all its overtones, particularly the high ones. Without them it sounds 'woolly' or 'muted'. You often can't even tell what instrument is playing, unless you can hear at least half a dozen harmonics. To detect subtleties of technique the player might be using requires many more. So to hear, say, the note A_5 (880 Hz) it might be important to hear up to 8 800 Hz. Likewise, in speech, in order to distinguish one vowel from another, you need to hear frequencies up to 3 000 Hz. To pick up nuances of intonation and accent, you want several times that. So, unless the recording process can cope equally with all these frequencies, the sound you hear won't have the same overtone structure as when it left the performer. It will be **distorted**. This kind of subtle distortion, obvious to careful listeners, was reduced by the change to 'ffrr'.

I have, thus far, been talking only about recordings; but the same sorts of things can be said about any device which processes audio signals. Take radio broadcasting for example; and go back to the simple diagram on page 272, with which I explained the radio telephone. Imagine that the carrier wave has a frequency of 500 000 Hz, and the audio signal 500 Hz. (It could be Caruso hitting a high C_5, just a little flat). Then the radio signal being transmitted might look like this (at least it might if what is called 100% modulation is used):

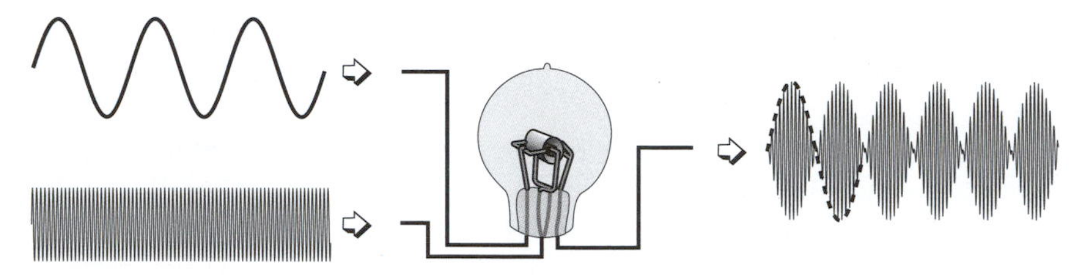

Now you've seen that shape in another context—it looks just like a **beat** between two pure sine curves (refer back to page 94). And this is an important piece of insight. From a mathematical point of view, it means that multiplying two sine curves together is exactly the same as adding two (different) sine curves. From a practical point of view, it means that it is possible to think of the modulated radio signal as equivalent in all respects to *two* pure radio waves of frequencies 500 500 and 499 500 Hz.

However singers don't produce pure sine curves. Their notes usually contain harmonics at least up to the twelfth (6 000 Hz in this case). Therefore an actual voice-modulated signal must be equivalent to a *couple of dozen* pure sine waves, with frequencies between (at least) 506 000 and 494 000 Hz. This makes great demands on the receiver (i.e. your home radio set). It must be able to resonate with all of these waves, more or less equally. And it can only do this if its **bandwidth** is considerably wider than the ±6 000 Hz in question. In other words, the frequency response curve for your radio, tuned to this station, must look more or less like this—which is just a **resonance curve**, like the one on page 226:

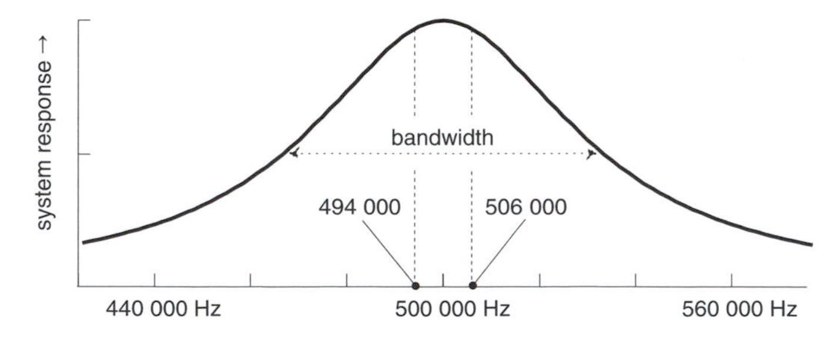

From this diagram, you can probably see the explanation of a few other related facts: why, for example, in any local area there are only a small number of radio 'channels'—because they can't be allowed to interfere with one another, and so each one must be tens of thousands of Hertz wide; or why, when you are not quite 'on the station', the sound is distorted—because the high harmonics are not being received properly.

But I hope you can see a more general point. Records and radios are both audio signal processors, and the equipment they use needs very definite frequency response characteristics. The same must be true for all communications systems. Even humble telephone cables must be carefully made so that they can carry frequencies up to about 4 000 Hz. The obvious fact that voices over the telephone are sometimes only just understandable, proves they haven't got a much wider frequency response than is absolutely necessary.

This has become increasingly obvious now that telephone lines are being used to carry information for computer communications. Sending this kind of information still relies on sending waves through your system. If you want to send a lot of information quickly, your medium (in this case the telephone cables) needs a lot bigger bandwidth than they normally have.

Of course you don't *always* need to have a wide frequency response range. Sometimes you actually want a circuit which will respond only to a particular range, and block out other frequencies. Such circuits are called **filters**; and it is easy to find examples in common use which allow through (or 'pass') only high frequencies or low frequencies, or some band in between. Their response curves might look schematically like this:

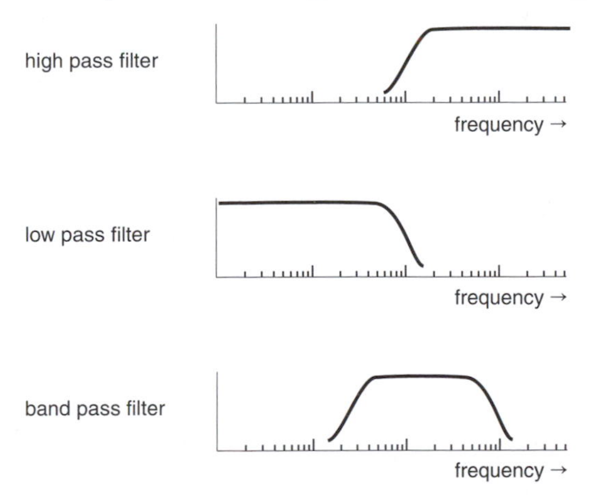

The 'tone control' on ordinary radios is really just a simple band pass filter, whose cut-off points can be adjusted. In fact, fiddling around with the tone control on your radio is an easy way to hear for yourself some of the effects I have been taking about.

The main idea that comes out of all this is that all communications systems are similar to one another in fundamental ways. It is the basis of what is called **information theory**. I will take this up again in Chapter 9; but for now, let me continue the story of the electronics communication industry.

The semi-conductor revolution

By the beginning of the 1950s it started to become clear that, despite its ubiquity, the radio valve had outlived its usefulness. That was when the first electronic computers were being built. There had been mechanical calculators for quite some time (cash registers for example) which performed additions and multiplications by ingenious arrangements of cog-wheels. But the experience gained during World War II of deciphering enemy messages by machine, convinced people that any computational problem could be reduced to a sequence of simple signal processing operations. This, and the wide experience gained in storing electrical information, made it natural to try to construct a device to do computing electronically.

But a problem quickly surfaced. Each one of these simple operations, and every single particle of information stored, needed its own individual valve. So a computer that could handle, say, a thousand numbers—and no one was interested in anything smaller—had to have some tens of thousands of valves. But valves consisted of fiddly bits of metal inside evacuated tubes, and were about the size of a light bulb. They were fragile and broke down with distressing regularity. Furthermore, they all had filaments that had to heat up before they would work. Therefore they used a lot of energy, were expensive to run, and needed an air-conditioning system to keep them from overheating.

One of the earliest of these machines was built at the University of *Illi*nois in 1958, the *a*utomatic *c*omputer which was christened ILLIAC. A second version of the same design (SILLIAC) was built at the University of Sydney, (which I became familiar with because I worked with it as a student). It was so big it filled a whole room, and it needed two full-time technicians and a roster of operators to keep it going. It cost over $20 000 (at 1950s prices!). Nowadays you can buy a pocket calculator which will do the same job for a few dollars. The explanation of this remarkable drop in price is traceable to quantum theory; and this is where the third critical piece of knowledge, which I referred to at the start of this chapter, comes in.

When valves were invented it was believed that there were only two kinds of materials (in regard to their electrical behaviour)—conductors and insulators. But quantum theory described how electrons would behave inside other sorts of matter, and pointed to ways their behaviour might be influenced just by changing slightly the chemical nature of the material. Such substances were called **semiconductors**. Incredibly, even though this knowledge was available within a year of publication of the Schrödinger wave equation, it was twenty years before it was put to use. But finally in 1948, a team from the Bell Telephone Laboratories constructed a device consisting of a judicious mixture of different kinds of semiconductors which could do exactly the same job as de Forest's valve. This was the world's earliest **transistor**.

At first it wasn't obvious that anything earth-shattering had happened. For a decade or so, transistors gradually replaced valves—but so what? Radios and televisions were a bit cheaper, a bit more compact, a bit more reliable. But then computers really began to develop. Because the critical feature of a transistor was the chemical nature of its semiconductors from which it was constructed, it didn't need to be the size of a light bulb. It didn't really have to be much bigger than atomic dimensions. So before long it was possible for a complete circuit, a whole array of transistors and other components (an **integrated circuit**), to be copied, photographically, onto a piece of semiconductor the size of a pin-head (a **silicon chip**). This miniaturization gave transistors their greatest advantage. They could do signal processing extremely rapidly, simply because they didn't have to wait for all those electrons to travel across the spaces inside valves. Suddenly computers became much faster, much more compact and much, much cheaper.

Because computers have become so powerful (and ones with over a million times more capacity than SILLIAC have been around for decades), there is a fundamental change in the kind of problems they can tackle. Modern theories of intelligence believe that the human brain is no different, in principle, from any other animal brain, except in its size and the complexity of its interconnections. Perhaps all that is necessary to create artificial intelligence then, might be a computer of sufficient size. That idea could never have been contemplated except for the miniaturization of individual components brought about by quantum theory and the understanding of the semiconductor. But what is relevant to us now, is the difference that computers have made to communications. By treating audio signals as the same kind of numbers they are used to handling (the process known as **digitizing**), they have improved enormously the reliability of all of sound recording.

That statement perhaps needs a little justification. All of the earlier sound storage devices are designed around some property which can vary *continuously*—they are known as **analogue** devices. For example, a record uses the depth of a groove cut in plastic to store a sound signal, and a tape uses the degree of magnetization of grains of ferrous material. These quantities can change by very small amounts; and any change at all will mean a difference in sound quality. Therefore they are sensitive to scratches, or nearby magnets, or less-than-gentle handling. But a compact disk, for example, is a **digital** device. It uses a series of tiny spots on a metal surface to store the sound; and each spot is either definitely *there* or definitely *not there*—there are no half measures. This makes them extremely reliable, because, although the stored information can be interfered with (some of the spots may be completely obliterated for instance), it is much less easy to do than with analogue devices.

Now this discreteness, this there-or-not-there property, is the very essence of computer operation. All computers store their signals in transistors that are either *on* or *off*, and what happens to this information is determined by whether or not other transistors are in the *on* or the *off* state. That is why things like compact disks only developed after the invention of the computer. So there is no question that the computer has made the recording and reproduction of sound much more reliable. But it is also starting to change the very *nature* of mass communication. In the 1990s there suddenly sprang into prominence the Internet (or if you prefer, the World Wide Web). It has become increasingly common for music (among much else) to be stored on web sites, in the form of computer files which you can download for little or no cost if you want to listen to it. It's an ideal way, for example, for up-and-coming music bands to display their talents and get their music noticed. So a lot of research is going into the problems of storing information efficiently, and moving it about quickly.

The trouble is that sound files can be very large indeed. Just how large depends on many things: the duration of the sound; the number of bits used to encode the data (by which I mean the range of numbers you use to represent the variation in pressure at any point in the sound wave); the **sampling rate**, that is the number of times every second that you measure that pressure; and whether the sound is monophonic or stereophonic. Just a minute or two of CD quality stereo, sampled at 16 bits and 44 000 Hz, gives you megabytes of data. That means you need hundreds of millions of *on/off* entities (magnetic grains or whatever) to store it all.

What you have to do is to **compress** sound files. You get rid of as much extraneous information as you can, for example all the frequencies above 20 000 Hz which the human ear can't hear. Right now (in the year 2001, when this is being written) one of the most successful solutions is a computer algorithm, a mathematical recipe, known by the acronym **MP3**. (Don't worry about the name, it doesn't stand for anything very imformative). It gives in real efficiencies and is more or less the industry standard.

It probably won't last. I'll bet in a few years something else will have come along. But don't let that worry you. Even if you're not 100% up-to-date on the very latest technology, the fundamental problem remains the same: how to communicate between musicians and their audience.

Has mass communication technology changed music?

I find it interesting that most of the standard histories of music, when they discuss the 20th century, make very little reference (if any) to the mass media. Yet the media *must* have had a huge effect which will become blindingly obvious from the perspective of time. After all, composers and their music don't exist in isolation from their audience; and audiences are unquestionably influenced by the media.

This is most obvious in pop music. At the turn of the 20th century, there were many different forms of Western popular music—English music-hall songs, French cabaret chansons, Viennese waltzes, American black-and-white minstrel music and so on; and I think it is possible to make the following observations about them all.

1. They were primarily active forms, meant to be played or sung, rather than just listened to. Their main mode of dissemination was by the printed sheet, which meant that they had to be able to be performed by people of limited skill. (An exception was ragtime, which was technically more difficult and disseminated mainly by piano rolls).

2. They had distinctive national styles, with great differences even between English-speaking cultures.

3. In the musical idioms they employed they were several generations behind 'serious' music: their almost universally diatonic melodies, for example, had little in common with the contemporary chromaticism of, say, Debussy or early Stravinsky.

But in the 20th century technology changed all that. Radio broadcasting popularized jazz in the 1920s and swing bands in the 30s. The 'golden age' movie musicals gave us the Tin Pan Alley ballads in the late 30s to mid 50s, and rock-and-roll in the late 50s. Television fostered the protest and folk-songs of the late 60s and the many kinds of 'rock' during the 70s and 80s, not only because of its use of light-show imagery, but also because of its coverage of those massive, semi-religious public happenings—pop concerts. And if you want tangible proof that technology does influence music, just ask yourself how much a good, clear amplification system is finally responsible for the ultra-loud, frenetic sound and the deep, almost infra-sonic, pulsating rhythms of modern hard rock.

Anyhow, if you want to generalize about pop music today, I think you would reach conclusions quite different from the previous three—and the changes are unquestionably traceable to technology.

1. The new popular music is passive; it is primarily meant to be listened to. Since the introduction of long playing records in the 50s and CDs

in the 80s, and the use of multi-tracking, a modern pop recording can be very complex, made up from as many as 24 different tracks, incorporating sounds from many different sources. It's ironic that one of the best known groups of the 60s, the Beatles, eventually had to stop touring, because their stage performances couldn't reproduce what their fans were used to, and wanted to hear again.

2. A sense of national identity has almost disappeared. American popular styles have found their way into Africa and Asia (especially the music written for their very extensive movie industry in India). Equally, music that originates elsewhere is very quickly absorbed by the American juggernaut—as happened to reggae music from Jamaica in the 80s, or rap music from their own Afro-American minority in the 90s.

   In the early 50s European recording companies started sponsoring popular music festivals in an effort to protect themselves from what they saw as American domination; but if you listen to the winners of the most widely known of these, the Eurovision Song Contest, they don't seem to have succeeded. Perhaps the time is coming when we should stop thinking of this style as 'American' and start considering it 'universal'.

3. It hasn't been true for many decades now that all (or even most) popular music is musically unsophisticated. Perhaps it is because financial rewards have lured talented composers into the field. Just think of the songs of George Gershwin or Leonard Bernstein or even the Beatles. The stylistic idioms of popular music may be simpler than those of the mainstream, but it is no longer true that they are as far behind in time as they used to be.

There was one even more interesting (to me) development, which is particularly relevant to the theme of this book, and that was the emergence of a new musical scale in American popular music. Starting with the end of the Civil War, the erstwhile negro slaves and their descendents spread through the country, particularly to centres like New Orleans, St Louis and Chicago. And they brought their own distinctive music with them. At first this was mainly gospel songs and spirituals, which typically were pentatonic melodies. But as time went on a new idiom emerged, which came to be called the **blues**.

There are many ingredients which go to make up this very characteristic new sound. Perhaps the most important is the special intervals which the music favours, which go to make up what is called the **blues scale**:

You will notice that it is basically just a minor pentatonic scale—the five black notes on the piano starting on E♭ (there is more information on this in Appendix 6 if you want to follow it up). But it has one extra note, the **blues note**, in this case the A♮. What gives this note its bite is the fact that the interval between it and the tonic is a tritone, which you will remember is an interval that ordinary diatonic melodies tend to avoid.

In case you don't know what this sounds like, here are a few bars from the *St Louis Blues*, written by W.C. Handy in 1914 and one of the earliest blues songs to be popularized by the new radio broadcasting.

I might say that there seems to be difference of opinion about whether the D♮ in the second bar and the F in the ninth are themselves blues notes or just accidentals denoting temporary change of key. Whichever it is, it seems to me to be a beautiful example of how the new technology has influenced popular music at its most basic level.

And what about serious music?

When we turn to what we might call 'serious music'—or perhaps 'art music' is a less judgmental term—matters are not so clear-cut. For one thing the whole field is exceedingly innovative anyway, and attributing any change to a particular cause is dangerous. But nevertheless it is still true that audiences have been changed by the mass media. I can think of at least two items of evidence to support this.

- The world-wide rebirth of opera. Before the advent of long-playing records, complete operas, on the old 78 rpm format, were unbelievably cumbersome to listen to. But after 1948 all that changed, and operatic recordings rolled off the production lines. Audiences could now become familiar with full scores and not just the standard arias (the kind of thing Puccini was good at writing to be just the right length to fit onto a 7 inch disk). They were then prepared to pay money to attend performances, and opera companies the world over started showing, if not a profit, then at least a smaller loss.

- The widespread rediscovery of old masters. A fascinating example is the modern career of the baroque composer Antonio Vivaldi. Before 1950 he was essentially unknown. There wasn't a single one

> of his compositions in the record catalogues. Then in 1950, one of the very earliest long-playing records was a performance of *The Four Seasons* under the auspices of the small American Concert Hall Society. It obviously struck a resonance with public taste because, since then, he has become standard fare in most concert repertoires (and has carried many of his contemporaries with him). Most of his 600 odd concertos are now in the catalogues, and *The Four Seasons* itself has been recorded well over a hundred times!

All this must be having an effect on music. It is impossible to believe that people who have had opportunities, as never before in history, to be acquainted with great works of the past and the music of other cultures, haven't become more knowledgeable and discerning; and that will reflect itself on the music they demand to listen to.

But so far as musicians, especially performers, are concerned, the benefits are not completely unalloyed. Perhaps the most worrying question is this. If, in a recording session, a performance can be corrected and edited (sometimes note by note) to a state of near perfection, what is left for the artist afterwards? One particular performer, the Canadian pianist Glenn Gould, who also enjoyed wide acclaim during the 60s, gave up his public career for exactly this reason. He decided that, having recorded his repertoire and being unable to reach the same standard in real life, he would devote the whole of his time to improving his recordings. And he spent much of the last ten years of his life cutting and splicing tape.

Such apprehension is understandable, but on the other hand, there may not be need for worry. The same pieces are still being recorded (and sold). There is, after all, no such thing as a definitive performance. Recordings themselves vary: each is only as good as the skill of the engineer in charge allows. And this introduces a new concept. A modern recording session is, in many ways, an act of musical interpretation in its own right, and demands artistic talents of the producer. There are musical purists who decry this fact, arguing that a recording which is extensively edited is, is some sense, not a *real* performance. But, as Glenn Gould has pointed out, the same thing was said, by stage actors and producers, when early cinema directors started constructing movies by editing strips of independently filmed action. Yet today, no one has any difficulty with the idea that a movie and a stage play are different, but equally valid, artistic events.

We know that the importance of *conductors* dates from the 19th century when symphony orchestras became popular; and the tradition of their being primarily responsible for the interpretation of the composer's intentions only arose with Wagner. Will the 21st century musical world accord similar status and honour to great virtuoso recording *producers*?

Technology in perspective

Let me now try to tie all this together. In this chapter I've talked exclusively about technological advances, rather than fundamental physics. I have argued that the very existence of this technology has changed the milieu in which the music of our culture exists and that, of necessity, shapes music itself. I will talk about the new instruments that have come out of this technology in the next Interlude; but for now let me give you an example of how the technology for processing musical sounds has transformed a whole area of theoretical music: **ethnomusicology**.

The idea that the history of music was also the history of the human race, and therefore that *all* peoples should be included, seems only to have arisen in the 19th century. Before that, since the early days of colonial expansion, Western scholars had made some studies of the music of other cultures; but the discipline, until quite recently known as 'comparative musicology', was limited because of the difficulty of getting access to the appropriate music. There had been some cross-fertilization of Western music with outside influences; 'Turkish' music, for example, was popular in Mozart's time, and clearly influenced some of his music. At the Paris Exhibition of 1889, Debussy first heard the music of the Javanese gamelan, and that (by all accounts) turned his interest to the use of non-Western scales. But those were isolated examples: by and large Western musicians remained largely ignorant of the music of the rest of the world.

Then the phonograph was invented in 1877 and all that changed. By as early as 1900, anthropologists had made collections on cylinders of the songs of American and Mexican Indians, and of Russian and Rumanian 'peasants'. The Hungarian composer, Béla Bartók, was particularly influenced by this music; and himself participated in the early field work which collected it. His extensive use of ancient musical scales led him to a style of composition which almost completely abandoned Western harmonies, and which virtually paralleled that of Schoenberg.

Nowadays of course, portable tape recorders (analogue or digital) are staple equipment for workers in the field, and video tape is being used more and more, especially where dance is important. By such means a mass of information has been assembled about details of performance, styles of singing, construction of instruments, and so on, and housed in recording archives in universities and museums throughout the world.

But that is only the start of the ethnomusicologists' work. Analysis and comparison of this music is difficult to carry out while it remains as sound on cylinder, disk or tape. It is much more useful when written down. But because little of it uses Western scales, it often cannot be written in standard musical notation. One important technical advance in this field was the invention, in the 1880s, of the **cent** system of pitch and interval measurement (see Appendix 3); but it was the new breed of electro-acoustical instruments which has provided the nearest thing to a solution to the problem. There was the oscilloscope of course, but more important was a device invented in 1958, known as a **melograph**, which can produce a simultaneous graphical record of the pitch, amplitude and overtone structure for a continuous melodic line. Such devices are limited—they can't, as yet, handle polyphony or even two voices in duet—but it can't be long before the next generation of computer controlled instruments solve that.

The main aims of this field of research have been enunciated by one of its foremost workers (in my paraphrase) as these:

- detailing the technical features of the music itself,
- exploring it as a form of communication, and
- relating it to the society it comes from.

Clearly the technological devices I have discussed apply directly only to the first of these; but without them the other two couldn't proceed. Therefore this whole area is fundamentally underpinned by technology; and it seems to me that, of all the inhabitants of this vast and varied world of music, ethnomusicologists are a singular breed in whom musical values are married particularly intimately with the scientific/technological frame of mind.

Interlude 7

Electronic instruments

In the last hundred years, electricity has replaced other forms of energy usage in most human activities. It isn't surprising therefore that it should have invaded musical instruments also. And just as, on the industrial scene, many of the new inventions caused resentment through the fear that workers would lose their livelihood; so also, in today's musical scene, there seems to be a nagging worry that current technological developments will eventually devalue the skill of the performer. We are too close to things to decide whether or not such fears are realistic: but at least it gives us a simple way to survey this class of instruments—and that is by focusing on just how large a role electricity plays in the operation of each.

The simplest level is illustrated by the electrically operated organ. In older, 'manual' organs, each key connected the wind-box to the appropriate pipe by means of a system of levers—I talked about that briefly in Interlude 5. But in many modern organs (though not all), each key closes a switch, which allows a current to flow in a circuit and make the same connection. It is very simple in principle, and rather less prone to malfunction than the older system. But even so, there are people who argue that it diminishes the player's control over the subtleties of shaping each note. So the worry I talked about is there, even at this level.

But electricity can do much more than that in the music-making process; and in this Interlude I want to talk about three classes of instruments;

- those in which electricity merely provides amplification for the sound from some 'natural' generator;
- those in which it also generates the initial vibration, as well as amplifying it; and lastly
- those in which it does all that, and actually plays the notes too.

Instruments with electrical amplification

Following the great increase in demand for loudness in the public performance of music that took place during the 19th century, the major skill (and expense) that went into the construction of many ordinary musical instruments came to be tied up with the problems of coupling the energy of vibration to the outside air. I pointed out in Interlude 2, that the piano developed into such a large and expensive instrument because it needed long, taut strings (and an iron frame to support them) in order to maximize the energy of vibration; and an extensive sound-board to convert this energy into a loud sound. Likewise in the making of plucked and bowed instruments (Interlude 3), the main skill is in the construction of the body, which projects each vibration into the air with as little distortion as possible. These are the precise functions which, once the technology of signal processing had been developed in the first quarter of this century, could be turned over to electronic amplifiers.

A very simplistic approach would be to position an ordinary microphone close to a conventional instrument. This has occasionally been done for special musical reasons—for example, some of the works of John Cage, demand a microphone to be placed inside a grand piano—but that doesn't make it a new instrument. What is more, it is plainly inefficient to have the energy of vibration of a string transformed into a vibration of the air, then to an electrical vibration in the microphone and then back to another acoustic vibration in the air. The trouble is that, with each transformation, you run the risk of introducing some distortion. It is more sensible to separate clearly the two functions of **picking up** the vibration of the string and transforming it into an electrical signal; and then of **amplifying** it and sending it out again as an acoustic vibration.

In most electrical instruments, the job of turning the string's vibration into a fluctuating current is done by a small magnet placed directly under the string. The string is made of steel and therefore capable of being magnetized. As it swings backwards and forwards, it distorts slightly the field near the magnet, and the rapidly changing distortion is detected by a small coil of copper wire.

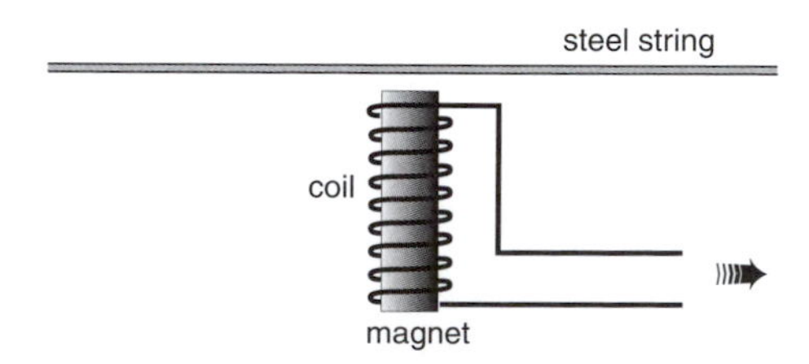

Copper is not magnetic, so the coil doesn't move, but the current carriers inside it—the electrons—feel a force as the magnetic field changes.

Thus tiny fluctuating currents are set up in the coil. The vibration has been picked up.

This is the most common such device, but there are others, different in detail, which do the same job. However they all have these important features in common:

- the strings are made of steel (or something else magnetizable), and
- the movement of only that part of the string directly over the pick-up is detected, rather than the movement of the string as a whole.

The piano was one of the first instruments to be modified in this way. The pioneering work was done in Germany by Walter Nernst, who had been a colleague of Max Planck's at the University of Berlin and a Nobel Prize-winner in his own right. In 1931, he designed the so-called 'Neo-Bechstein'. It looked like an ordinary grand piano, though, having no sound-board, it was much lighter. Inside, its strings were arranged to converge on eighteen pick-ups. In addition, it had several features which would outlast the instrument itself. The loudness of the notes was regulated by a 'volume control' pedal; and the timbre could be adjusted to the taste of the performer by a 'tone control' knob—essentially just a variable filter which increased or reduced the strengths of various harmonics.

Several similar instruments, with only quite minor differences between them, came out at about the same time, on both sides of the Atlantic. The Vierling company in Berlin produced their own version of an electric grand piano, as well as electrical violins and cellos. The **radiotone**, an instrument of French design, was a kind of half-way house, which used a piano keyboard to sound a string by a circular bowing motion. None of these achieved lasting popularity.

Today the name '**electric piano**' is usually reserved for a more recent member of the class. It doesn't use strings as the basic vibrators at all. Instead it has a set of tuned metal reeds—similar to the tines of a music box—which are struck by small, felt-covered hammers. It has completely lost the traditional grand piano shape, and looks just like a keyboard on legs.

These were developed mainly in the 1960s, and had a brief time in the sun during the 70s, where their characteristic sound—often described as 'gleaming'—proved popular in jazz and the milder versions of rock music.

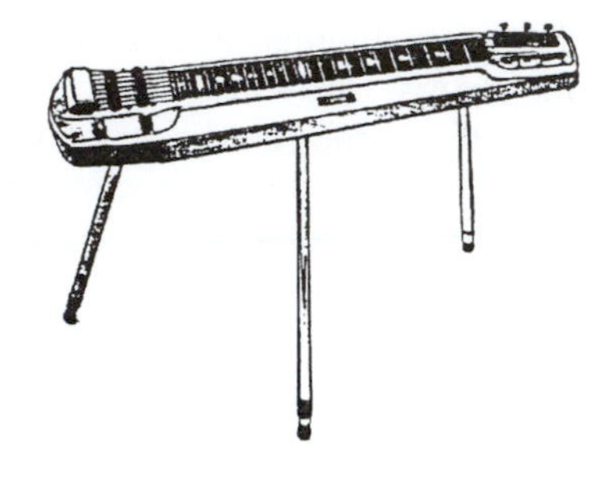

But one instrument which did catch on was the electric guitar. The earliest design was the **electric Hawaiian guitar**, released in 1936. It was played horizontally by moving a metal slide along the (steel) strings, and strumming the free part. It proved immediately popular in jazz groups and swing bands in the 40s with its twanging sound and characteristic sliding notes. But then it seemed to acquire a reputation for producing only schmaltzy tourist music and faded from the scene.

The more conventional **electric guitar**, played vertically rather than horizontally, was produced shortly after. It came into its own in the multi-tracking recording studios in the 50s, mainly because it was easy for different instruments playing together in a group to be recorded separately. So by the time that rock music had taken over the pop scene in the 60s, it was firmly established as the central instrument.

Today the typical electric guitar has no sound box, merely a whimsically shaped flat board which has no acoustic function, but merely carries the strings and the electrical controls. It usually has six strings (steel or nickel alloy) and is plucked with a plectrum. There are nearly always two sets of pick-ups (sometimes three). One set is placed towards the middle of the strings where they respond most strongly to the fundamental modes of vibration; and the other set is closer to the bridge where they pick up better the high harmonics. Either or both sets can be brought into use by means of a selector switch; and each set has its own volume and tone control knobs.

The amplifier is not an integral part of the instrument, but is usually connected by a length of cable. In the early days, it was placed close to the speakers, well apart from the player; but as greater loudness became important to pop music during the 70s, the amplifiers became more significant in their own right. Originally there were problems in trying to prevent them distorting the sound as they worked closer and closer to their design limits; but musicians are adaptable people, and it was soon realized that distortion, too, had musical possibilities.

Today any good electric guitarist will employ several devices which deliberately distort the sound, usually operated by foot controls. A common one is the **fuzz box**, which flattens the tops of all signal waveforms, in the process adding extra harmonics. The resulting sound is described as 'fuzzy'; it gives the impression of loudness, even when it isn't particularly loud. Another device is the **wah-wah**. This modulates the high harmonics, sliding them periodically up and down in amplitude. It produces a wailing effect, much as its name suggests.

The world of art music has not yet taken much notice of the electric guitar; but it seems to me that, in this device, the last generations of popular musicians have produced a valid new instrument—robust and relatively cheap, and one whose enormous musical possibilities have yet to be completely explored.

Instruments with completely electrical generation

In all the instruments I've just been describing it's clear that, because of all those tone controls, the timbre of the final sound they produce bears very little relation to that given out by the original generator. So the question arises: why do they need a 'natural' vibrator at all?

I've already said that it was possible, from the early days of radio telegraphy, for electrical circuits to produce a signal which oscillated like a simple sine wave, with a single pure frequency. It wasn't long before oscillations with more complex shapes were also produced—square waves, sawtooth waves, pulses; all characterized by a great number of overtones. . Of themselves, such specialized wave forms may not be very useful musically, but they make splendid raw material for filtering circuits to work with. Perhaps their key feature is that all their overtones are absolutely, accurately harmonic. Indeed one of the reasons why electronic tones often sound 'artificial' to musical ears is that they have none of the subtle inharmonicities that all natural instruments possess.

When musical instruments first started using such circuits, the designers had to decide how the frequency was to be changed in order to produce the *scale* of notes from which its music was constructed. Conceptually the simplest procedure was to have a different circuit for each note, and a set of switches to turn each one on and off, constructed like a keyboard.

This concept was the basis for a whole class of instruments known generically as **electric organs**. Initially the switches were simple on/off toggles like the ones you are familiar with in domestic situations; later they became more like the so-called 'dimmers' which turn up and down the brightness of lamps. Nowadays they are usually pressure sensitive: as you push the key down you *bend* a wire contact, which changes its resistance slightly, and hence the strength of the current. So the player has direct control over the loudness of the notes, just as in a piano.

The trouble with having a separate circuit for each note was that the electrical properties of circuit elements could change slightly with time and with temperature; and the whole instrument would go out of tune. However, the different frequencies could also be produced by signal processing operations, and as time went on, most of the commercially available electronic organs exploited these.

The simplest such system is a **frequency halver**. A pair of transistors (or valves) is set up in such a way that, if the current coming into one of them suddenly increases, it will turn itself *off* and the other one *on*. It is easy to see that if you input a square wave signal, each transistor will be turned alternately *off* and *on*.

Engineers call these things **flip-flop**circuits. Now if you arrange for either transistor to output a signal only when it is *on*, you will end up with a square wave of exactly half the frequency.

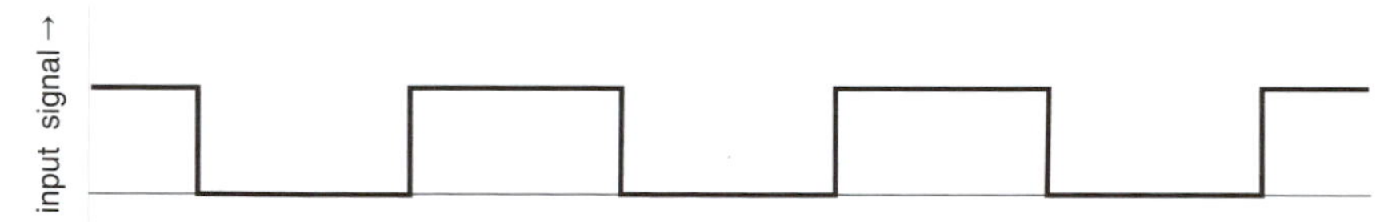

In principle it is possible to do the same sort of thing with more elements, and end up with a more general **frequency divider**. And there are a vast number of related procedures that usually go under the generic name of **heterodyning**—if ever you should meet that word in outside reading.

One of the earliest of the instruments using these techniques, which came onto the market in 1935, was the **Hammond organ**. Its frequency generating arrangements were an interesting compromise. It used a rotating shaft fitted with a number of disks, each with circular rows of evenly spaced teeth around the rim. As the shaft rotated these teeth moved past a magnet, periodically distorting its field and thereby generating fluctuating electrical currents

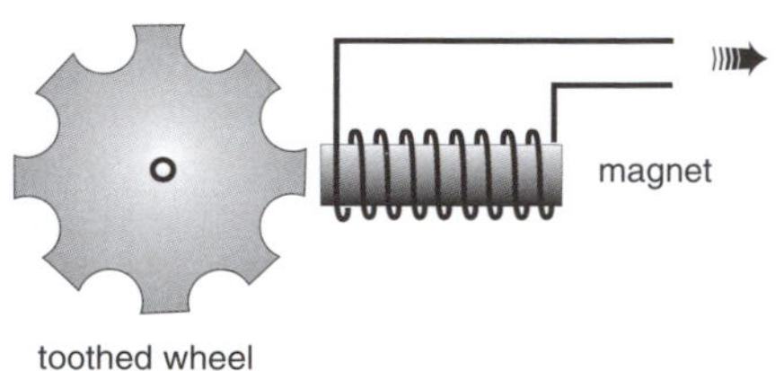

There were over ninety different frequencies produced this way, corresponding to the fundamentals and seven overtones of the twelve notes of the highest equal tempered octave on the keyboard. The other necessary octave ranges were filled out by these same notes transposed down by frequency-halving circuits. The Hammond organ therefore was controlled by a single master frequency—that of the rotating shaft—and so, no matter how much things changed, it could never go out of tune with itself.

Later electronic organs use an oscillating circuit as master, and an elaborate system of signal processing to produce all the other notes and their overtones (and it needs more than simple frequency halving to get the frequencies of the equal tempered scale). They also need some fancy circuitry because the business of constructing the timbres for different 'stops' is complicated. Simple filtering isn't enough; and it is also important to control the way each note varies with time. As one straightforward example of what is possible, it is standard practice to include in the circuits, capacitors which discharge relatively slowly—thus allowing the simulation of any desired reverberation time, and enabling the instruments to overcome unsatisfactory acoustical features of the halls in which they are played.

Although electronic organs were the most lastingly popular of this class of instrument, during the amazingly creative period of the 20s and 30s, close contact between musicians and scientists came up with much more imaginative solutions to the problem of giving the player direct control over what they were playing. In 1930 for example, there was an instrument called the **trautonium**. It was played by pressing down along a long steel wire—the effect of which was to change the length of the wire and its resistance, and hence the frequency at which the circuit oscillated.

An even more unusual instrument was the **theremin**. It was invented in 1920 by the Russian radio engineer and cellist, Leon Theremin, who originally called it the **etherophone**—a reference to the imaginary medium, disproved by Albert Michelson, through which electromagnetic waves were once thought to propagate.

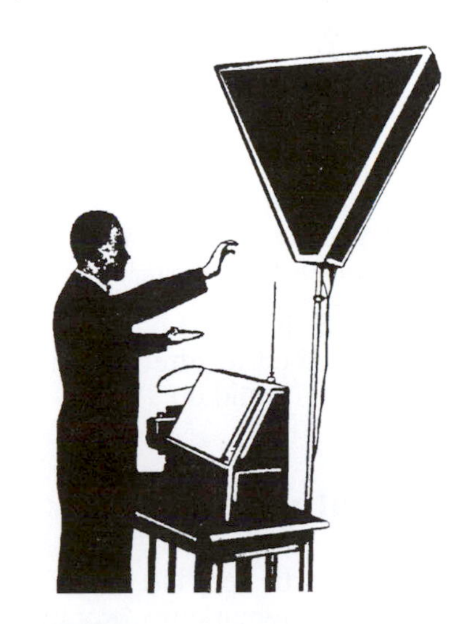

To play it, you positioned your hand in the space *above* the device, which affected its electromagnetic field sufficiently to be detected and fed back into the electronics. In practice you used both hands. One controlled the pitch and the other loudness. The sound came out of a large triangular loud-speaker and was characterized by a thin timbre and lots of swoops and slides between notes. It was often described as spooky. Indeed in the 1940s it was frequently used to supply background music for suspense movies—most notably in the Academy Award winning score by Miklos Rozsa for Alfred Hitchcock's *Spellbound* (1945), and in Bernard Herrman's score for the cult science fiction movie *The Day the Earth Stood Still* (1951).

Just by the by, I might add that Theremin himself had quite a suspenseful life. He emigrated to the USA in the 30s, and gained fame when his invention was featured in performances at Carnegie Hall. But he mysteriously disappeared and was assumed dead. Then in the 90s a television documentary producer discovered him alive and well in Russia. He told how he had been kidnapped by Russian agents for his technical expertise, and put to work in government laboratories and universities to help fight the Cold War. He even got a medal from Stalin. He died a few years after the documentary was made. If you ever get a chance to see it, it was called *Theremin: An Electronic Odyssey*.

But I digress. There was also an instrument, similar to the theremin, called the **ondes martenot**, which sometimes had a dummy keyboard against which the player's hand moved, to assist in precision. All these instruments were rather idiosyncratic devices, and somewhat limited, in that they were strictly melodic (they could only play one note at a time). But they opened up exciting possibilities, and serious composers were fascinated. Among those who contributed to their quite extensive concert repertoire were: Hindemith, Stokowski, Honegger, Messiaen and Milhaud.

However, they remained novelties and never attained the status of mainstream orchestral instruments. Probably the reason was that they needed enormous skill to play them musically, and the next generation of players preferred to wait and see which survived before devoting all that time to mastering them. As the composer Pierre Boulez has said:

> "The history of musical instruments is littered with corpses: superfluous or over-complicated inventions, incapable of being integrated into the context demanded by the musical ideas of the age which produced them."

And as it turned out, in the late 50s, the **synthesizer** was developed, which seemed to offer, in one device, everything that all of these early instruments could do.

The futuristic fantasy

In 1910, in the very last paragraphs of his *Theory of Harmony*, Schoenberg wrote these remarkably prophetic words;

> "If it is possible to create patterns out of tone colours that are differentiated according to pitch, patterns we call 'melodies', then it must also be possible to make such progressions out of that which we simply call 'tone colour', progressions whose relations with one another work with the same kind of logic entirely equivalent to that logic which satisfies us in the melodies of pitches.
>
> That has the appearance of a futuristic fantasy, and is probably just that. But it is one which, I firmly believe, will be realized. Tone colour melodies! How acute the senses that would be able to perceive them! How high the development of spirit that could find pleasure in such subtle things!"

It is difficult to believe that he was thinking about the electronics industry when he wrote this; but within forty years his 'futuristic fantasy' had indeed become technically feasible. Electronic computers suddenly made it possible to apply the knowledge that Fourier had developed centuries before, and construct musical tones of any imaginable harmonic content. Up till then, the timbre of notes on electronic instruments was controlled either by filtering out harmonics in certain ranges, or adding them together in more or less inflexible ratios. But computers can shape individual sounds, note by note. It should be possible, therefore, to synthesize the tones of any instrument so perfectly as to be indistinguishable from the real thing.

But when this was tried, experimenters came in for a rude shock. No matter how accurately they duplicated the harmonic content of ordinary instruments, their elaborately constructed notes just didn't sound right. They were all too obviously 'synthetic'. Scientists and musicians were forced to a conclusion that went against everything they had believed up till then. *The human ear does not recognize the sound of different musical instruments simply by listening to harmonic structure.*

So far in this book I have made no effort to prepare you for this observation. Nothing I said was actually untrue—oboe notes, for example, do tend to have strong second and fourth harmonics, and clarinet notes do have mainly odd ones; and our ears do notice that. But that is not *all* we use to enable us to say definitely: "That's an oboe playing, and that's a clarinet". And it is worth stressing how absolutely unexpected this was when it was discovered in the 1950s.

Success came to the early experimenters only when they realized that musical notes *change* all the time, and the way they change is absolutely characteristic of the instrument which makes them. It is not enough to treat a musical note as a *steady* collection of harmonics: you have to pay attention to the *transient* ones too. Therefore, before I go on to talk about the new instruments that emerged, I want to say something about this behaviour.

Transients

When thinking about how a musical note changes with time, it is useful to identify three distinct phases: the **attack** (or **starting transient**) when the sound rises from nothing to its highest value; the **steady state**, when it remains more of less at the same level; and the **decay**, when it falls away again to silence. It is usually easy to identify (roughly) these phases on an oscilloscope trace, and to abstract from that the structure of the whole note, by drawing what is called the **envelope**:

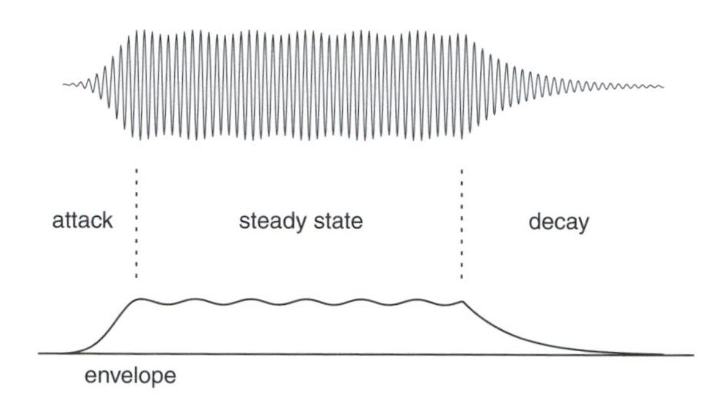

It isn't difficult to guess what this envelope might look like for certain instruments. The note of an organ, for example, usually has a relatively long attack, lasting some 400–600 ms (the symbol 'ms' means 'milliseconds' or thousandths of a second). The steady state is perfectly constant, because the sound is mechanically produced; and its length depends on the time value of the note as written on the printed sheet. The decay period is of similar length to the attack. So its envelope for different length notes might look like this.

On the other hand, a violin note usually has quite a short attack (50 ms is about average), and a rather longer decay, 100–300 ms. The 'steady' state is often not constant at all, depending on the intention and the skill of the player: usually it has at least some *vibrato*—i.e. rapid fluctuations of pitch in a small range around the main note.

A piano note is different again. Since it comes from a hammer hitting a string, its attack is extremely rapid, around 20 ms. Strictly speaking, it has no steady state at all, since nothing is done to maintain the note: instead it simply has a very long decay—which can last for up to 10 seconds. But when the pianist's finger is lifted and the damper touches the string, this slow decay is turned into a much more rapidly falling, 'real' decay phase.

Incidentally, careful observation of piano notes shows that there are often two distinct decay rates even when they are completely undamped. This comes about because pianos have more than one string for each note, and their energy of vibration actually moves around between the strings.

The reasons why attack and decay times vary so much lie in the instruments themselves, and are determined by the physical processes which make the sound. Hitting something with a hammer is a much faster way of giving it energy than stroking it with a bow. Likewise, a string rubbing against the felt pad of a damper will lose energy much more quickly than one which is vibrating freely. That much is straightforward. But it also happens that these processes vary in effectiveness as the *frequency* changes: rubbing against the damper rapidly, for example, will dissipate energy more efficiently than rubbing slowly. This means that different harmonics usually build up or die away at different rates. Therefore the *timbre* of a note is often quite different during the transient phases than during the steady state.

But there is more. In both of the transient phases, particularly the starting one, there always occur overtones which are not harmonics of the fundamental. The explanation is this. You will recall that, when I talked about the Fourier theorem in Chapter 4, I showed that any perfectly periodic vibration must consist of pure components (sine curves) of harmonically related frequencies (i.e. once, twice, three times, etc., the fundamental). The corollary to this, which I didn't dwell upon, was that a fluctuation which is *not* periodic must therefore consist of frequency components which are not harmonically related.

The upshot of all this is that the starting transient of any note is very complex acoustically; and it seems pretty clear that, when we learn to identify the sounds various instruments make, we rely on the starting transient every bit as much as we do the timbre of the steady state. Certainly that must be how we pick out different instruments in an ensemble, where the same harmonics may be coming from many sources. To convince yourself of the importance of starting transients, the following simple experiment is revealing. Record a series of piano notes on magnetic tape, or in your computer, and then play them backwards. The long decay of each note turns into a long attack, and the notes don't sound like a piano at all, but almost as though they're being played on an organ.

Composers and performers have always been aware of the particular importance of the starting transients. It is standard practice for string players to emphasize certain notes with a sharp bow attack (or even by semi-plucking at the same time as they bow, a playing style known as *quasi-pizzicato*); and wind players can accentuate notes by flicking their tongues as they blow. In each case the rapid attack causes high overtones to come in early; as opposed to a more gently graded approach which eases the low ones in first.

It can even be argued that composers are striving for a similar effect when they use **grace notes**—short notes immediately preceding the main note, played so quickly that they almost lose separate identity. By way of example, in his musical tale for choldren, *Peter and the Wolf*, Prokofiev represents the duck by this melody, played on an oboe.

Notice how the combination of notes, and the tone colour of the instrument, are used to suggest the kind of noise a duck makes. The grace notes have the same sort of effect as the 'kw' sound at the start of the word 'quack'. So, if you like, you can think of transients as serving much the same function as *consonants* on the start and end of a syllable of speech.

Indeed, in playing the bagpipes, the *only* way you can detach notes from one another is to introduce grace notes into the melody. As an example, here are the opening bars of *Highland Laddie* as printed in the 1923 edition of a handbook for playing Scottish pipes:

Scientific knowledge of these matters didn't only begin in the 1950s of course. Sound engineers with the recording industry had been studying the subject for most of the century; and it was one of the main reasons for the overhaul in recording techniques in the late 40s. All signal processing devices used in music recording need to be able to handle transients properly; and, since a transient is equivalent to some particular collection of frequencies, they can only do this if their frequency response is adequate. If they can't respond to high frequencies properly then they will distort any note with a very sharp attack. This is where you really notice the difference between pre-war and post-war recordings. The transients are crisper; and the separation between notes, especially in fast passages, is cleaner. But it is no coincidence that all this happened just before it was realized how important are the transients in the actual process of recognizing musical notes.

The instrument that can play its own notes

The first synthesizers appeared in laboratories and sound studios all over the world in the middle 1950s. Right from the start they were **modular devices**, essentially just collections of specialized bits of circuitry, each one chosen to do a particular job. A typical system consisted of (at least):

- **oscillators** to produce the basic pitch of each note,
- **amplifiers** to control their loudness,
- **mixers** to combine different signals and build up timbres,
- **filters** to change the harmonic balance, and
- **noise generators** to introduce percussive effects.

All of these were designed so that their output could be controlled by applying particular voltages at key points in the circuits (and remember this is just the kind of thing that transistors are ideal for). The devices that performed this modifying function were known generically as **voltage controllers**. Typical ones were:

- **envelope generators** which caused the amplitude of each note to rise, stay steady and decay for the appropriate times;
- **triggers** to ensure the starting transients began at exactly the right time;
- **frequency shifters** which were capable, among much else, of gliding the pitch gracefully from one note to the next (which musicians refer to as *portamento*);
- **frequency modulators** which allowed slight fluctuations in pitch characteristic of *vibrato*;
- **amplitude modulators** which did the same thing for loudness and produce *tremolo*;
- **reverberators** to produce artificial reverberation and echoes;
- and many, many more.

All of these devices then had knobs or buttons that had to be regulated by the 'performer'. There was usually at least one **keyboard**, though it didn't necessarily control only the pitch as on an electronic organ—it is perfectly possible to 'play' loudness, say, on a keyboard. All other regulatory settings had to be done manually by turning knobs or dials, or by plugging and unplugging connecting cables (the technical term is **patching**).

As these systems increased in sophistication, more and more of this regulation was done by computer: and this is where the unique capabilities of computers came in. They could be *pre-programmed* to perform the often intricate sequence of jobs at the right time during 'performance'. In the early computers the programming 'instructions' were fed in by punched paper tape; later they too were entered by keyboard (of the typewriter kind).

When the name 'synthesizer' was first coined (by the RCA company in 1955) it referred therefore, not so much to a single instrument in the old-fashioned sense of the word, as to a *system* of interconnected electric circuits and their controls. One of the earliest systems (known as a **Moog**, after the designer, Robert Moog, who put it together) covered a whole laboratory wall with dials, control knobs, keyboards, loud speakers and punched-tape readers. In a smaller, commercial form (circa 1970), it looked like this:

For the next decade or so there were many similar synthesizers on the market, of which the keyboard was the most obvious part; and many people tended to think the keyboard was the synthesizer. Nothing could be further from the truth. When you learned to play them, acquiring keyboard skills was only part of the problem: a good performer had to be able to 'play' the electronics as well.

It is probably true to say that synthesizers burst onto the public consciousness in 1968, when Walter Carlos released an album of the music of J.S. Bach, performed on a Moog synthesizer, whimsically entitled *Switched On Bach*. Technical descriptions of the project, written at the time, give evidence of how tedious was the production of such music, as late as 1968. It had to be constructed note by note, and the simplest phrasing could involve hours of production for mere seconds of music. Only tone envelopes were pre-programmed, all the sequencing and timing of notes were regulated manually, mostly using keyboards. No more than one melodic line could be constructed at a time: and each line had to be recorded separately on magnetic tape, to be mixed together at the end by multi-tracking procedures.

But the final sound was worth the trouble. It presented the music in a mixture of conventional musical sounds and frankly electronic tones: gurglings, poppings, beepings, contrasted with blatant harpsichord imitations. The effect on the listening public was electric. *Switched On Bach* became, within six weeks, the most popular Bach recording of all time, and one of the fastest moving classical records in history. Many argue that its popularity opened the ears of pop musicians, and that it was largely responsible for the key place these instruments were to gain in the modern music recording industry.

But they weren't there yet. While the production of music needed such a vast amount of technical expertise, synthesizers remained stuck in the acoustics laboratory. It was true that computers were gradually taking over all the technical functions, but what ordinary musician could possibly afford to own a computer?

The personal computer revolution

As I've said before, the original computers were very large, expensive devices, usually referred to as **mainframes**. Even by the mid 70s, when transistors had been around for twenty years, the newly produced "mini computers" were still way beyond the reach of most individuals. But then in 1977, manufacturers like Apple and later IBM started producing **microcomputers**—or as we prefer to call them these days, **personal computers**. Within a decade or so it became clear that it would soon be possible to have a computer in every home and on every desk—a development, I might say, that even science fiction writers had not foreseen when the computer was first invented.

The effect on the electronic music industry was immediate. All those specialized components, those voltage controllers and what have you, could now be miniaturized and packaged inside compact cases. Synthesizers started to be mass produced, and as a result their design and appearance became standardized. They were now much more like ordinary, stand-alone musical instruments and in order that they could be played by amateurs, much of their functionality was hidden. Two different kinds emerged. The first were the canonical synthesizers, devices that directly generate the timbre of the notes they play by adding together overtones according to some pre-determined set of rules. The second were **samplers** (sometimes called **wavetable synthesizers**). A sample is a 'snapshot', or detailed recording, of the sound of some acoustic instrument. All the information about that sound—its harmonic content, its transients and so on—are stored in computer memory and used as a model to recreate notes as they would be played by that instrument. As you might imagine, sampling involves encoding and storing enormous amounts of data, and it requires huge quantities of computer memory, much more than any other type of computer based instrument needs. Furthermore, samplers must be able to manipulate all this data instantly in order to produce a note a soon as the performer strikes the keyboard. They are remarkable achievements of computer technology, and are getting better all the time. The sounds they make seem to come closer to their acoustic models with each passing year. For that very reason, they concern some professional musicians. There can be no doubt that these instruments are largely used exclusively to save money—to avoid hiring traditional players. And that is a real worry.

Anyway, that aside, it is worth remembering that all these instruments are still modular devices, and *connectivity* is critical to their performance. In the old synthesizers, being analogue devices, the different modules were physically patched together by electric cables. But with the personal computer revolution, these devices became digital. All the electric currents flowing through different parts of the system were digitized (turned into patterns of on/off pulses), and what was now being shunted around was information. So the problem of connectivity was how to move information, efficiently and reliably, around networks.

It is basically a commercial problem. All the different components that go into these music producing systems must have what is called a **digital control interface**—if you like, a metaphorical front panel which you plug input and output leads into. And to continue the metaphor, it's not going to work unless your plugs fit into the sockets that the manufacturer supplies. In the late 70s, when the industry started going digital, each manufacturer made control interfaces to their own design. But then in 1983 an industry-wide standard was announced, the **Musical Instrument Digital Interface (MIDI)**. This was not a piece of equipment or anything, it was basically a *protocol* which lays down both a hardware interface and a language for passing musically meaningful messages between different musical electronic components.

It is helpful to think of MIDI as a kind of musical language, in many ways analogous to the notation we use to represent a piece of music on paper. When you play one of these instruments, what you are doing, deep inside the instrument, is generating and using a sequence of MIDI commands, which describe, among other things, the pitch of each note, its velocity (that is, how hard it was hit) and its duration. If you like, you are generating a kind of electronic 'score'. And that is what runs everything.

So in order to appreciate how one of these instruments works, you should continue to think of it as an interconnected system of modules:

- a **sequencer** which receives information from the player about what musical notes are to be played and when (it might, for example, take this information from a keyboard that the user is 'playing'), and encodes this information into a MIDI message;

- a **controller** which decodes this message and translates it into the sequence of voltages and currents, or other digital messages, that the sound maker can understand; and

- the actual **sound synthesis engine** which consists of the electronics that actually generate the sound.

Well that is what the electronic instrument scene is like today. Synthesizers are firmly established in the world of pop music (as are electric guitars). Their position in the art music world seems to be more equivocal. Much experimental music has being written for them, but very often it appears as a finished product—a tape or a disk, ready to be listened to. It seems that not very many composers in this area are prepared to write for commercially available instruments, arguing that they are not flexible enough to accommodate their ideas. But there are a few willing to give the new instruments a go: Morton Subotnick, Steve Reich and Ingram Marshall to name but three. Time will tell if change is on the way.

Time will also very quickly make obsolete much of what I have just been saying. If I ever get round to writing a third edition of this book, I'm sure this interlude will have to be extensively rewritten (again). There's no way I can guess what will be important then, so all I can say is:

Watch this space.

Chapter 9

I think, therefore I am

Much of this book has been about attempts by scientists to answer the question : "What is music?"; and when you clear away all the side issues, you can see a trend in their answers. Pythagoras believed we should try to understand music in terms of arithmetic. Galileo and Mersenne argued that the arithmetic wasn't important of itself, but was merely a reflection of the physical motions of sound sources. Then Sauveur and Rameau shifted attention from these motions to the properties of the sounds they gave rise to, and argued that therein was to be found an understanding of the structure of music. Lastly, Helmholtz showed, again, that it wasn't the sound itself which was important but the way our ears respond to that sound. All the time the centre of attention is getting closer to the human brain.

Ultimately music is how our brains *interpret* the arithmetic, or the sounds, or the nerve impulses—and how our interpretation matches what the composers and performers thought they were doing when they were making the music in the first place. At the most fundamental level, any music making is an act of *communication* between musician and audience.

Now, although that statement seems a bit vague and pompous, it does offer a clue about where we should go next. Scientists have, after all, been studying communication for at least a century. And in that time, it has been realized that there are certain abstract features, very obvious in the context of electronics, which are common to all kinds of communication. In 1948 a completely new branch of learning was invented, called **information theory**; and that's what I want to talk about in this last chapter.

But before I do, let me stress that there will be no answers to fundamental questions here, merely hints about where we might look for answers. The whole subject is still quite new and speculative—but that doesn't mean that we shouldn't talk about it.

The theory of communication

Let me start off with a simple example of how an electrical engineer might think about a problem in communication. Let's say we want to send a message on an early Morse code telegraph. That message will be made up of words, and each word of letters of the alphabet. The average message might be, say, four sentences long—perhaps a hundred words; and each word about five letters, plus spaces or punctuation marks. So the average message must be something like 600 characters.

Now in Morse code each letter is a number of dots and dashes, and spaces separating them—on average perhaps 3 *ons* and 3 *offs*. It is important to distinguish clearly between dots (which are short) and dashes (long); and your detecting device must be able to tell the difference even when there are unpredictable changes going on—surges in the power supply, variations in the rate at which the keys can move, peculiarities of your operators, and so on. (In the jargon of the trade, all such unpredictable effects are referred to as **noise**). This imposes minimum standards on your equipment, and in all probability, each burst of current (each *on*) couldn't really last for much less than 1/10 second.

Now that's a severe limitation. It means that each letter of the alphabet will take something like 1/2 second; and the average message 300 seconds, or 5 minutes! If you were thinking of a commercial operation you're limited to, at maximum, about 300 messages per day.

Now compare that with a telephone. It carries sound; and a 100-word message would take maybe 30 seconds to read. So a telephone system could handle some 3 000 messages a day. It is clearly at least ten times more efficient a means of communicating information. But it's even better than that. The spoken word consists of a large collection of individual sounds—certainly more than the 26 letters of the alphabet. Furthermore, when you hear a voice you get more than just the words. You hear intonation, accent, stress. You may even be able to tell who is speaking. Clearly there is much, much more information in the spoken word than in its Morse code equivalent. The telephone must be many hundreds of times more efficient than a telegraph.

Mind you, you have to pay for it. In order to transmit speech you must deal with sounds whose fundamental frequencies are in the range 200–800 Hz, as well as many harmonics of these. Unless your telephone lines have the appropriate bandwidth, you're going to start losing the very information you want to communicate. (I talked about this briefly on page 277.) And you've got to shield your line so that it doesn't pick up signals going along other lines, or from the outside air space. Opposed to this, the telegraph cable is simplicity itself. It only has to be able to carry a current which turns on and off; and so long as it doesn't actually have a break in it, it will probably do that adequately.

The purpose behind that simple discussion was to introduce you to a few key ideas:

- that quite different communication systems can be discussed in the same kinds of terms;
- that the efficiency of communication depends on physical properties—both what the system is intrinsically capable of doing, and how easy it is to protect it from outside 'noise'; and
- that there is a concept of 'average amount of information', which is quite independent of what actual message is being sent, but depends only on the *form* that the message takes.

Out of just such a way of looking at things did the abstract theory of communications develop. In 1948 an American engineer from the Bell Telephone Laboratories, Claude Shannon, proposed the following 'universal model' as a means of visualizing how an act of communication can be broken up into distinct units:

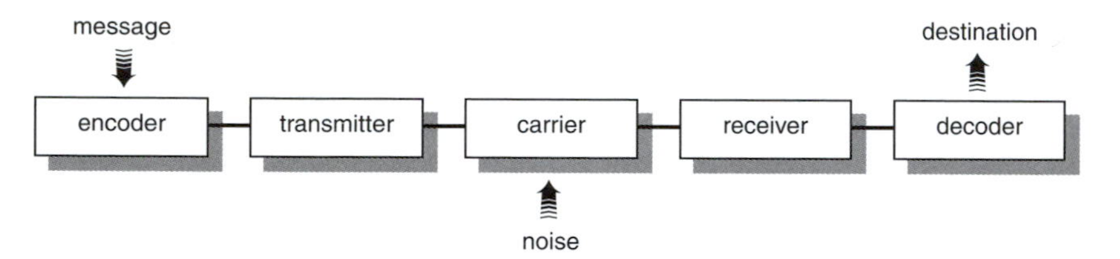

It says that in any communication, the sequence is more or less the same. The sender first **encodes** the message; and then the code is **transmitted** through some particular channel (or on some **carrier**)—in the process being exposed to the possibility of picking up **noise**. At the other end it is **received** and has to be **decoded** before it can be understood.

To fix the ideas in your mind, think about radio broadcasting. The message (already an audio signal) is used to modulate a high frequency electrical current. This is the carrier, and the modulation process is the encoding. The carrier wave is then broadcast (transmitted) over the 'air waves'; and while it is in this form it can get mixed up with other radiations from thunderstorms, motor cars, lawn mowors and so on, which all ends up as 'static' (or if you like, noise). Eventually the aerial of your radio (the receiver) will pick it up and demodulate it (decoding), and turn the energy back again to a fluctuating electrical signal which can go directly to your loudspeaker.

Though all this started off in the context of electrical engineering, it very soon came to be realized that the same ideas can be carried over into other fields. One of the earliest, interestingly, was the way animals communicate territoriality to other members of the same species by means of scent; and the way bees convey information about the location and direction of a source of nectar to other members of the hive by means of an elaborate dance ritual. In each case the complexity of the message communicated is limited by the versatility of the encoding process. Exactly the same parallels can be seen in spoken language; and of course, as is important to us here, in music.

In music there are (at least) two distinct phases of the communication: from the composer to the performer by means of the printed page, and from the performer to the audience by means of sound waves. We will try to focus attention on the possibility of quantifying the amount of information transmitted in each phase.

Information theory

After a message has been encoded it consists of a number of what are conventionally called **signs** (which can be letters of the alphabet, bursts of electricity, hand movements, or whatever). Consider, for the sake of simplicity, that the signs are the ten digits, 0 to 9; and the message is made up of just three of these digits—it might, for example, be an automobile number plate (in a very small country). There are clearly a thousand different three-digit numbers (that is $10 \times 10 \times 10$); or if you like, 1 000 messages that can be sent with this code. This number then, may be taken to be one measure of the 'amount of information' in each message.

On the other hand, if there were six digits—a telephone number perhaps—a million combinations are possible. The new 'amount of information' is 1 000 000. But this is counter-intuitive. A message with 6 signs is twice as long as a message with only 3; it seems reasonable to expect there to be twice as much information in the first message as in the second. This intuitive feeling is telling us that the natural way to measure information is not to accept the numbers 1 000 or 1 000 000 directly, but rather to use the *logarithm* of those numbers. (And, if you still unfamiliar with logarithms, read Appendix 2).

Now no real code is as simple as that. Any transmission is susceptible to being degraded by noise, so it is very seldom you find a code in which every single possible arrangement of the available signs corresponds to a completely different message. Ideally, codes are designed so that, even if a few of the signs get distorted in transmission, the general drift of the message is guessable. Business firms, for example, like to have telephone numbers with repeated digits, because they are easy to remember; and that is also the reason why so many people like personalized number plates. In technical jargon, you talk about there being some **redundancy** built into the code. But it means that there aren't actually as many different messages that can be sent. In such a code the amount of information is reduced.

The crucial idea being developed here is that the term 'information' is used not to specify what any particular message *does* say—I suppose a better word for that would be 'meaning'—but instead, what all such messages *could* say. In scientific contexts the word often used instead is **entropy**, which actually means 'disorder'. That is perhaps a more descriptive term, because any attempt to incorporate redundancy introduces a pattern and decreases the disorder. If you want a code that is able to convey many subtle meanings, then it must be capable of lots of different arrangements, of much possible disorder. But to make sure that the meaning doesn't get lost and is understood by the recipient, you need some redundancy, some predictability, some order. It is the balance between these two which specifies what the code is capable of.

A useful analogy occurs in painting. The canvas of an *avant garde* 'modern' work might have paint splattered all over it, apparently at random. There is lots of detail, so there ought to be much opportunity for the artist to communicate subtle messages; yet somehow it is hard to believe it actually 'means' much. On the other hand a piece of wallpaper has lots of order, perfectly repetitive geometrical patterns. It may be pretty, but it doesn't 'mean' anything either. In a sense, good art is a balance between the two extremes.

The science of information theory says nothing at all about the 'meaning' of a painting or a piece of music, but it enables the amount of information to be quantified, and gives a way to measure the balance between order and disorder. And in so doing it helps (just *helps* mind) understand the process of artistic communication.

Applying information theory to music

All this theoretical background was developed during the early 1950s; and music was an obvious field in which it could be tried out. A melody is made up of notes, which are chosen from a limited set of possible frequencies (a scale). Rhythm also is made up from a small set of allowable note values; and harmonic structure from a discrete set of chords. Any of these can be thought of as the 'signs' in a particular kind of coding.

The most obvious way you can apply mathematical analysis to these codes is to calculate their degree of order (or redundancy). But the trouble is that there isn't a unique way of defining what you mean by order—it can be done in many ways. One simple thing to do is to go through all the notes of a piece of music and count the number of times each is followed by another particular note. In other words, you ask the question: what is the probability of getting each possible pair of notes in succession?

The way you turn this into a measurement of order is as follows. You add together some combination of the logarithms of these probabilities (I don't think the exact form of the equation is particularly relevant), and come up with a figure for the 'information content'—though of course it is only one particular kind of information. Then the complementary quantity to this, the redundancy, is defined by this equation (which probably is worth writing out):

$$\text{redundancy} = 1 - \left(\frac{\text{information content of this particular sample}}{\text{maximum information possible with this code}} \right)$$

You interpret the resulting number (which is usually expressed as a percentage) like this. If there is absolutely no restriction on which notes may occur in succession (if for example there are no such things as standard cadences) then all possible combinations are likely to occur and your piece, if it is long enough, will have the maximum possible information (disorder). Its redundancy will be zero. On the other hand your piece may be absolutely predictable. It might, in the absurd limit, consist of just a single note, endlessly repeated (shades of the old Judy Garland song *Johnny One Note*). In that case the probability of that note occurring is 1 and its information content is zero (the logarithm of 1 is 0 remember). Hence the redundancy is 100%. In most real music, of course, this number will fall somewhere in between. If the notes tend to jump around a lot and there isn't much structure, the information content will be high and the redundancy low. But if there are very clear rules being obeyed about note progression and so on, the redundancy will be correspondingly higher.

Incidentally this demonstrates what an unfortunate choice of terminology is the word 'redundancy'. The last chord of any Beethoven symphony, for example, is probably entirely predictable. Given what has gone before, there is only one thing it can be. Hence in that chord there is zero information, 100% redundancy. But although it may be predictable, it isn't 'redundant' in the ordinary sense of that word. It certainly couldn't be omitted in performance. However I'm afraid that's the word that everybody uses, and we're stuck with it.

I'm sure at this stage you can see why this kind of work had to wait for the invention of computers. Imagine for example that you were trying to analyse in this way, a piece of music for string quartet. The basic signs of this 'code' are four-note chords. Each of the instruments can play some 50 notes (four octaves); therefore there are about 6 250 000 different signs in the code (that is $50 \times 50 \times 50 \times 50$). The job of looking for all possible pairs of these chords in succession would be horrendous (the number of pairs you would have to keep track of would be the square of this number). That is why most of the early work done in the 1950s was restricted to single notes in a melody (or at best pairs of notes). Even so, on the first computers, the calculations were exceedingly time consuming.

The details of what those early research projects found are not really relevant here: if you're interested you can read about them in references in the bibliography. By and large they yielded no great new insights. Studies were made, for example, of some well known examples of Gregorian chant. The redundancy values calculated depended, not unexpectedly, on what the investigator assumed to be the basic 'code'. If this was taken to be the 7 notes of the diatonic scale, then the redundancy came out to be very low. But if the basic code was taken to be the 12 notes of the chromatic scale, it was calculated to be much higher, for the simple reason that these extra notes were just never used, and their absence was therefore predictable. I repeat, there is little new in that. It simply says that, in their own stylistic context, those pieces of plainchant were remarkably free and innovative; but to ears used to post-19th century music, they seem much more formalized and predictable.

Calculations done on the music of Romantic composers showed up similar differences. Some music of Mendelssohn was found to be more redundant than some Schumann. Again it was no big deal. What was being detected was the fact that Mendelssohn's use of chromatic notes (in those pieces) was less frequent than Schumann's. Similar analyses on examples of rock songs of the period showed, not unexpectedly, even greater redundancy—much less melodic or rhythmic inventiveness.

Please don't get me wrong. I am not decrying those early research calculations, nor ones that were done later. It's just that so much had been hoped for. When physicists in the late 19th century had applied exactly the same kind of probability calculations to the theory of heat—giving rise to the subject we now call **statistical mechanics**—the increase in understanding was breath-taking. (This in fact is when the word 'entropy' was first invented). But music proved, as ever, more elusive. Still, some gains were made, and the fact that no blinding insights were forthcoming was, in the words of one of the researchers, less an impeachment of the techniques than an over-simplistic view of how they should be applied.

Since those early days, this whole approach to musical analysis—consisting, I need hardly say, of much more diverse methods than the simple ones I have singled out to describe—is a valuable tool for musicologists. And it has been used in many other musical contexts. Recently, for example, researchers in the field of marine biology have analysed the songs of humpbacked whales and have shown that these songs follow a definite 'grammar'. They have also shown however that, if these songs are used as a means of communication, then their rate of information transfer is quite low. These findings are interesting and potentially important. This field of investigation is still developing, and may take off in unexpected directions. It may yet yield insights that will help us in understanding what music is all about. But I just don't think we are there yet.

The computer as composer

Of more immediate interest, for those of us who are not professional musicologists, is not just to have our computers analyse music, but to see if they can make new music. It was very soon after computing machines were first mooted that suggestions along these lines were made. This was in the middle of the 19th century when the English mathematician and astronomer, Charles Babbage, constructed a steam powered 'analytical engine' to aid in the calculation of astronomical tables. He personally hated music, but a colleague, the mathematician Ada Lovelace, was farsighted enough to realize that Babbage's engine could handle other things besides numbers, and might one day manipulate tones and harmonies to put together "elaborate and scientific pieces of music".

One of the earliest suggestions about how this might actually be done appeared in 1950, strangely enough, in the magazine *Astounding Science Fiction*. It was made by an occasional writer of scientific articles and science fiction stories, one J.J. Coupling. He suggested that because computers could make decisions quickly, based on the results of calculations, they could be programmed to choose sequential notes at random and discard those which didn't obey some predetermined rules, eventually to build up a complete melody. (It is interesting to note that the writer was in fact a leading scientist in the Bell Telephone Laboratories, John Pierce, who had long been interested in the scientific aspects of music, and who is today the author of one of the best text books on the subject. His strange pen-name, incidentally, is a groan-worthy pun, taken from an obscure procedure in quantum theory for combining the rotations of electrons inside atoms, for which physicists traditionally use the symbol **J**.)

Actually the idea that music might be written by procedures of chance was much older than this. In the 18th century it was a popular diversion, involving, among others, Joseph Haydn and Carl Philipp Emanuel Bach. Mozart himself wrote a *Musical Dice-Game* consisting of a set of alternative groups of notes for each measure of a standard form minuet. The particular notes to be played in each measure were determined by the throw of a dice. I believe that, though it gets a bit repetitive, there is a surprising amount of variety in the resulting 'compositions'. But what was new in the 1950s of course was that there was a theoretical basis to guide the process, and computers to do the choosing in real time.

Starting in 1951, the man who would later invent the RCA Sound Synthesizer, Harry Olsen, built a specially designed 'composing machine'. He and a colleague painstakingly analysed the songs of Stephen Foster, in just the way I described earlier, and set their machine to see if it could come up with something similar. The results were equivocal. Some of them sound vaguely Stephen Fosterish, in small snatches, and occasionally moderately tuneful; but that is all.

You can get an idea of the outcome of this kind of experiment from a project which was published in 1956 by one Richard Pinkerton. It is simple enough that you may be able to reconstruct it for yourself. What he did was to analyse a whole collection of nursery rhyme jingles, just on the basis of what was the probability of any one note being followed by any of the other notes of the scale. Then he simplified his findings by discarding all but absolutely the most common note pairings; and constructed what he called a *banal tune maker* (because it was "scarcely more inventive than a music box").

It will compose four bars in 6/8 time in the key of C. At each point there are at most two choices of where to go next—the choice being made by throwing a dice, or getting a computer to pick a random number. I reproduce his scheme on the page opposite in the form of a computer flow diagram. You start at the top and follow the arrows. The notes (or rests) to be played are indicated directly; and where there is a choice to be made, you can take either path. By the time you reach the bottom of the page you will have composed one bar of six beats; you then return to the top and go through it again. You exit after the third bar (or earlier or later if you wish). If you are competent at programming a personal computer you might like to translate it into code and run it; otherwise I think you can see, just by reading through, what it does.

When you run this program it gives you a different snatch of melody each time. They're rather repetitive, and in fact when you calculate their redundancy you get a very high figure indeed, much higher than the original set of nursery tunes they were based on. But that isn't surprising: after all, Pinkerton did simplify savagely in order to get a scheme that was easy to operate, and it uses only four different notes. The important thing is that some of melodies it gives you don't sound bad at all. Here is one I quite like. I could easily imagine a nursery rhyme being sung to a tune based on this:

Many such experiments were done throughout the 1950s. Their importance didn't lie in the music they produced: they never claimed to be writing real music. Rather it was in the insights they were giving about the process of composition. Then, toward the end of the decade, a project was undertaken which came very close to crossing the gap. It was done at the University of Illinois under the direction of a professor of music (who had actually started his career as a research chemist), Lejaren Hiller, using the ILLIAC computer I talked about in the last chapter. In its program, the computer was instructed to choose one note at a time, again by random

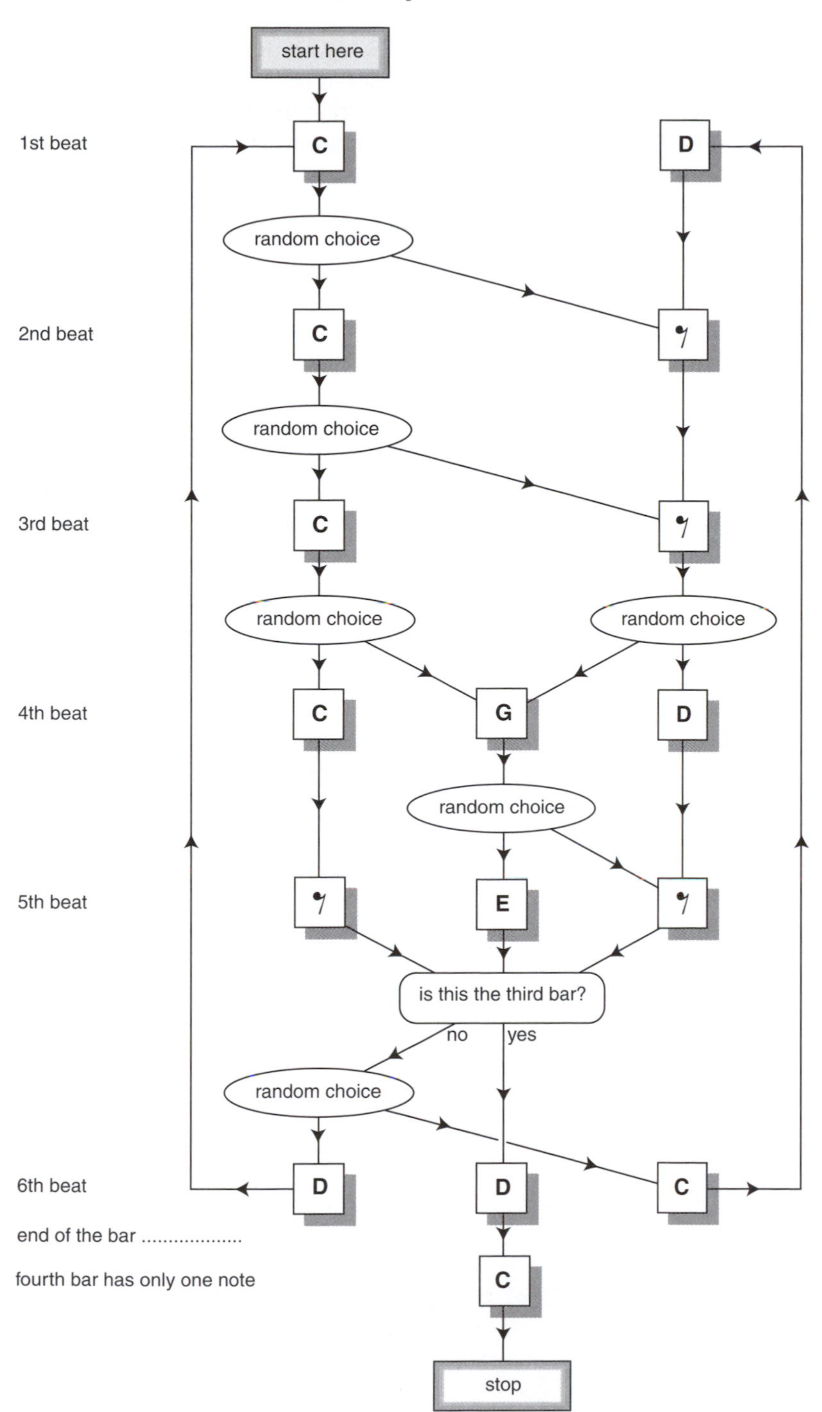
start here
1st beat
C
D
random choice
2nd beat
C
random choice
3rd beat
C
random choice
random choice
4th beat
C
G
D
random choice
5th beat
E
is this the third bar?
no
yes
random choice
6th beat
D
D
C
end of the bar ....
fourth bar has only one note
C
stop

selection; but this time the criterion it used to decide if the note should be retained was based on whether of not it accorded with the classical rules of composition—those rules which, you will remember, were worked out during the late Baroque period. In other words, this was an attempt to program into the computer, the accumulated wisdom of generations of musicians.

The results of this project were published in 1959—as the now famous *Illiac Suite for String Quartet*. They were first given public performance a year later, though to no great critical acclaim. One reviewer said they aroused "much curiosity but not much envy from flesh-and-blood composers".

But in a way, their reception as music was irrelevant. They showed that it could be done. And immediately the idea that a computer could be used as an aid in music composition was taken away from the physics laboratories and into the studios of composers like Iannis Xenakis and John Cage (to name but two). At the same time, interested scientists turned their attention in a different direction—away from the ordered side of musical composition, the redundancy, and back to the disordered part, the entropy.

Noise and music

The procedures used in the kind of composition I have been talking about are usually known by the jargon word **stochastic**—a word which actually means something which moves by steps which are unpredictable, but nevertheless controlled by well-defined probabilities. They incorporate both redundancy and entropy in an obvious way—the former by specifying probabilities or rules which guide the making of choices, and the latter by making those choices in a random manner.

But there are different kinds of randomness. If you throw a dice, or pick from a pack of cards, the numbers you get fluctuate wildly. They bear absolutely no relation one to the next. This kind of randomness is called in the trade, **white noise**. The word 'white' is meant to signify that no particular frequencies are present in the fluctuations: it's very like the 'snow' you see on your TV screen when it isn't working properly. It's the kind of randomness I have been talking about up till now.

Here is a short example of 'white music' constructed completely by white noise randomness.

There are no rules of composition here. The pitch of each note is chosen completely arbitrarily from a set of 15—the two octaves of the diatonic scale from C_4 to C_6; and the duration is chosen, again randomly, from six possible values. If you play that to yourself on a piano, I think you'll agree that it is not what you would class as interesting music.

All the early computer compositions used this kind of randomness to make choices. When you listen to them they don't usually sound any good: they jump around too much. Even when they have composition rules built in, they sound all right in small bits; but in longer segments they seem uninteresting, they never seem to 'get anywhere'. And that makes sense of course. When you remember how they were composed, it's obvious that individual notes would be influenced with those which have just gone before, because of those rules; so there will be short range structure. But that isn't true of the larger segments of the work; these have absolutely no correlation with one another.

But there are other kinds of randomness. Think about a drunk staggering about in the middle of an empty parking lot, each step he takes being in a completely unpredictable direction. After some time he could end up anywhere at all in the lot. That is quite random. Yet at any one time his position is strongly correlated with where he was a few minutes back; and, what is more, it is possible to calculate how far, on average, he will be from his starting point at any time. It's randomness of a different character, and it's known as **brown noise**. The name here is a rather weak pun, based on the fact that this kind of fluctuation was first described by the Scottish botanist Robert Brown, in the early 19th century.

There have been some experiments in creating music using randomness of the 'brown noise' kind. Here's an example.

By and large, 'brown music' tends to sound dull and rambling; it just seems to go on and on (which is, of course, exactly the character of drunken meanderings). Clearly, if these ideas are to be useful in music composition, we want something 'in between'.

As it happens, there is another kind of randomness, shown, interestingly enough, by many natural phenomena. Clouds are a very good example. Their shapes are absolutely unpredictable. Yet each single cloud has a definite shape, and the various bits of the cloud fit together to make that shape. And no matter whether you look close up or far away you get the same impression: each bit does its own thing, yet in a way it is also correlated with all the other bits. Now that's a different kind of randomness. It is characterized by overall unpredictability, but at the same time by long and short range correlations. Its technical name is **1/f noise** (and let's not bother about where that name comes from).

The key feature of this behaviour seems to be this **self-similarity**, this looking much the same over a whole range of scales. It is also the property of what are called **fractal images**. Look at this picture.

It is constructed from a whole lot of 'Y' shapes, with two smaller Ys joined on to the 'branches' of each larger Y, but all with slightly different angles between the branches. But there's no doubt it looks like a tree. That's the kind of shape some trees grow in. And it's quite attractive: whether *because* it looks like a tree, or whether our brains sense something of the pattern of its structure is not clear.

As I said, this kind of randomness occurs all through the natural world: not only in clouds and trees, but in coastlines, mountain ranges, star clusters, all sorts of shapes. And for some unknown reason we find most of them aesthetically pleasing. Artists have always enjoyed painting clouds for instance, and most people like finding patterns in a flickering fire, or on the face of the moon. Is this kind of randomness just what we are looking for to aid in the composition of music?

A lot of work has been done on this question in the 1970s and 80s, largely by a professor of mathematics at the Massachusetts Institute of Technology, Richard Voss. Here is an example he has published of what '1/f music', generated by a computer program in the same manner as the two previous examples (which were also his), is like:

I don't know what you think, but that example, while by no means a show-stopping tune, sounds quite pleasant to me. And if you look at it carefully you should be able to see that the way it varies from note to note, is much the same as the way that chunks of the melody (say, groups of ten notes at a time) vary. If you like, the melody doesn't forget where it has been, there is some correlation with all that has gone before.

I should say that composers didn't have to wait for electronic engineers to develop the theory of 1/f noise before they started using these same basic ideas. In 1938, Prokofiev was asked to compose the background music for Sergei Eisenstein's *Alexander Nevsky*. In parts of the score, he used the silhouette of the landscapes from the movie itself as a pattern for notes on the staff, and orchestrated around them. And since we know that the profile of mountain ranges typically follow patterns of 1/f fluctuation, what he was doing was writing 1/f music.

The meaning of computer composition

Computer composition started being seriously regarded by composers in the 1960s. And since that time several important centres of learning have been established, devoted to the musical rather than the scientific aspects of the work. Among these, particularly prestigious examples are the *School of Mathematical and Automated Music* set up in Paris in 1965 by Iannis Xenakis, and the *Institut de Recherche et Coordination Acoustique/Musique* set up in the same city a decade later by Pierre Boulez.

So, the computer as an aid in composition has been absorbed into the musical scene; but again it is only a tool. Its role is more active than, say, that of a piano on which Romantic composers tried out ideas they had already conceived in their minds. But equally well, it has certainly not automated the process completely. Composers I talk to seem to have a semi-interactive relationship with their computer. They tend to let it play around with bits of program and see what it comes up with; and they will accept the interesting sounding bits and discard what they don't like. But they reserve for themselves the job of incorporating these 'ideas' into a finished composition. In that part of the composing process the most they will delegate to the computer are a few straightforward orchestrations or tedious transpositions.

From a musical point of view, that's obviously how things must go. If the time ever arrives where the computer can be trusted with the whole job, then composers must be very sure how to write the original programs; and that will only come after a lot of experience (if at all). But from the scientific point of view the unanswered question still rankles. Is it really possible?

Again, most of the composers I know believe that it may be—at least in principle. For some time now, computers have been able to write low grade TV jingles or musak (the kind of thing they play in lifts and on factory floors). After all computers can draw pretty pictures easily enough, and can easily produce pleasant geometrical designs for wall paper; so why not 'wall paper music'? But they also argue that even the simplest pop songs have some creativity in them and there are plenty of unemployed composers who want to earn their bread and butter; so why not let them write the TV jingles?

But as a scientist I still want to know. You see, if it were possible, then that tells us something about the working of the composer's mind. There's an old story that used to be told about himself by the 19th century physicist, Nicholai Tesla; the man who invented, among much else, the system of distributing electrical power as alternating current (thereby making his rival, Thomas Edison, an enemy for life). He was reputed to have a truly phenomenal memory, and he claimed that he could test any new machine he invented without building a working model. He would design it completely in his head, and then set it going. It would continue, entirely subconsciously; and days later, he would know which particular bits wore out first. Now that may be a bit hard to swallow, but you can't help wondering if that's what composers do. Perhaps they start some sort of program in their heads, for all the world like the original ILLIAC experiments, choosing notes and discarding them if they don't fit in with their overall scheme, until at last the whole thing is finished. Maybe that's what Mozart meant when he said melodies just popped into his mind, "whence and how they came he knew not".

One of the most exciting fields of scientific research in the last decades of the 20th century has been into **artificial intelligence**. It is a truly massive inter-disciplinary undertaking, involving physicists, engineers, computer scientists, psychologists, linguists ... and musicians. Much important work in this field is done by a group at the Massachusetts Institute of Technology in Boston, led by one of the so-called 'fathers of artificial intelligence', Marvin Minsky, who has based much of his research on the logic behind computer composition. His argument has been that one of the great unsolved problems about the brain is why it likes something so abstract as music. It is just possible therefore that if someone could write a program to compose music that genuinely sounds acceptable, then the logical steps in that program might reveal a lot about the complex network of nervous reflexes that go to make up the human brain.

May I just make the final comment that this is another example, to add to the many from the last 2500 years, of science in the process of changing music, and music in the process of changing science.

Interlude 8

Sublimest of instruments, the voice

The voice is the oldest of all instruments, but in many ways the slowest to be understood. That the relevant scientific knowledge should have only relatively recently been discovered is not difficult to believe—after all it was only with the invention of electronic equipment that the voice became completely accessible to scientific study. But even knowledge of a musical nature about how the voice was used in earlier times is hard to come by. So let me lay the groundwork for this last interlude with a brief description of what we know about what singing styles were like up till now.

A short history of singing

From the very earliest writings we know that people have always sung, and that singing has always had religious associations. Yet even the ancient Greeks, for whom music was an important part of every child's education, left very little written instructions about what you should actually do when you sing. Plato wrote about the need to suit the style of singing to the words being sung, but that doesn't really tell us much. (Of course we know very little about any facet of the performance of Greek music; for them, the greatest importance was always in its theory.)

But we don't know much more about early Christian singing either. In 367 A.D., the Council of Laodicia banned all forms of congregational singing in church, so presumably there must have been a need to train choirs. Certainly in about 600, Pope Gregory set up schools to teach church singing, and his name has come to be associated with the chant of the period. The ideal voice for that style of singing, with its strong heritage from the synagogue, was probably a high, flexible (male) voice capable of elegant delivery of long fluid lines of melody. Writings of the day declared the best voice to be "high, sweet and strong".

By about the 13th century, early polyphony had developed to a point where the ornate alleluias, tracts and graduals required genuine virtuoso singing, particularly in the upper parts. Most of the well known singers were also composers—very few people made a living singing someone else's music; and the likes of de Vitry, Dufay and Josquin were, in their own times, as famous for their voices as for the music they wrote. But the timbre of the high male voices that dominated this kind of music can only be guessed at. It has often been observed that singing styles tend to parallel the instrumental sounds which are currently popular. So, given that many early medieval instruments had a strong Moorish connection, it is probable that the typical vocal music was a bit reedy or nasal in tone. Indeed, in *The Canterbury Tales*, Chaucer has these often quoted lines:

> "Ful weel she soong the service dyvyne,
> Entuned in hir nose ful semely."

In the heyday of classical polyphony, in the 14th and 15th centuries, the vocal range of singers expanded. Most of that music had four well differentiated lines: bassus, tenor, altus and descant. To supply the lowest part, the bass voice suddenly came into prominence. Up till that time vocal music rarely went below C_3; but then the Flemish composer Johannes Ockeghem, himself reputed to be a particularly fine bass singer, wrote masses which regularly got as low as F_2, and on occasions, C_2. At the other end of the scale, the descant lines got so high they had to be sung by young boys; although there was also a tradition, particularly strong in Spain, of using men singing *falsetto* (being more reliable and not subjected to the inconvenience of having their voices break just when a lot of time and effort had been invested in their training). It is from this period that the first written references can be found to the observation that the human voice has several different 'registers' (though that word wasn't used till much later). Men singing falsetto were said to be using a 'head voice': the usual range was called the 'chest voice'.

In the late Renaissance, one of the most popular forms of secular music was madrigal singing. Its sound, which once again reflected the lute and viol ensembles which were also fashionable, demanded several high lines of melody—and women's voices were suddenly discovered. Why half the human race should have been overlooked up till then is not clear; perhaps mothers taught their daughters that only "certain kinds of women" sang in public. Whatever the reason, a few strong minded characters, like Isabella d'Este, Marchioness of Mantua, became seriously interested in musical accomplishments and dragged the female voice out of the closet.

Initially most music written for women's voices was low, not really very different from the high male range. But as ornamentation became more fashionable, and as the newly established form of recitative put more stress on the ability to improvise, the pitch went up. By the end of the 16th century, sopranos were all the rage; and, for example, the composer Giulio Caccini advocated this kind of voice as his ideal sound (possibly influenced by the fact that his own daughter was to gain great renown as an operatic soprano).

From church music, of course, women were still excluded; so the new musical sound had to be supplied by other means. It isn't clear just when the gruesome habit began of preventing adolescent boys from entering puberty (and therefore from having their voices break) by surgically removing or otherwise destroying their testicles. But in Western music, the *castrati*, as they were called, came into their own in the early 17th century. Both in church music, and in the newly invented operas, a voice with the high notes of a boy and the power of a man was just what was needed. A completely new profession arose, that of the professional opera star, and the castrati were on top. Their feats of vocal virtuosity became legendary.

For at the beginning of the 18th century, virtuosity was the name of the game in the new singing style. And so another profession also blossomed—that of the singing teacher. Countless books were written containing scales and exercises; and great emphasis was put on training the head voice. It is in these books that you find arguments about just how many registers the human voice has, and the idea that the way to achieve these different registers is to 'place' your voice in certain parts of the body. The fundamental aim was to teach the singer to perform on the voice almost as on a flute, striving always for purity of tone and technical brilliance of the highest calibre. It was a style of singing that came to be called *bel canto*.

These changes in vocal music, again, paralleled what was happening to instrumental music. There too the ideal sound was clear, bright and elegant. And, as the 18th century progressed, a succession of new (or newly reworked) instruments extended the tone colour range of orchestral sounds. So too with singers. Sopranos started getting higher and higher, rivalling the castrati in popularity. Up till then singers were seldom asked by composers to sing beyond a range of about one and a half octaves, and rarely higher than C_6. But there are plenty of reports of sopranos whose range was much greater. One Lucrezia Agujari (according to Mozart writing in 1770) could get as high as C_7! Since these extreme notes didn't appear in any of the scores, they must have been used only in improvised passages or cadenzas. Gradually however, composers started writing them in, and by the end of the century, in the Queen of the Night's aria in *The Magic Flute*, Mozart asks his soprano, on four different occasions, to sing the note F_6.

In the 19th century, as I've pointed out many times, audiences for music increased and halls got bigger—halls that were primarily designed for the music of orchestras and choirs. Solo singers found themselves at a severe disadvantage. If their art was to survive they had to find a way to be heard. It wasn't just extra loudness they needed, but a quality which is usually called 'projection'—the ability to be heard in competition with the great volume of sound coming from an orchestra. As a result, there was a definite shift in singing technique to the use of the chest voice for volume; and singing teachers started exploring certain internal 'resonances'. In many ways it was the high male voice which possessed these qualities naturally, and especially in the Wagnerian tradition of constant sustained singing, the tenor voice gained prominence. The age of bel canto faded, and a new one took its place—one which is sometimes disparagingly referred to as *can belto*.

The 20th century has seen some new ways of using the voice. Works like Schoenberg's *Pierrot Lunaire* have used a stylized form, halfway between speaking and singing. But, in general, operatic requirements haven't really changed; aspiring opera singers are still put through the same kind of training as they were a century ago. However other forms of singing have undergone a radical shift, due entirely to the invention of the microphone.

Art musicians, by and large, have ignored this device and have steadfastly refused to allow it to change their public performance techniques. But they do rely on it to make records, and I think it can't be denied that it is having a subtle effect because of that. If you listen to old recordings of the great singers of the early part of the 20th century—Melba, Caruso, Chaliapin and so on—they seem much more eccentric and florid than is the accepted style today. It is almost as though the ability of the public constantly to make comparisons between singers has imposed a degree of conformity, a kind of regression towards the norm.

It must also be true that modern recording techniques underlie the current trend towards recreating historically accurate performances, one result of which has been the resurrection of the counter-tenor, the use of the male falsetto voice in a style of singing not much heard since English music of the 17th century. Another has been renewed interest in the soprano voice with a pure treble sound, as opposed to the heavy *vibrato* of the older operatic tradition.

However it is in popular singing that the effect of the microphone is most noticeable. Prior to 20th century there wasn't much difference between popular and 'classical' singing styles; although in lighter musical theatre, like Offenbach or Gilbert and Sullivan, in which the words were just as important as the music, the extreme operatic style of singing was toned down a bit. You see, voice 'projection' usually tends to distort vowel sounds: it is often extremely difficult to understand the words being sung by an opera singer at full blast, especially a soprano. (And if you are tempted to dispute that statement, think how often today opera companies use 'surtitles'—that is, the text projected onto a screen above the stage—even when the work is being sung in the language of the audience). But once electrical amplification had become widespread, it was possible to reach a large audience without the need for projection, concentrating instead on articulation of the words being sung. This is the road that popular music chose.

To my mind, there are distinct parallels between what happened in the 16th century following the invention of recitative, and what happened in the 20th. Afro-American singing styles which came to the forefront after the First World War, were full of the cadences of everyday speech, just the kind of sound that electronic systems had been designed to handle. A new ideal singing voice arose; one which was light, mellow and intimate. This sytle was never so well exemplified as with the 'crooners', the best known of whom was Bing Crosby who had an amazingly successful recording career in the 1940s and 50s.

Today, virtually all popular singers sing in the middle ranges. There are very few deep basses or high tenors, and essentially no sopranos at all. Yet, particularly in jazz singing, the same vocal tricks can be heard as were popular in the early days of recitative—subtle variations of tempo, turns and slurs, 'leaning notes' which displace the stress from the first beat of the bar—even when these aren't specified in the score. I wonder if it is fanciful to suggest that, once the current fashion for extreme loudness passes, popular singing styles may become very ornate, just as did early Italian opera; and perhaps, when historians look back over the 20th century, they will judge that pop music led the way that the rest of Western vocal music followed.

Science and the voice

The background knowledge necessary to understand the voice as a musical instrument comes from three distinct fields of study. Firstly there is the basic anatomy. The systematic study of the organs of the body which produce the voice start (like so much else in anatomy) with the 2nd century Greek physician, Galen. He had (by our lights) a strange understanding of the nature and function of the breath, but he nonetheless left a fair description of the lungs and the windpipe through which the breath entered and left the body.

In the 16th century, mainly under the influence of the great Italian anatomist, Andreas Vesalius, the first really accurate drawings appeared, and the organs of speech were given most of the names we still use today. Since I will be referring to these many times in what follows, let me include at this point a diagram indicating what and where the most important of these organs are.

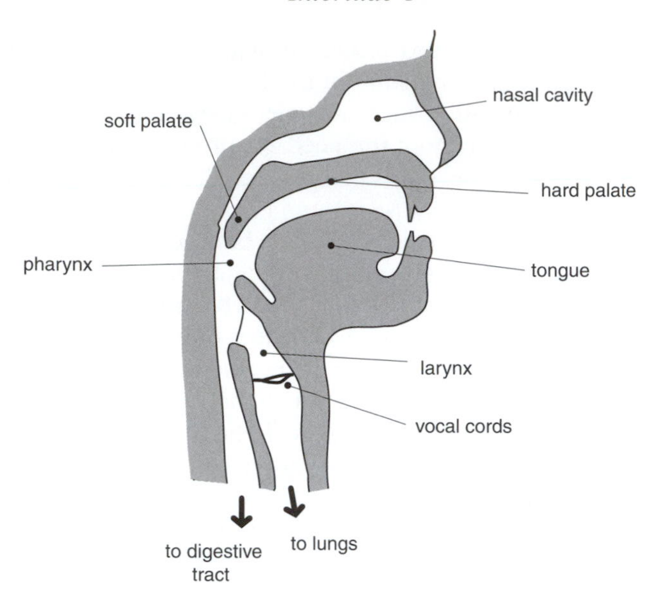

At the top of the windpipe coming from the lungs there is the **larynx**, or voice box, a short tube which houses the vocal cords. This opens into the **pharynx**, a longer tube, the bottom of which also connects to the digestive tract; and this opening can be closed off during swallowing by a special valve. The top of the pharynx then connects to the oral cavity—the mouth—whose internal shape is controlled by the tongue, teeth and lips. The roof of the mouth is mainly rigid (the **hard palate**) except at the rear (the **soft palate**) where it acts as another valve to open or close a connection to the nasal cavity.

The early anatomists weren't much interested in the vocal mechanism as a source of musical sound. That was the preserve of a special breed who applied this growing body of knowledge to the practical business of training singers. The earliest known treatise on this was in the 16th century by one Giovanni Maffei, a trained physiologist and practising music teacher. It is interesting because it was written in the form of a long letter to his patron, and it was intended for those seeking to learn without a teacher—which is perhaps why he felt he had to go into so much detail about the vocal mechanism.

But after that, despite the plethora of books written in the 17th and 18th centuries about methods of voice training, surprisingly few of them were concerned with voice production as such. It wasn't really till the 19th century, with its pressures to produce a new kind of singing, that writers started using a more 'scientific' approach and treating the voice as just another machine. And no one epitomized better the new approach than one who was reputedly the greatest singing teacher of the century, Manuel Garcia.

Born in Spain in 1805, the son of the famous and colourful operatic baritone (with the same name) who was the first Almaviva in Rossini's *Barber of Seville*, he started life as a singer himself—a bass—singing Figaro with his father in American productions of the same opera. But he soon retired from the stage and took up the position of Professor of Singing at the Paris Conservatoire, and later at the London Royal Academy of Music. He invented a remarkable device which he named a **laryngoscope**—essentially a kind of periscope for looking down his pupils' throats while they sang. It must have been fiendishly uncomfortable, but it worked. Garcia was able, for the first time, to see what the whole vocal mechanism—vocal cords, throat, palate, tongue—was doing while it was producing a musical note. He wrote an enormously influential treatise on the art of singing, based on a clear understanding of how the voice works. But he always kept his scientific knowledge subservient to his musical needs. As one of his most famous students said of him:

> "It must always be remembered, he taught singing, not surgery."

The third branch of learning which contributes to our study is that which deals with speech sounds—the science of **phonetics**. This was very popular during the 18th century, attracting many big names from mathematical fields. Then in 1828 the English mathematician, archaeologist and amateur musician, Robert Willis discovered the interesting fact that every vowel sound is connected with a particular frequency related to a resonance in the oral cavity. In 1860 Helmholtz, by virtue of the remarkably sophisticated equipment he had devised, showed that certain vowels were associated with *two* such resonances, which later came to be called **formants**. Then in 1924, Richard Paget, barrister, physicist and sometime composer, proved that this was true for every vowel; and actually constructed models of the vocal cavities which would produce artificial vowel sounds when you blew into them.

This particular branch of science has dramatically increased in importance in the last half century. Again it has to do with the invention of computers. There is today a pressing need to communicate with computers by means of voice—both to get data into them, and to retrieve information from them. To this end, laboratories all over the world have analysed human speech into the smallest possible units of sound (the technical word is **phoneme**), and computer programs are being written to assign meaning to the code that these sounds represent. Not all the problems have been solved, but we are very close. In 1968, when the movie *2001, A Space Odyssey* was released, the idea was still considered fantasy, that a computer could talk and be talked to. Today you can buy in any toyshop for a couple of dollars, any number of children's toys or computer games. And they all talk.

The vocal cords

Turning now to the physics of the voice, the obvious place to start is with the generators of the sound—the vocal cords. Actually that name, though widely used, is misleading. They are not cords or strings, they are two folds of tissue stretched across the windpipe; so from here on I will always refer to them as the **vocal folds**. You can see their structure in the following diagram, which is a horizontal section taken across the front of the throat.

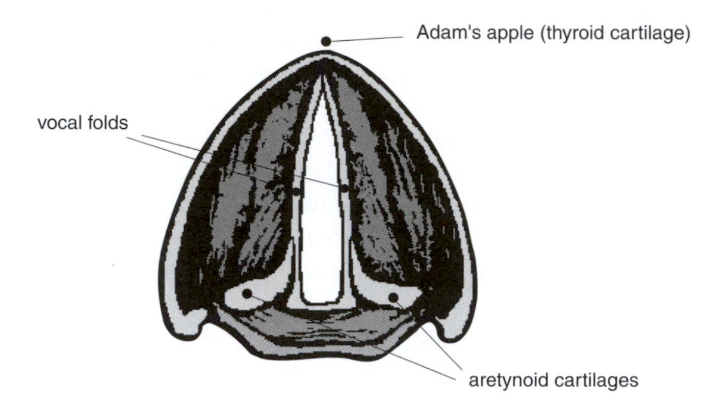

These two folds meet at the front where they are attached to the Adam's apple, a short band of cartilage which forms the front wall of the voice box (at the top of the diagram). The back ends are attached to two extremely mobile pieces of cartilage (called the **aretynoid cartilages**). The whole assemblage is controlled by an elaborate system of muscles, which can separate or bring together, as well as stretch or loosen, the two folds.

To help you understand how the vocal folds produce their sound, the simplest picture to hold in your mind is a rubber balloon. If you hold the neck nearly closed between your thumb and forefinger of each hand, so that the air comes out slowly, you can produce a satisfying squealing sound. As you pull your hands apart you stretch the neck and the pitch of the sound goes up: as you bring them together, you loosen it, and the pitch goes down. That, crudely speaking, is what happens in the voice box. Air from the lungs makes the vocal folds vibrate, just like the rubber of the balloon, and the pitch is controlled by the tension in the various muscles. (An actual description of the forces involved gets a bit hairy, but in essence it just involves a dynamic balance between the forces of the air trying to push the folds apart, and the elastic force of the tissues tending to pull them back together. We've met enough of those examples all through this book to know that under these conditions a vibration will result, and its frequency will be determined by the mechanical properties of the tissues.)

Your first temptation therefore would be to think that the voice, as a musical instrument, is just like, say, a clarinet or a trumpet. In both of those you have, at the bottom of a long column of air, something which vibrates under the influence of an air-stream—a reed or the player's lips. But there is one critical difference. In those instruments, you will remember that there is a feedback mechanism which controls the pitch of the note. The air column can only vibrate easily at one of its natural frequencies; and the fluctuating air pressure, associated with the standing wave that is set up, forces the reed or the lips to vibrate in sympathy.

That couldn't possibly be true for the voice: the wavelengths are all wrong. Think of a bass singing a very low note—say one of those C_2's that Ockeghem could obviously get down to. That frequency is 65 Hz, so its wavelength is over 5 m! Now for a column of air to have that as one of its natural modes, its length must be at least a quarter of 5 m (refer back to Chapter 6, page 186 if you don't remember where that result comes from). But there aren't any columns of air inside the human body anything like a metre long. Therefore there cannot be any aerodynamic feedback to keep the vocal folds vibrating at the right frequency. They must be acting like 'free reeds', and in that sense the vocal mechanism is more like a mouth-organ or a concertina than a clarinet.

Let us go back therefore and think about the vibrations of the vocal folds in more detail. As with the balloon, pulling on the folds makes them tighter and thinner, hence—by an extension of Mersenne's laws—the frequency goes up. But it also makes them longer, which tends to lower the frequency. So the effect isn't as strong as you might expect, and you can't produce a wide range of frequencies with such a simple mechanism. The actual system is considerably more complicated. In order to get higher notes, the vibrating length of the folds has to be made shorter *as well as* tighter. In simple diagrammatic form this is how it is achieved:

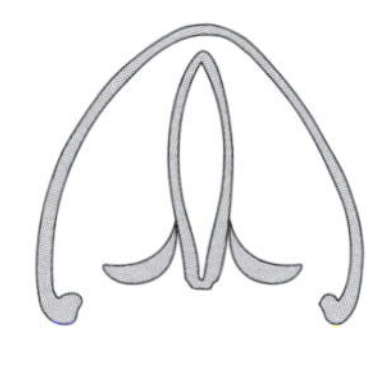
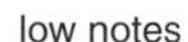
low notes

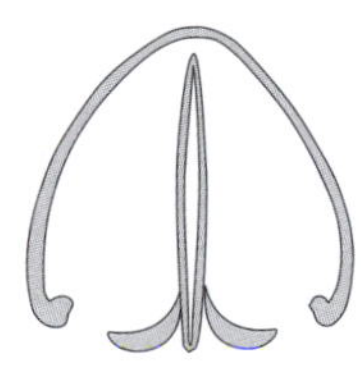
middle notes

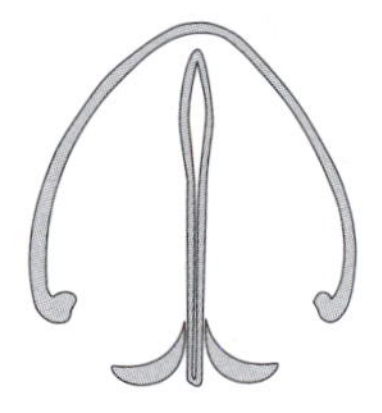
high notes

Obviously those diagrams are only an approximation to what really happens, and hide the complexity of the muscles which produce all these different configurations. Nevertheless they do explain a lot of what you know about your own voice. If you place your fingers against your Adam's apple while you sing you can feel, not only the vibrations, but also the muscular movement. It also explains the difference between men's and women's voices. In a man the whole voice box is bigger, the Adam's apple is more prominent. The folds are thicker and heavier, and therefore their sound is lower, often by as much as an octave.

There is, however, a difficulty concerning the different voice 'registers'. The simplest idea, presumably behind the thoughts of those who gave them that name in the first place, is that they are like the registers of a woodwind instrument—different internal modes of vibration of the air column which lead to different ranges of notes and timbres. But we have seen that the voice isn't like a woodwind in this respect. So when the pitch goes up with a change to *falsetto* that change can only be caused by something in the vocal folds themselves.

What happens is that they start vibrating in another mode (and recall that stretched surfaces can vibrate in different modes, just like stretched strings). The difference in motion is easiest to visualize from these two diagrams of their up and down motion, looking *edge-on* at one of the folds:

This explains a lot about the falsetto voice. As the two folds flap backwards and forwards against one another, because the motion of each is so complicated, they never completely close off the windpipe, whereas in the normal singing mode they do. This means that there is less variation between the highest and the lowest pressures of the vibration than in the normal mode. The sound will be softer and poorer in high harmonics. You will recall (see page 198 in Interlude 5) that much of the difference between the sound from the air reed of a flute, and from the cane reed of an oboe arises from a similar effect.

You can understand too why the voice of an adolescent boy behaves as it does while his voice is breaking. He has to learn again how to work the new instrument which has appeared so suddenly in his throat—to work out how instinctively to adjust the different muscles to achieve the pitch he wants. It's just like a young violinist learning to bow. For any given bow pressure there are a number of vibration modes the string can respond with, so, until she learns exactly what to do to select one mode rather than the others, the resulting sound can jump between them, resulting in all sorts of squeaks and squeals. So too with the boy. Until he knows what to do with his voice, it can unexpectedly, and embarrassingly, shift from the normal register (the chest voice) up an octave into falsetto and back again.

There is still the question of whether the voice really does have any more 'registers' than these two. Many singers claim they can detect another point at which their voice changes noticeably in quality. I confess I don't know whether there is a physical basis for this or not. They could, I suppose, be detecting the point at which the new set of muscles start to shorten the vocal folds while still increasing the tension (as symbolized in the diagram on the page 327); but, to the best of my knowledge, there is no general agreement on this.

Anyhow, we now have a fair understanding of the voice box as the generator of a sung note, and the next question to be addressed is: if the air column in the throat and mouth doesn't determine the pitch of the note, what exactly does it do?

The air column

The channel between the voice box and the lips is called the **vocal tract**. It is a slim tube of air, made up of the pharynx and the oral cavity (the mouth). Its cross-section is almost infinitely variable, being controlled by the angle of opening of the jaw, the position of the tongue, the degree of closure of the lips, the position of the soft palate and the tightness of the throat. An average value for the length of this tube is about 14 cm in adult women, and about 17 cm in men.

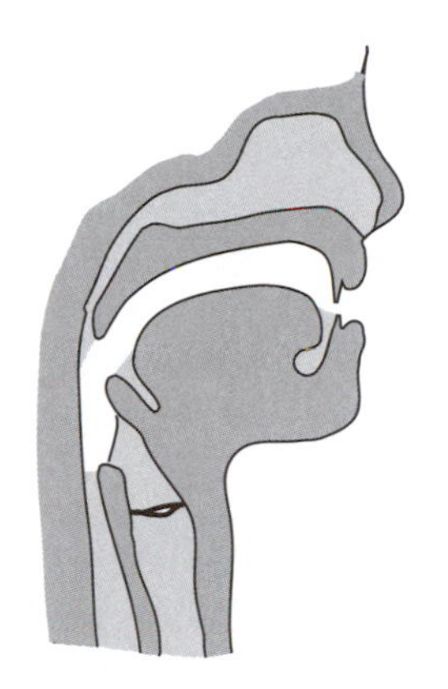

If its cross-section were reasonably uniform, then we would expect it to behave acoustically like a cylinder closed at one end. Taking a representative length to be 15.5 cm (halfway between women and men) we would predict it should have natural frequencies at about 550 Hz, 1650 Hz, 2750 Hz and so on (again refer to Chapter 6). Now of course, the vocal tract isn't a perfect cylinder; but we know enough about the behaviour of air columns to be pretty sure that, whatever its actual shape, provided its length is much greater than its diameter, it will still have a set of resonances not too far away from those frequencies.

So now we can see what is the function of the vocal tract. It acts in some ways like the box of a violin, in which the enclosed air vibrates in sympathy with the strings. The strings' vibrations have a complex harmonic structure; but the box has its own set of resonances, and that determines which of the frequencies—fundamentals or overtones—will couple efficiently to the outside air.

Likewise the vocal folds also produce signals with a rich harmonic structure. As they open and close they let through puffs of air and the flow rate changes periodically. Your first guess might be that it would vary more or less as the area of the opening between the folds. But air, or indeed any fluid, has the property that it flows much more easily through a wide opening than a narrow one. Therefore the difference between the highest and lowest flow rate is accentuated. The resulting variation can be represented schematically in the following diagram.

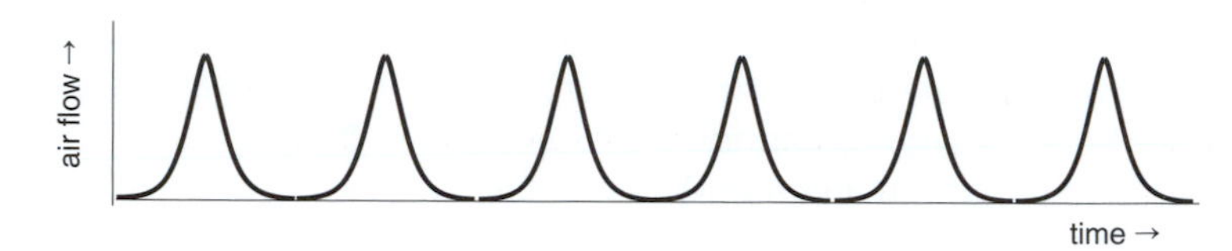

You will recall, from our discussion of the Fourier theorem in Chapter 4, that the sharpness of this repetitive pattern means that, when you analyse it, you will find it has a lot of high frequency Fourier components. And that means that when you hear the sound, you will hear a lot of high harmonics. I could indicate this by drawing an imaginary frequency spectrum for the sound as it leaves the vocal folds.

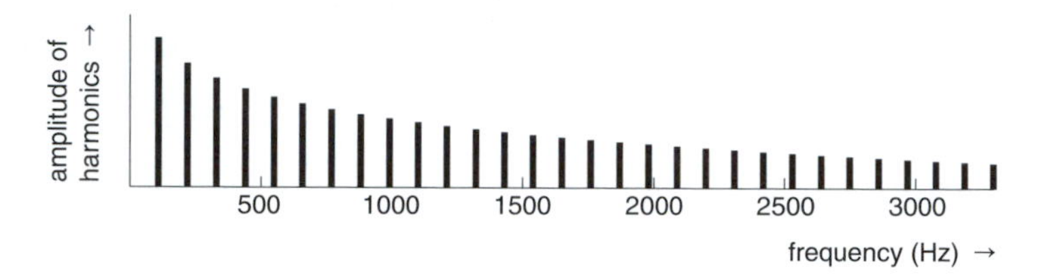

These frequency components must be communicated to the outside world through the vocal tract. Those which happen to match the natural frequencies of the tract will come through strongly; those which do not, will be weaker. If you want to think of it like in a different way, the signal will be filtered according to the frequency response of the vocal tract, which, for a more or less cylindrical column of air, might look like this:

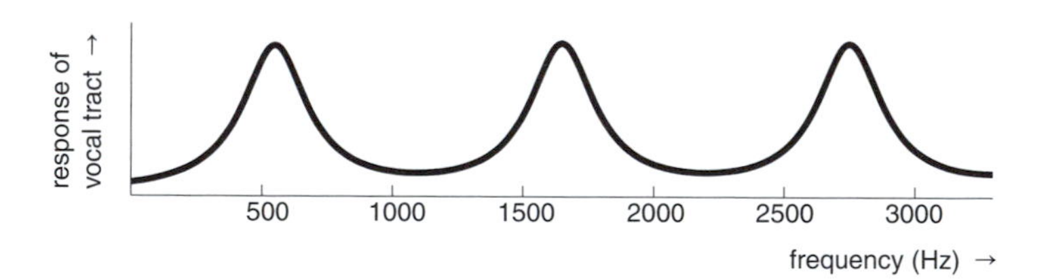

Therefore what emerges after passing through such a 'filter' will have this kind of frequency spectrum:

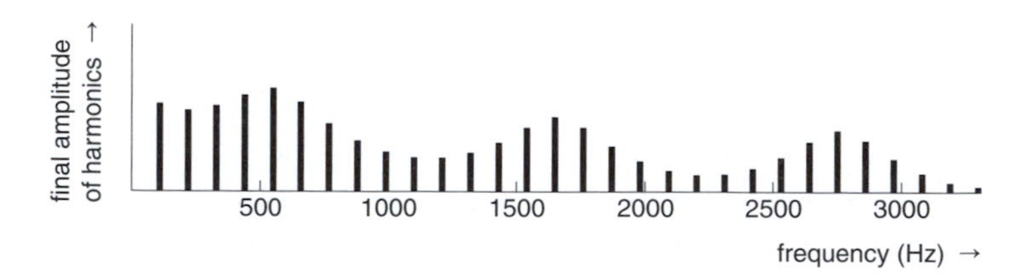

From the point of view of music, this looks like a very strange timbre indeed. And so it is. When you listen to it, it will sound, not just like an ordinary musical tone, but like a sung vowel sound. The peaks in the frequency spectrum are exactly what the early workers in the science of phonetics had identified as the formants of articulated speech.

Speech formants

It is worthwhile at this point saying a little more about the results of those experiments in phonetics, just to check that we have correctly recognized the role that the vocal tract plays. I said earlier that the findings were that each vowel sound was characterized by a different pair of peaks, or **formants**, in their frequency spectra. Actually there are more than this number for each vowel, but I will just concentrate on the first two. A simplified summary of the results of many different measurements done over the years is contained in the following graph.

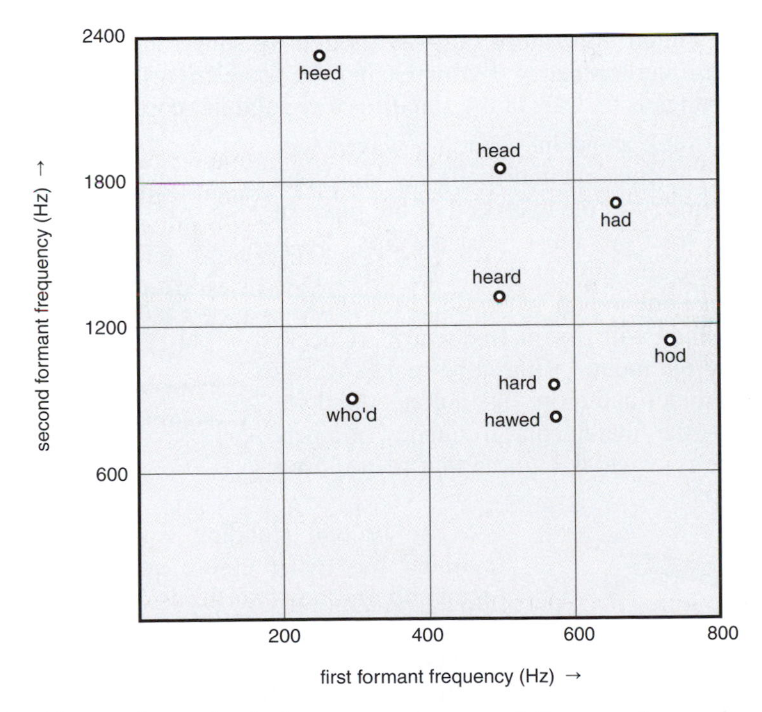

In this graph, each circle represents one of the vowel sounds, as in the words *heed*, *hard*, *who'd* etc. The frequency of the first formant is measured along the horizontal axis, while that of the second is along the vertical axis. So, for example, you can see from this graph that the formant frequencies of the vowel sounds *ee*, *aa* and *oo* are as in this next table.

	ee	**aa**	**oo**
First formant frequency	290	570	340
Second formant frequency	2340	1100	910

These figures are only rough, of course. The real values for women will be a little higher, and for men a little lower—there is usually about 10% to 20% difference between them.

The question then is: how does the vocal tract manage to produce all these different resonant frequencies? Well, each resonance of the air column corresponds to a standing wave, and certain points in the standing waves are particularly sensitive to a change in the width of the tube. This is especially true of the *pressure nodes* where the back and forward movement of the air is greatest. A change here will affect the whole pattern and alter the resonant frequency. In general, a contraction at these points tends to stretch the pattern out—as though by squeezing it—and hence to decrease the frequency (by increasing the wavelength). Conversely, an expansion tends to draw in the standing wave and increase the frequency.

Now think about the standing waves separately. The fundamental mode has only one pressure node—at the open end of the pipe. In the vocal tract this must be at the lips. So if you expand the tube at this point—that is if you open your mouth wider—the fundamental resonance will rise in frequency. (Check: widening the mouth of the tube makes it less like a cylinder and more like a cone—and we know that the fundamental frequency of a perfect cone is twice as high as that of a closed cylinder.)

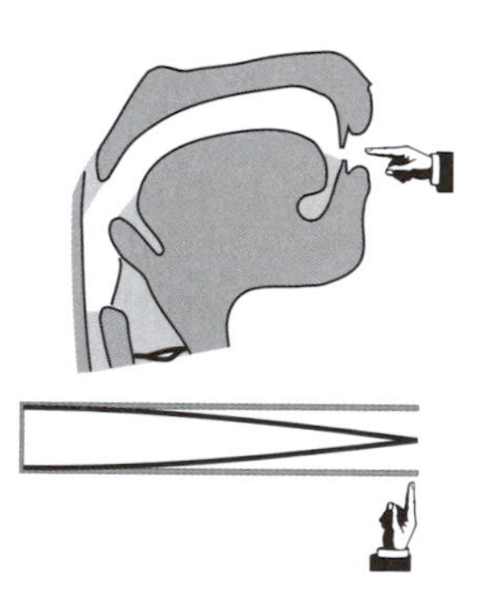

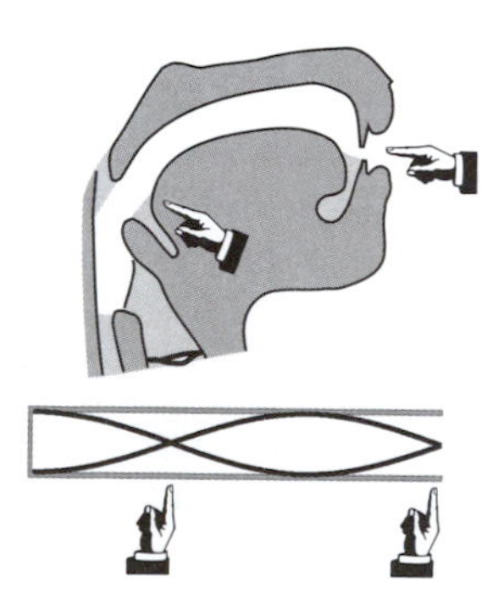

The second standing wave of a closed cylinder has two pressure nodes, one at the open end and one two thirds of the way back. So if the tube is widened at either of these points, this resonance will rise in frequency. In the vocal tract these points are at the lips (again) and at the back of the tongue. The third and higher resonances will also be sensitive to other locations in the mouth, but there isn't much point in going into more detail. We already have enough understanding to make the simple observations that the mouth opening affects the first formant most, and that the back of the tongue has most effect on the second.

Think about the three vowels *ee*, *aa* and *oo*. For *aa* you open your mouth wide, its first formant is high: for the other two you bring your lips together, their first formants are a lot lower. On the other hand, when you say *ee* the whole front part of your tongue is very high in your mouth, and the back is relatively open, so the second formant is very high: whereas it is much lower for the other two, because the main tract is more uniform. X-ray photographs bear out these observations, and the following drawings are schematic representations of what they show:

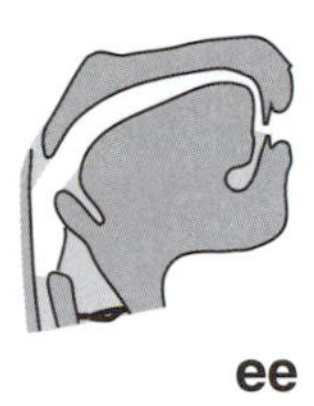

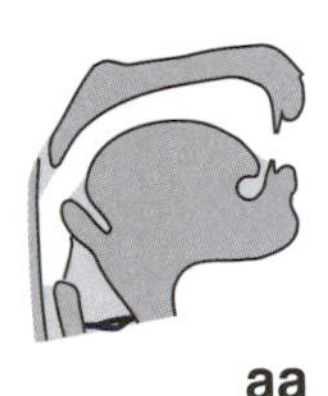

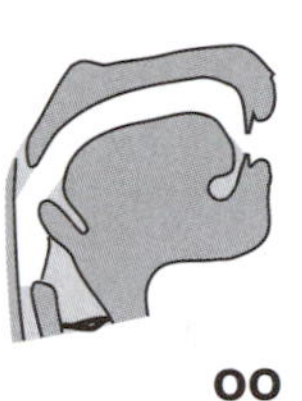

What I've been saying applies only to vowel sounds of course. Consonants are different. They are transient noises made in a number of quite different ways. But even with them some of the same considerations apply. For example, think about the three so-called 'plosive' consonants *p*, *t* and *k*. The first is made by closing your lips and suddenly forcing them apart with a puff of air. The sound it gives is an incoherent mixture of low frequencies—the same kind of mixture you might get by tapping a cardboard cylinder about 15 cm long. The second (*t*) is made in the same way, only here the constriction is made by the tip of your tongue against the roof of your mouth. It's much the same sort of sound, except that the frequencies are a bit higher, as you'd expect because the tube is that much shorter. And the third (*k*) is higher again, because the constriction here is made by the back of your tongue.

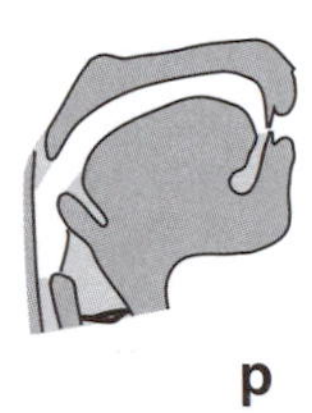

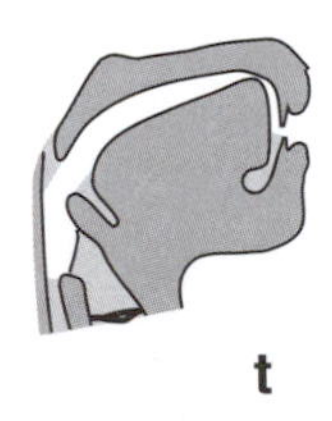

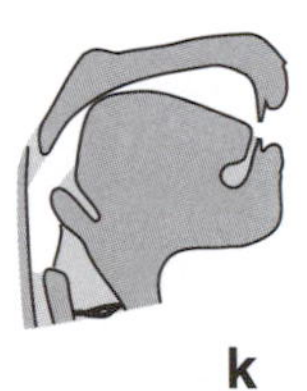

These then are some of the basic acoustic features that come into the sounds of speech. The question that is of most interest to us is: what is the effect of all this on singing?

Singing and articulation

The basic difference between singing and speaking is in the mixture of frequencies present in the sound. Speech leaves the voice box as an incoherent collection of all sorts of frequencies (i.e. noise). When these are filtered by the resonances of the vocal tract, the resulting signal will look just like its frequency response curve, because there are plenty of frequency components to 'fill out' all the formants. Therefore your ear will have no difficulty in recognizing them. But with singing you only have a special set of frequency components in any one note—a fundamental and its harmonics—and, especially if the pitch is high, these can be quite widely separated.

In the diagram I drew on page 330 the first formant was just over 500 Hz, and the fundamental of the note was 110 Hz. It could have been a bass singing A_2. But if I were to draw a similar diagram for higher notes, keeping exactly the same articulation so that the formants are at the same frequencies, it would look different. Here, for example, is the same vowel sound at 220 Hz (a tenor singing A_3 perhaps):

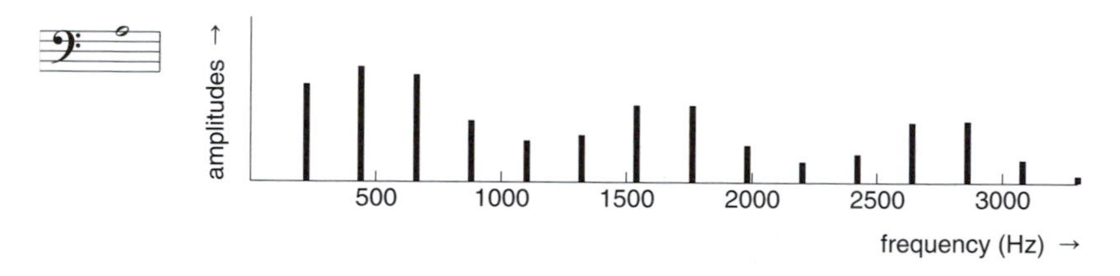

440 Hz (alto A_4):

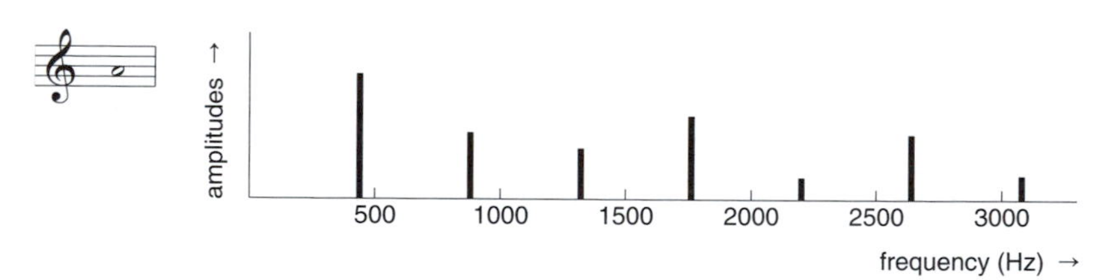

and 880 Hz (soprano A_5):

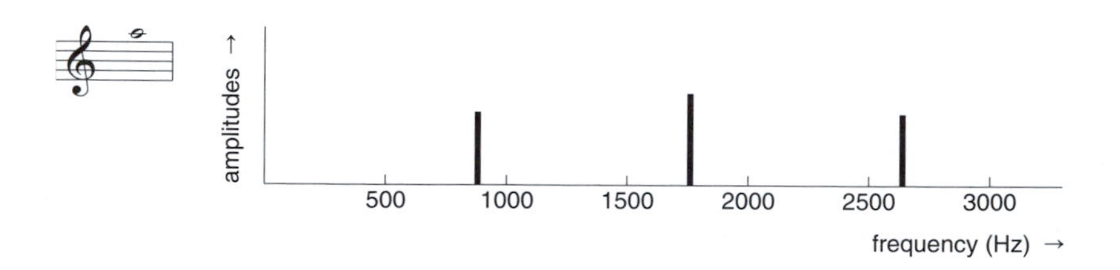

Actually, had I really wanted to represent the various singing voices more accurately, I would have to increase the frequencies of the formants a bit. Since women's vocal tracts are on average about 20% shorter than men's, their formants, for the same articulation, should be about 20% higher. And remember that an increase of 20% corresponds to a musical interval of a major third. So a tenor and an alto both singing E_4 (330 Hz) might be represented like this,

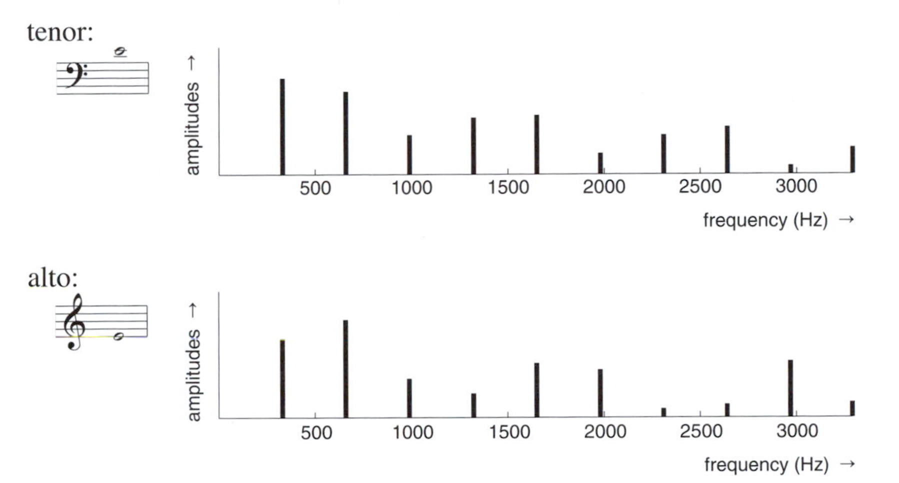

That would explain the very common observation that when you hear altos and tenors singing together (as they often do in choirs, because tenors are often hard to find) the men seem to be higher than the women, even though you know they're on the same note. Our ears have learned to compensate for the differences in formant frequencies between men and women, and there is a strong tendency to measure pitch subjectively against the formants.

However that is only a small effect. The main point of the diagrams on the previous page was to show that, on very high notes, the fundamental can *miss* the first formant altogether—by considerably more than 20%. This will have two very serious consequences. Firstly the listener's ear won't be able to recognize the first formant, so it won't know which particular vowel is being sung. Secondly, since the fundamental frequency, which usually carries most of the energy, is coupled to the outside air with such a low amplitude, the note will sound very soft even with a considerable amount of effort on the part of the singer.

Therefore whenever a soprano sings a high note, she must try to raise the pitch of the first formant; and the easiest way to do this is by widening the vocal tract at the pressure node. That is why the stereotype of the operatic soprano, popular with cartoonists, is a large lady with her mouth wide open. That is also why choir conductors spend so much time (with depressingly little result) telling their members to open their mouths when they sing. It's the only way to get volume on the high notes.

But there's a price. Changing the frequencies of the formants distorts the vowel sounds. The instinctive tendency, when singing with an unnaturally wide mouth, is to keep the back of the throat the same, to preserve the upper formants. The result is that pretty soon, for example, the sound *oo* becomes indistinguishable from *aa*. And as the note gets even higher, eventually all vowels sound much the same.

Most composers recognize this of course. I've said many times that they have a thorough knowledge of this kind of acoustics; and you'll notice that a good operatic composer never gives an important line of text to a high soprano, unless it's repeated at a lower pitch. Of course careful articulation of the consonants helps a lot. Being an incoherent collection of frequencies, they are much less influenced by the basic pitch of a note, so even if the listener can't understand a single vowel, it's often possible to recognize the word from the consonants alone. That's why articulation of consonants is so important, especially for choirs, where the members will be singing their vowels with different degrees of distortion at any one time.

The singing formant

I still haven't said anything about 'projection'. Choir conductors, those treasure-chests of traditional wisdom, continually tell their members to "throw their voices out", or to "sing to the back of the hall". In many cases what they're saying is quite straightforward. There are probably many choirs like the one I sing in, in which most of the time the singers are looking down at their music, desperately trying to read what notes come next. And because sound from the mouth is reasonably directional, it will tend to get absorbed in the sheet music, if that's what it's pointed at. So if you look up, the sound will go more directly forward. But there is another effect.

Human beings have a reflex throat action which allows them make a very loud sound on occasions. You realize it when somebody yells "Look out!" when they see an accident occurring. That sound has a very distinctive timbre, and will cut through a remarkably high level of background noise to be heard by whomever it was intended for. It's pretty obvious why such a property of the voice should have evolved—in a Darwinian sense it's a valuable survival characteristic. It's also clear why it evolved as a reflex—you wouldn't want to talk like that all the time. That's the mechanism that the singers of the early 19th century learned to bring under conscious control.

Only quite recently however, have people come to understand what gives this sound its carrying quality. Experiments have been done to measure the average kind of frequency spectrum of typical sounds from orchestras and singers. They vary a lot of course, but, as a general rule, the overtone frequencies of most instruments and untrained singers decrease fairly steadily in amplitude as the frequency increases above about 2000 Hz. Except trained opera singers. Their voices have a very pronounced peak somewhere between 2000 and 3000 Hz. This gives their sound a very distinctive quality which the ear can fix upon (and remember from Chapter 7 that the ear is particularly sensitive around these frequencies); and therefore it allows them to be heard even in competition with a full symphony orchestra. This peak, which is of course just another formant, is so typical that it has come to be considered one of the hallmarks of a highly trained voice. It is known as the **singing formant**.

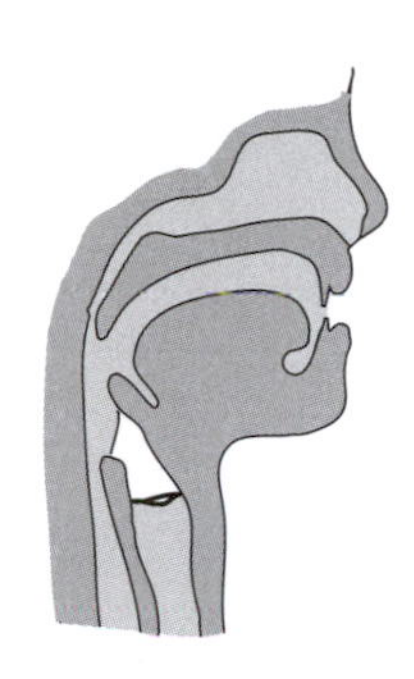

The physical reason for its existence is bound up with the way the vocal tract is put together. The air column whose properties we have been talking about, extends to the bottom of the pharynx where the tube going down into the digestive tract narrows. However at the top of the larynx, above the vocal folds, there is a very small cylindrical volume. At the junction between this and the pharynx proper there is a very severe impedance mismatch, and normally it takes no part in setting up any kind of standing waves. In this respect it is not unlike the mouthpiece of a brass instrument.

The length of this small volume is usually less than a couple of centimetres, so if it could establish its own standing wave, it would have a frequency over 4000 Hz. But trained opera singers have learned how to lengthen it by consciously relaxing the muscles in their throat, and dropping their voice box slightly, and bringing its fundamental frequency down to between 2000 and 3000 Hz. It is the resonance of this tube that produces the singing formant. And, again, that is why choir conductors always give their members those 'jaw relaxing' exercises that everybody hates.

Looked at as a piece of physics, it is a remarkably efficient way to increase loudness. It simply relies on resonance to increase the coupling of the vibrational energy into a sound wave; so it doesn't cost the singer any more effort. It does however distort vowel sounds even further. Its frequency is round about where the third and fourth formants of most of the vowels are normally located. So by the time a wide open mouth has shifted the first formant, and a lowered larynx has overridden the third and fourth, the vowels are essentially distorted beyond recognition.

The early 20th century marked the parting of the ways between operatic and popular singing styles. Human beings are very adaptable, and obviously 19th century audiences learned to cope with this kind of singing. They knew that loud singing had to sound like that. But with the invention of the microphone they could turn *down* the volume; and then it became glaringly obvious just how much distortion there was. So popular music went another way.

Loose ends

The voice is an enormously ancient instrument, and over the eons there have grown up a multitude of mythologies and old wives' tales, mixed in with good sound observations. I cannot in this book cover more than a tiny fraction of all that might be talked about.

For example, there is the question of whether the other cavities in the head (for example the nasal cavity) resonate and contribute to the sound. Research in this area is still going on. Some singers open their soft palates and establish a connection between the vocal tract and the nasal cavity. Others don't. The two configurations certainly add different timbres to the sound—the difference you get by 'talking through your nose' (which actually means closing the connection and *not* letting the sound come through your nose). But it doesn't seem to affect the formants or the general loudness. However I don't want to be too dogmatic about that; further research could quite possibly prove me wrong.

There is also the business about 'placing your voice'. You often hear of singing teachers telling their students very strange things like "to sing through the top of their heads". And what they mean is that this is the way to achieve a pure 'head tone', but in physical terms this is nonsense. Sound cannot possibly come out of the top of your head, no matter which register you're singing in.

There is an amusing story told by the well-known film actor, Peter Ustinov, in his autobiography, *Dear Me*. He was preparing to play the emperor Nero in the 1951 movie, *Quo Vadis*. Since he was expected to sing as well as fiddle while Rome burned, the studio paid for singing lessons with a professor of music at the Rome Opera House. But it was all very rushed and there wasn't enough time to do any actual singing. Instead the professor simply gave him three maxims: "Breathe with the forehead, think with the diaphragm and sing with the eye".

Now I really don't know what the professor was trying to say, but it must be remembered that singing teachers have a very difficult job to do. They have to communicate to their students things which are entirely internal—subtle subjective impressions. So when they talk about 'placing the voice' in a certain way, they could well be drawing attention to some very real response in that part of the body—a resonance in some air cavity for instance. I don't know.

But then life would be very barren if there were nothing left to discover, wouldn't it?

Epilogue

Well, I've come to the end of this book and it's time to look back on where I've got to. I said right at the beginning that I find the connection between physics and music a source of endless fascination; and that's what I've tried to communicate to you. If you're primarily interested in music, I don't know if I've convinced you that it could be worthwhile knowing a little more physics. But if science is your main concern, I hope you can see why I believe that physics needs music. Today, more and more it's coming to be realized that science is a human activity, and scientists must understand what it is they are doing and why they do it. May I give you one last example—the theory of sub-nuclear matter.

At the end of the 1920s, when Schrödinger and Heisenberg had developed the quantum theory, it was believed that all matter consisted of atoms, and each atom consisted of a number of electrons orbiting round an electrically charged nucleus. The nucleus was made up of smaller entities—positive particles (or **protons**) which were responsible for the electric charge, and neutral ones (or **neutrons**) which gave the extra mass. So, when you remember that light also consists of particles, photons, there were just four different 'building blocks' from which the whole universe was constructed. It was a beautifully elegant picture of the world.

But as time went on, and more sophisticated experiments were done—with cosmic radiation from outer space, and with enormous particle accelerators—other 'building blocks' were found. They were given names like: **muons**, **pions**, **neutrinos**, **hadrons**. By about 1960, some hundred were known. But then it was noticed that all these particles occur in groupings with similar properties—'families': and there is a recognizable pattern in how they're arranged within these families. This pattern was interpreted as suggesting that the particles themselves are made up of a small number of much more fundamental things—which were given the name **quarks** (from an obscure line in James Joyce's *Finnegan's Wake*).

Once again, a beautiful simplicity had returned to our view of the universe. And I think the parallel with the ancient Greek idea of the cosmos (that word means 'order': as opposed to chaos, or 'disorder') is very striking. In that case, a theory of the way nature worked was put forward on the strength of the recognition of a common pattern—between the arrangement of planets in the heavens and the notes in a musical scale. If the patterns were the same, then so might be the underlying causes.

When I wrote the first edition of this book, in 1989, that was more or less where the theory of sub-nuclear matter stood. Most physicists believed that quark theory was probably valid, even though, despite long and expensive experiments, no one has ever found a quark. But physics changes very quickly sometimes. It had long been a holy grail of physicists to find a 'grand unified theory' which would explain the very large (the universe, controlled by gravitation) as well as the very small (sub-nuclear particles, controlled by other forces). Einstein spent the last forty years of his life looking for it, without success. Only later did it come to be realized that (for reasons which are far too complicated to go into here) the two could be described by the same mathematical model if the basic entities of nature were not infinitely small points, but rather very small one-dimensional objects, which came to be called **strings**.

Today—that is in 2001—a commonly accepted view of the fundamental nature of things is described by **superstring theory**. It says that, if you could look deep inside an electron or a quark or whatever, you'd find a tiny little loop of energy, rather like a loop of string. And, just as the strings in a violin vibrate in different modes, sounding different notes, so these loops of energy can vibrate in different patterns, producing different particle properties. In other words, an electron is simply a string vibrating one way, and a quark is a string vibrating differently.

Many researchers writing about this field have drawn attention to the fact that all this is closely parallel to the ancient doctrine of the music of the spheres. Certainly in superstring theory, musical metaphors take on an amazing sense of reality. The basic building blocks of the universe are vibrating strings, and the laws of nature are an expression of the harmony binding these together.

All of this may turn out to be nonsense of course. Superstring theory may have some fundamental flaw, yet to be discovered. It doesn't matter. It's the method that counts. We believe that order exists, and we look for it. In that respect the aims of science and of music are identical—the desire to find harmony. And surely, without that very human desire, science would be a cold and sterile undertaking. As Shakespeare said:

> "The man who hath not music in himself,
> Nor is not moved by concord of sweet sounds,
> Is fit for treasons, stratagems and spoils;
> The motions of his spirit are dull as night,
> And his affections dark as Erebus:
> Let no such man be trusted."

Appendix 1

Musical notation

A convention for writing music down so that it can be performed by others is a surprisingly recent invention. Even today, although music exists in all cultures throughout the world, many of them don't bother to put it on paper (or whatever is the equivalent). The system that is used in Western music settled down into its present form some 500 years ago. It is essentially a system of pictographs—small symbols, each representing one idea; and if you want to read the music you've simply got to learn what each pictograph means. I will restrict this brief survey to the minimum number required to enable you to interpret the musical examples I have used throughout this book, in case you can't read them already.

Specification of the duration of notes

The rhythmic nature of music makes it possible to define the basic unit of duration as a **beat**. The most commonly used symbols for different numbers of beats appear in the following table; together with the symbols for periods of silence (or **rests**) of corresponding lengths of time. The basic unit is called the *crotchet*—which originally meant 'crooked' because it used to be drawn with a crook attached to its stem.

In those days, the beat was a bit faster than we take it today, and the next note up, the minim, was considered the most sensible counting unit. But in the 15th century there was a general tendency to slow down, and that was when they introduced the system of drawing the shorter ones 'filled', to distinguish them from the longer notes which were 'void'.

DURATION	NOTE	REST	NAME (English)	NAME (American)
4 beats	𝅝	𝄻	semibreve	whole note
2 beats	𝅗𝅥	𝄼	minim	half note
1 beat	♩	𝄽	crotchet	quarter note
1/2 beat	♪	𝄾	quaver	eighth note
1/4 beat	𝅘𝅥𝅯	𝄿	semiquaver	sixteenth note
1/8 beat	𝅘𝅥𝅰	𝅀	demisemiquaver	1/32 note

The English names appearing in this table are all based upon the old Latin words in use during the early Middle Ages. The American names, on the other hand, are a translation of those used in Germany. They are undoubtedly more logical and easier to remember; and are slowly coming into common use throughout the world today.

Notes whose durations are composites of the above are designated either by connecting those symbols with a **tie**, or by using a **dot** to signify that the length is increased by 50%. Thus both of these examples denote a note whose duration is one and a half beats.

The *absolute* measurement of duration is seldom of great importance in music. Performers are usually allowed wide latitude; and (not very precise) instructions are conveyed by the words 'fast', 'slow', etc (usually in Italian). Where the composer wants a little more definiteness, a symbol like this is written above the start of the first line of the music. This particular example is to be interpreted as saying that (approximately) 90 crotchets take up 1 minute.

♩= 90

Specification of the pitch of notes

The system in most common use in Western music today can be thought of as being built around the white keys on an ordinary piano keyboard. Each of these keys is named by a letter, thus:

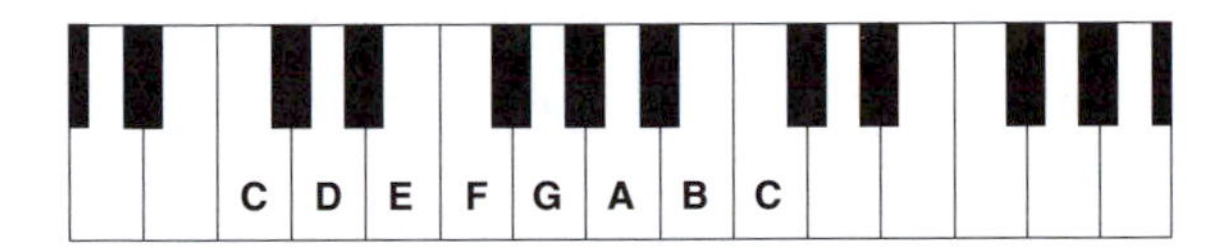

The sequence of letters repeats itself, as does the pattern of keys (at least five times on the normal keyboard). The particular C key which is closest to the middle of the keyboard is called **middle C**.

Which actual notes are to be played or sung are denoted by drawing the pictographs representing the length of the notes, at specified vertical positions on a set of five horizontal lines, called a **stave**. The 'specified vertical positions' are either *on* a line, or *between* two lines. If the range of possible positions represented by the stave is not great enough, it can be extended by drawing short extra lines—known as **leger lines**.

Then the eight white notes starting on middle C—that is to say, one complete octave of a diatonic scale in the key of C—is represented thus:

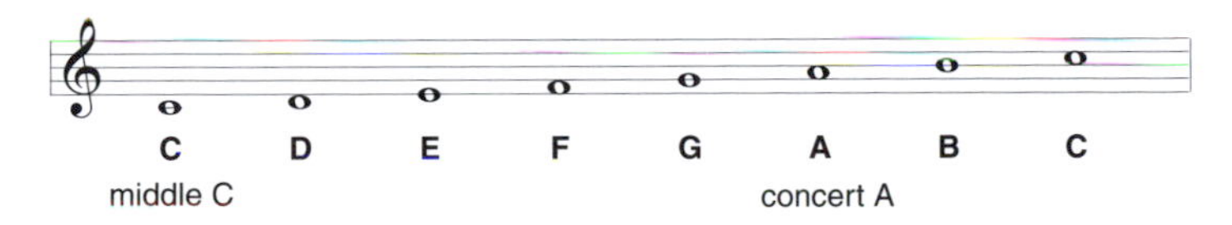

The strange symbol on the far left of the stave is known as a **treble clef**. It is actually a stylized script G and was originally intended to curl around the line representing that note. However today it is best merely to think of it as being a roundabout way of indicating the location of middle C (on the first leger line below the stave).

The notes of the diatonic scale one octave down from the last diagram are represented like this

The **bass clef** used here can likewise be interpreted as specifying the position of middle C to be on the first leger line *above* the stave. You will occasionally see other clefs, denoting other positions of middle C, but these two are by far the most commonly used.

The black notes on the keyboard are one semitone up (or down) from their neighbouring white notes. They are represented in printed music by drawing, along side the pictograph representing the note, a **sharp sign** ♯ (to denote an increase in pitch), or a **flat sign** ♭ (to denote a decrease). Thus the diatonic scale starting on concert A is represented as

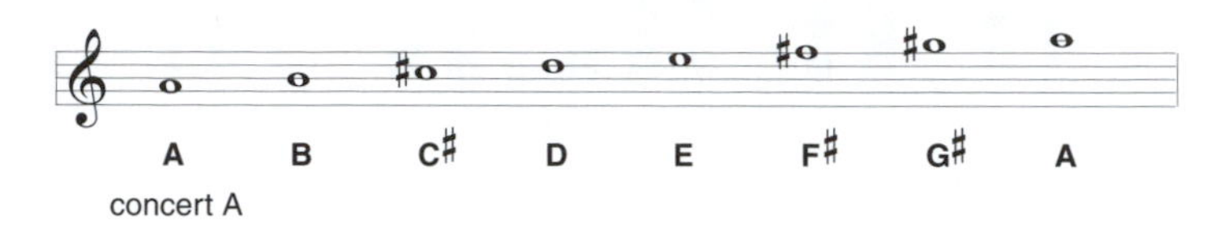

In the Western tradition of tonality, a piece of music is constructed (more or less) from the notes of one particular diatonic scale, though not necessarily the one which starts on C. However, music in another key would need many sharp or flat signs to be written throughout the piece. A convention to get round this is to print these at the *start* of each stave—making up what is called the **key signature**. So for example, the following is to be taken to mean that the music is in the key of A; and *all* F's, G's and C's are to be sharpened unless otherwise specified (by the **natural sign** ♮):

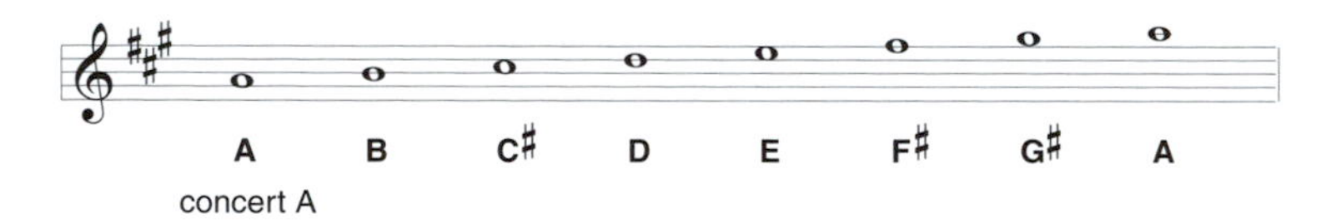

It is important to realize that this whole system of writing notes is not *absolute*, so far as the frequency is concerned. No distinction is made between various tunings—there is no way of telling whether the music should be played in a Pythagorean, or a just, or an equal tempered scale. Nor is there any attempt made to designate the exact frequency of Concert A at the time the piece was written.

Indeed there is sometimes an ambiguity about which *octave* the notes belong to. For example, in choral writing, the tenor line is often written on a stave with a treble clef, even though the notes are to be sung an octave lower (though on careful manuscripts the clef may have a small figure 8 underneath it to remove this ambiguity).

Matters are even more bizarre on the orchestral scores of 'transposing' instruments. Here a key signature will be used which is completely different from that which the instrument actually plays. The reason for this lies in the history of those instruments, and usually means that at one time they normally played in a different key. It really only tells the performer what to do to the instrument (what fingering to use or what keys to press) rather than what pitch to sound.

Conventions for writing about music

Although many cultures haven't bothered with writing conventions to standardize the performance of music, civilizations which have developed an intellectual life generally wanted to leave a permanent description of what their music was like, and therefore had conventions for writing *about* music. The ancient Greeks had at least two systems, one of which (like many which followed it) used letters of the alphabet to stand for the series of sounds. The Romans did likewise, and the influence of Boethius made sure that all of Western Christendom followed suit.

In scientific and musicological writing today there are (at least) two quite different systems in common use, as given by the table on the following page, which covers the complete range of the piano keyboard.

	American system	German system
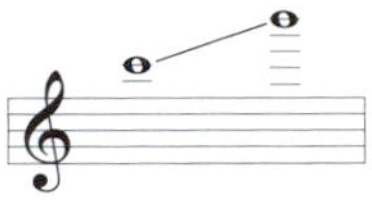	C_6 to B_6	c‴ to b‴
	C_5 to B_5	c″ to b″
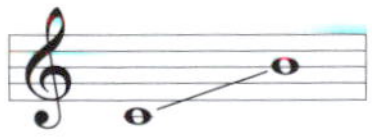	C_4 to B_4	c′ to b′
	C_3 to B_3	c to b
	C_2 to B_2	C to B
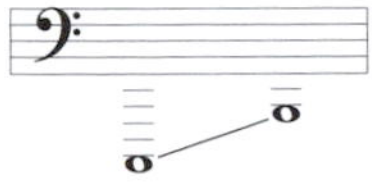	C_1 to B_1	C′ to B′

I think there is no question but that the American system is the more sensible, and that is the one I have used consistently throughout this book.

Appendix 2

Logarithms

A logarithm is a mathematical device which may be understood by thinking about the convention of writing large numbers in terms of powers of 10. For example,

the number 100 is equal to 10×10, and can be written 10^2;
the number 1000 is equal to $10 \times 10 \times 10$, and can be written 10^3.

Consideration of the logic involved in this convention will convince you of the validity of this table:

1 000 000	10^6
100 000	10^5
10 000	10^4
1000	10^3
100	10^2
10	10^1
1	10^0

The table is by no means complete. It is possible, for example, to have *negative* powers if you want—you just get them by continuing to divide by 10.

The main usefulness of this system, from a computational point of view, is that, when you multiply two numbers together, you *add* the corresponding powers. For example,

$$100 \times 10\,000 = 1\,000\,000$$

is equivalent to:

$$10^2 \times 10^4 = 10^6$$

It is simple enough to check others for yourself.

In about the year 1600, the Scottish mathematician, John Napier, introduced the idea of concentrating attention on the number written in the superscript position, and he invented a new word for it—**logarithm**. So instead of writing, for example,

$$10^3 = 1000,$$

the same information was conveyed by writing

$$3 = \log 1000.$$

Then the previous table could be rewritten as

NUMBER	LOGARITHM
1 000 000	6
100 000	5
10 000	4
1000	3
100	2
10	1
1	0

Next came the idea of logarithms (or powers) which were not whole numbers. One simple way to appreciate this concept is with square roots. We know that $\sqrt{10}$ is equal to 3.1623, or in other words,

$$3.1623 \times 3.1623 = 10.$$

Now because it makes sense to write

$$10^{0.5} \times 10^{0.5} = 10$$

then it must follow that,

$$\log 3.1623 = 0.5.$$

Here's a sample table. Don't worry about how they're calculated: that's what Napier spent half his life doing, and even today it's rather messy.

NUMBER	LOGARITHM
1	0
2	.3010
3	.4771
4	.6051
5	.6990
6	.7781
7	.8451
8	.9031
9	.9542
10	1.000

It isn't so long since school students (and others) regularly used logarithms to perform complicated long multiplications and divisions; and books consisting of nothing but columns of numbers and their logarithms—called **log tables** —were laboriously calculated and published. You see, they still have the important property that, when you multiply two numbers together, you add their logarithms. So, for example, you can check for yourself that

$$\log 2 + \log 3 = \log 6,$$

or,

$$.3010 + .4771 = .7781.$$

A mechanical device which did exactly the same job was the **slide rule**. This consisted of two scales which could be lined up along side one another (which performs the function of adding two lengths together). What was special about these scales was this. In an ordinary ruler, for example, the numbers 1, 2, 3 ... are arranged with equal spacing between them, like so:

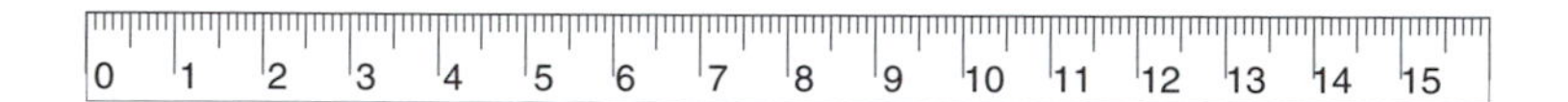

This is what is called a **linear scale**

But it is also possible to arrange the numbers so that their spacing is proportional to their *logarithms*. This kind of scale is called is a **logarithmic scale**.

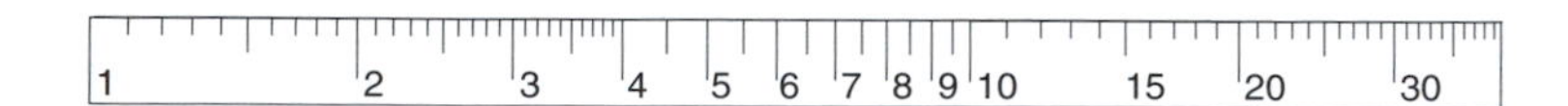

This is what was used in the slide rule, and when a length from one scale was 'added' to the length from the other, this effectively *multiplied* the two numbers on the scales.

Nowadays pocket calculators have made log tables and slide rules obsolete, but the logarithmic scale still has not disappeared; and this is because it is useful for measuring quantities which typically increase multiplicatively rather than additively (particularly psychological ones). The most obvious of course, which I have been talking about throughout this book, is musical pitch.

It is an interesting exercise to draw lines on a sheet of logarithmic graph paper (paper in which the lines are spaced logarithmically rather than linearly), corresponding to the notes which define the lines of the ordinary musical stave. What you can see clearly is that the ordinary stave is in fact a good approximation to the 'natural' arithmetical scale against which you should measure pitch.

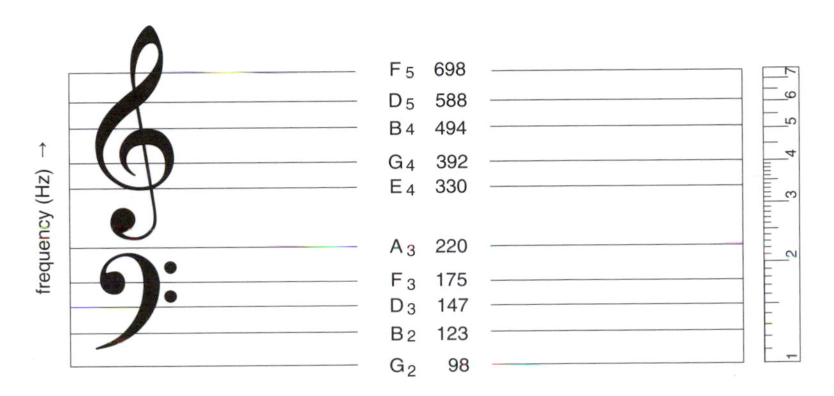

Appendix 3

Measurement of pitch intervals

Any system which attempts to quantify the perceptual entity, pitch—as opposed to the physical quantity, frequency—must be based on the fundamental observation (discussed at length in Chapters 1 and 2) that intervals of pitch which are judged to be equal to one another, have frequencies which are in the same numerical ratios.

The ratios corresponding to the most important pitch intervals in music may be represented on a bar graph like this:

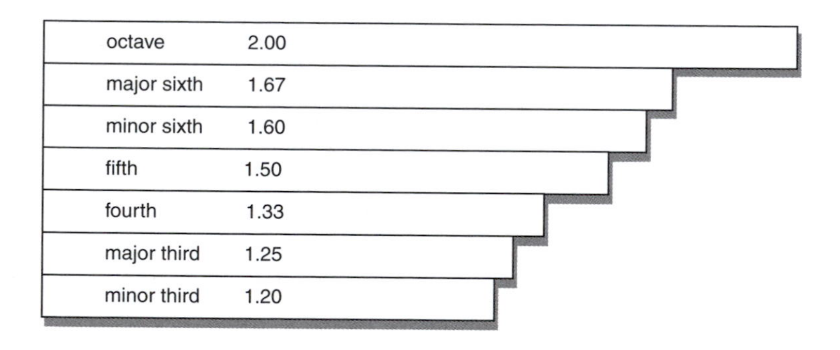

In this diagram a *linear scale* is used to represent the ratios; and clearly that is not a *useful* representation. All your experience suggests that the minor third, for example, is a much smaller interval than the octave. So far as your perception is concerned, the octave is in fact equal to four minor thirds. My diagram, which uses a linear scale, does not make that that clear at all.

Therefore it is much more sensible to use a *logarithmic* scale (see Appendix 2). If I do this, I would be representing the same information by this diagram:

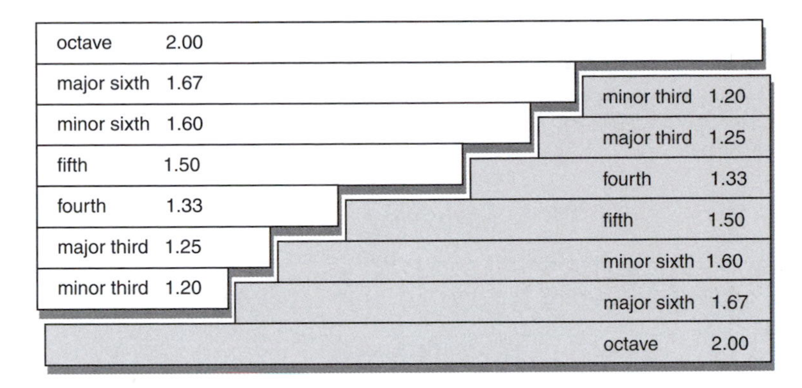

The convenient thing about this diagram is that the 'lengths' associated with the different intervals show clearly the known results of 'adding' certain pitch intervals together:

fourth	+	fifth	=	octave
major third	+	minor sixth	=	octave
minor third	+	major sixth	=	octave
minor third	+	minor third	=	fifth
...etc.				

This means that it is possible to use the 'length' of the ratios along the logarithmic scale (i.e. the logarithm of the numerical ratio) as a direct *measure* of the size of the intervals. But to make such a measurement useful, we need a *unit* of measurement that everyone agrees on. Such a unit was suggested in 1895, by the English musicologist (and translator of the works of Helmholtz), Alexander Ellis. He chose the hundredth part of an equal tempered semitone, and called it the **cent**.

Recall, by referring back to page 71, that the ratio corresponding to the equal tempered semitone is 1.0595 (actually 1.059463 is a more accurate value); and the logarithm of this is .025086, (which you can check with a pocket calculator). Then the cent is a hundredth of this,

$$1 \text{ cent} \quad = \quad .00025086$$

Therefore, if you have two frequencies, f_1 and f_2 (where f_2 is the higher), you can calculate the size of the pitch interval between them by this important formula:

$$\boxed{\text{number of cents} \quad = \quad 3986.3 \times \log(f_2/f_1)}$$

The number 3986.3 appearing here is simply the reciprocal of .025086. Some sample calculations based on this formula, showing the number of cents in various intervals are:

PYTHAGOREAN TUNING

semitone	tone	fourth	fifth	octave
90	204	498	702	1200

JUST TUNING

semitone	minor tone	major tone	minor third	major third
112	182	204	316	386
fourth	fifth	minor sixth	major sixth	octave
498	702	814	884	1200

EQUAL TEMPERED TUNING

semitone	second	minor third	major third	fourth
100	200	300	400	500
fifth	minor sixth	major sixth	seventh	octave
700	800	900	1100	1200

Appendix 4

Measurement of loudness

The loudness of a sound is the psychological perception that our brain ascribes to that property of an acoustic wave, for which the most obvious corresponding physical quantity is the rate at which energy is carried by the wave into our ears.

Any freely vibrating body loses energy as its vibration decays, and the rate at which this energy escapes can be measured in so many joules/second (the **joule** is the unit of energy in the S.I. system). Technically this quantity is called **power** and its unit of measurement is given the name of the **watt** (after guess who). To give you some feel for numbers, here are a few generally accepted measured values for the acoustic power output of some musical instruments.

INSTRUMENT	ACOUSTIC POWER (W)
Clarinet	0.05
Double bass	0.16
Trumpet	0.31
Piano	0.44
Cymbals	9.5
Bass drum	25
Orchestra	67

By way of comparison, light bulbs which give out radiant energy (heat as well as light) at a rate of 100 W are common. But in the case of most acoustic systems, the processes by which this escaping energy is transformed into sound are, in general, remarkably inefficient. A good average figure is that, of the power lost by a freely vibrating mechanical system, at most 1% will end up as acoustic energy.

Acoustic energy travelling through ordinary space spreads out, so that at some distance away, the amplitude of the wave depends not only on the power output of the source, but also the *area* over which it has spread. Any 'measuring' instrument, like an ear for example, will respond only to that energy which crosses the area of its detecting surface. The physical quantity being 'measured' is called the **intensity** of the wave, and its units are watts/square metre. Again typical values, at various distances away from some of the same instruments, are given in this table.

INSTRUMENT	INTENSITY (watts/m^2)	
	at 1 m	at 10 m
Clarinet	0.004	.00004
Trumpet	0.012	.00012
Cymbals	0.75	.0075
Orchestra	5.3	.053

These figures are only average values of course. Sound from any source will not necessarily spread out equally in *all* directions. Indeed a careful series of measurements around these instruments shows surprising variations of intensity in different directions for various frequencies. Such work is obviously important in determining the appropriate seating arrangements for players in an orchestra; but for the present we can ignore this kind of fine detail.

In general, the softest sound that most ears will respond to has an intensity of about 10^{-12} watts/m^2—an almost unbelievably small figure (which just goes to show how sensitive the ear is). As a comparison, a 100 W light bulb emits an intensity of radiant energy at a distance of 10 m of just under 0.1 watts/m^2. The acoustic intensity at the threshold of hearing is therefore a factor of 10^{11} times less than this.

Now we know our ears do not respond *linearly* to acoustic intensity. Instead they seem to compare sounds with one another in terms of the *ratios* of the energy collected (as do most of our perceptual organs). Therefore the sensible scale on which to measure the psychological quantity **loudness** is a logarithmic one; and the minimum discernible intensity provides a convenient reference point from which to start the scale. The earliest agreed system was one in which the loudness, measured in the unit of a **bel**, was simply defined as the logarithm of the ratio between the measured intensity and the minimum discernible level—that is,

$$\text{loudness (in bels)} = \log\left(\frac{\text{intensity}}{10^{-12}\ \text{W/m}^2}\right)$$

The greatest intensity that a sound can reach before it becomes physically painful, and therefore cannot really be considered a 'sound' any more, is about 1 W/m^2—which corresponds to a loudness of 12 bels. In order to have a scale that wasn't limited to 12 levels only, it was decided that in general it was sensible to use as the basic unit the tenth part of the bel, or the **decibel**. Hence the fundamental formula for working out the loudness of a sound, once you know the intensity in W/m^2, can be rewritten as

$$\boxed{\text{loudness (in dB)} = 120 + 10 \times \log\text{(intensity)}}$$

So the table on page 238 can be completed by including the absolute intensities, thus:

	INTENSITY (W/m^2)	LOUDNESS (dB)
Pain threshold	1.0	120
fff	.01	100
fortissimo	.001	90
forte	.0001	80
mezzo-forte	.00001	70
piano	.000001	60
pianissimo	.0000001	50
ppp	.00000001	40
Hearing threshold	.000000000001	0

It is worth pointing out, in case it isn't obvious, that the standard musical loudness markings are anything but absolute. Clearly a symphony orchestra playing *pp* is still much louder than, say, a single clavichord playing *pp*. The correspondences on the preceding table can be taken only as suggestive.

There is another way in which this description of loudness is oversimplified. The human ear is not equally sensitive at all frequencies. The graph on page 233 shows clearly that it responds most readily to sounds at frequencies between 2000 and 4000 Hz (largely because of the dimensions of the auditory canal). This means that it will perceive a sound of fixed intensity as being much louder than otherwise, if the sound has a frequency in this range. Therefore the definition of the decibel is only a crude approximation to true subjective loudness, because it assumes an average sort of frequency for whatever is being measured.

There are in existence several systems which attempt to make allowance for this variation with frequency. Series of measurements have been conducted with many listeners to determine what intensities of *pure* tones are judged to be equally loud. The average results of many such experiments are then used to produce a plot of contours of equal loudness against frequency, and the range between the thresholds of hearing and pain can again be divided up into 120 different levels. These levels are measured in a unit called a **phon**, and the scale is calibrated by making the loudness level in phons of any contour equal to the sound level measured in dB at 1000 Hz. Needless to say this unit is rather subjective.

There is yet another unit you may come across, a **sone**, which is defined to be the subjective response equal to that of a 1000 Hz tone at a sound level of 40 dB. Both these systems of measurement are useful to experts working in the field of noise control and such like, but for the purposes of most discussion in this book, the decibel is really quite good enough.

Appendix 5

Acoustic impedance

The concept of acoustic impedance was introduced in Chapter 6, and I don't intend to define it more carefully here. That would require the use of calculus quite beyond the scope and purpose of this book. However there are some points that can only be made by arithmetical calculations, and therefore I want to say something about the important formula I quoted on page 181.

Calculation of energy reflected

Let me start by again using the analogy of the Newton's cradle, and consider a collision between two balls

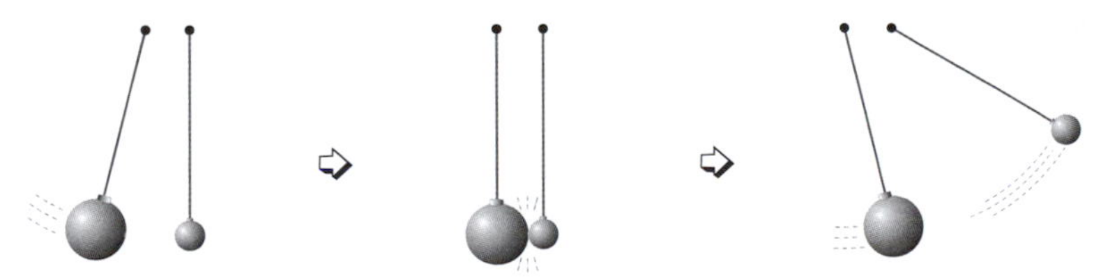

The way you would work out how much energy is given by the first ball to the second in the collision proceeds as follows. First you note that, because of the law of conservation of energy, you can say something about the initial and final kinetic energies (KE) of the two balls.

$$\text{Initial KE of ball (1)} \quad = \quad \text{Final KE of ball (1)} \quad + \quad \text{Final KE of ball (2)}$$

Secondly, because the same force acts (though in opposite directions) on the two balls, for the same length of time (the time they are in contact), the change in the velocity of each will be in inverse proportion to their inertias. So you can say something else about the initial and final velocities of the two balls.

$$\text{mass (1)} \times (\text{initial velocity (1)} - \text{final velocity (1)}) = \text{mass (2)} \times \text{final velocity (2)}$$

(This expression, incidentally, is called the **law of conservation of momentum**. Though I haven't had occasion to mention it before, it is an extremely important law in other contexts, rivalling energy conservation in overall importance to physics.)

If you are good at algebra you might like to try to express kinetic energy in terms of velocity, and then to solve these two equations. If not, just take my word for it that they are equivalent to this single equation for the *ratio* of the final KE of the first ball to its initial KE—i.e. the *fraction* of the energy which stays with the first ball (and can therefore be thought of as having been 'reflected').

$$\text{fraction of energy reflected} = \left(\frac{\text{mass of first ball} - \text{mass of second ball}}{\text{mass of first ball} + \text{mass of second ball}}\right)^2$$

This formula straightforwardly expresses the important point that has carried all through my development of the ideas in Chapter 6. If the two masses are different, irrespective of which is the heavier, the first ball will retain a fraction of its initial energy; and the bigger the mass difference, the greater this fraction. Only when the two masses are equal will *none* of the energy be retained and will all of it be passed on.

The more complicated, but completely analogous, problem of a sound wave travelling down a tube of air is analysed in exactly the same way. You consider the 'collision' of one slice of air in the tube with the next one. You note again that energy has to be conserved, and that the same forces act on the 'slices' for the same length of time (or, if you like, that momentum is also conserved). The only difference is that the resulting equations have to use calculus (they are called **differential equations**). But the *form* of the logical conclusion is still the same, with the role of inertia being taken by **acoustic impedance**.

$$\text{fraction of energy reflected} = \left(\frac{\text{impedance (1)} - \text{impedance (2)}}{\text{impedance (1)} + \text{impedance (2)}}\right)^2$$

If you put numbers into the formula you get a clearer idea of how the fraction of energy which is reflected depends on the *ratio* of the two impedances.

$\frac{\text{impedance (1)}}{\text{impedance (2)}}$	% energy reflected	% energy transmitted
0.01	96	4
0.02	92	4
0.05	82	18
0.1	67	33
0.2	44	56
0.5	11	89
1.0	0	100
2	11	89
5	44	56
10	67	33
20	82	18
50	92	18
100	96	4

From this table you will appreciate the important point that if the two impedances don't match exactly (that is, if their ratio isn't exactly 1) then some energy will be reflected, and therefore not all will be transmitted. And it doesn't matter if the first impedance is greater than the second or the second greater than the first.

There is however one difference. The quantity inside parenthesis actually represents the *amplitude* of the reflected wave. It is *positive* if the second impedance is smaller than the first, and is *negative* if they are reversed. Of course, when you calculate the energy, you square this factor, so its sign doesn't matter. But there are circumstances when the sign of the amplitude is important, in particular if the impedance mismatch we are thinking about is that which occurs *at the end of a pipe*. An open organ pipe represents a case when the second impedance is smaller than the first, and a closed organ pipe a case when the second impedance is greater than the first. We noted that these two cases gave rise to different standing waves because the amplitude of the reflected waves were of opposite signs, and that is exactly what this analysis predicts.

Impedance matching

The fraction of energy *transmitted* from one medium to another can be improved by putting a small length of another medium with intermediate impedance between the two. A simple example will serve to demonstrate this.

Imagine you have two pipes, one half the area of the other, so that their impedances are in the ratio 1:2.

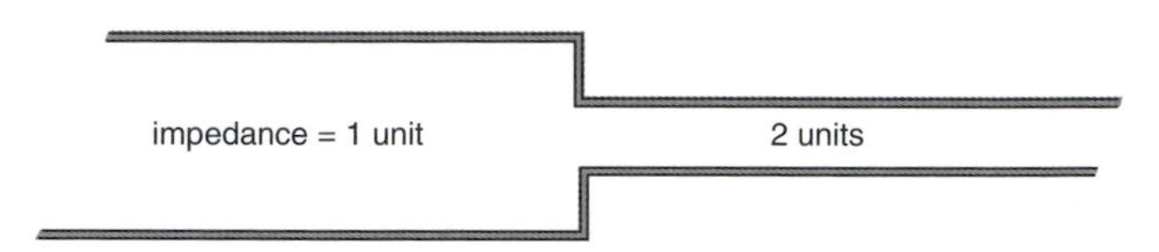

You can calculate easily that 11% of the energy of any wave hitting this junction will be reflected; and therefore only 89% gets through.

But now consider what happens if you insert another piece of pipe between the two, whose area is 2/3 that of the first pipe, so that the three impedances are in the ratios 2:3:4.

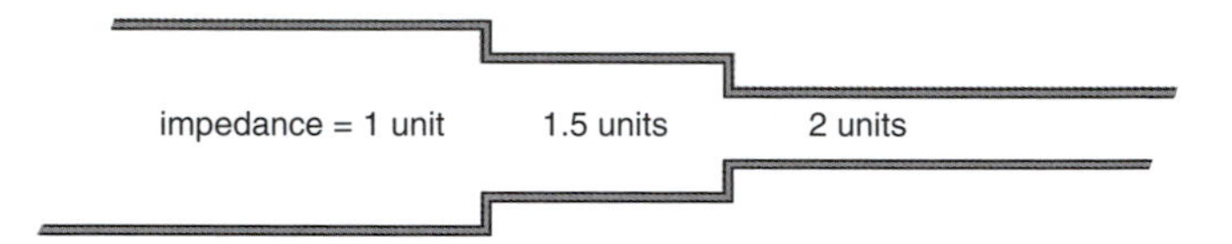

If you do the sums—and please check my calculations for yourself—you will see that 4% of the energy is reflected at the first junction, and 96% let through. Then at the beginning of the next pipe the fraction reflected is only 2%, so that 98% gets past this junction—that is, 98% of whatever falls on it. So the total fraction that gets from the first pipe all the way to the end, is 98% of 96%—which is 94%. And this is a considerably larger fraction than the 89% I calculated earlier.

I hope now you will be able to accept that this result is quite general. Any value of the acoustic impedance of the intermediate pipe, provided it is between the two values on either side of it, will have the effect of allowing a greater fraction of the energy than the original 89% to get through. It relies on the fact (which the formula confirms) that if the impedance mismatch is not too great, then the fraction of energy transmitted is quite close to unity.

The corollary to this should also be plausible without my proving it rigorously. More than one intervening piece of pipe will improve the transmission still further; and, taking the argument to its logical limit, it should be possible to get *all* the energy through by using an infinitely graded set of 'steps'—that is, a gentle flare.

Brass instruments revisited

This whole subject is obviously very important for any wind instrument whose bore differs greatly from geometric simplicity—in particular brass instruments. When I discussed these in Interlude 1, I hadn't yet established the physical concept of acoustic impedance, so I couldn't say much about the important question of how the makers design the shapes of their instruments to play the notes they want. Let me do that now.

When you look at a trumpet, open at one end, but covered by the player's lips at the other, your first expectation might be that it should behave like a closed organ pipe—that it should play only odd harmonics. Yet you know that it plays a *complete* harmonic series. The explanation is that the makers have tampered with its shape so much that they have completely changed its modes of vibration. And they have done this by careful shaping of the bell and the mouthpiece.

The **mouthpiece** consists of a cup shaped bowl joined on to the narrow end of a short conical tube (which is designed to fit into the main tube of the instrument). There is a severe impedance mismatch at the constriction. Therefore waves will suffer a sharp dislocation at this point—not quite a complete reflection, but close. So this point will have some of the effect of an 'open' end.

The **bell** on the other hand, changes its diameter much more smoothly. It starts off like a gentle cone, but soon curves away from the cone shape more and more.

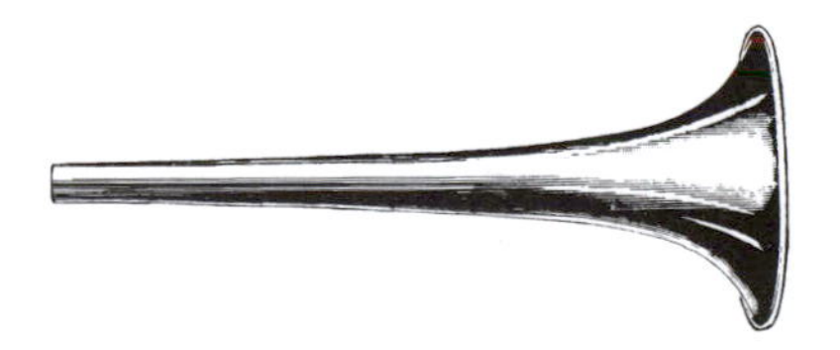

The effect of this is dramatic. Whether or not the flare of an expanding tube seems 'gentle' depends on the wavelength of the standing wave and how it compares with the radius of curvature of the pipe (that is the length over which the bore changes appreciably). At the start of the bell the flare is very mild, the radius of curvature is large. But even so, very long wavelength standing waves (the fundamental and the first overtone) find it too rapidly changing for their taste. So they reflect well down inside the instrument. But the higher overtones (having shorter wavelengths) will reach further out until each one comes to a point where the curvature is much tighter and the variation of the bore is too 'abrupt' for it, Then they reflect—producing a pressure node at that point. So, by the time the bell and the mouthpiece have had their effects, the different standing waves therefore look more or less like this, and you end up with almost a full harmonic series.

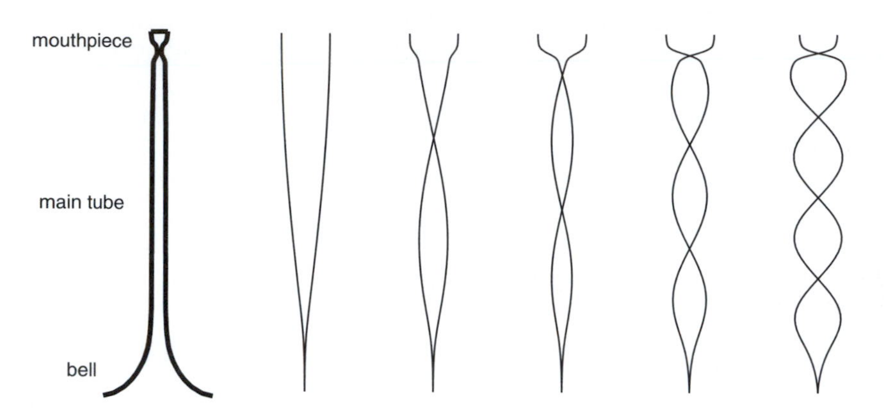

I say 'almost' a full series because the lowest mode is such a distorted standing wave that its frequency is usually quite out of tune with the others and musically unusable.

As one last example of these ideas, I think you could appreciate what happens if you rest your hand lightly inside the bell, as a French horn player does. You constrict the air channel at the point of reflection, increasing its impedance. Therefore you *lessen* the mismatch between the inside of the tube and the outside air, and the pitch is lowered slightly—for exactly the same reason as I outlined when I was explaining the 'end correction' on page 190. As you possibly know, this is exactly what happens.

Appendix 6

Pentatonic scales

I pointed out in Chapter 1 that many different musical cultures developed a five-note, or **pentatonic**, scale, apparently independently of one another. It was important to my story because the musicians who invented them were, in some sense, thinking along parallel lines to the philosophers of ancient Greece. I didn't pursue that topic, but there is a lot more that can be said about pentatonic scales that shows how widespread is the musical/mathematical way of thinking about nature. So I'll talk about them here.

As I said earlier, you construct a pentatonic scale (if you're a mathematician) by twice raising the starting note by a fifth and then twice lowering it by a fifth. Alternatively you can raise it by a fifth four times (bringing it back within the octave each time if necessary). Or by doing other equivalent operations. Whatever you do, you'll end up with the set of five notes (six counting the octave of the starting note) which have the same intervals between them as the five black notes on a piano.

Musical theorists have developed a pictorial way of representing what I'm talking about here. They call it the **circle of fifths**.

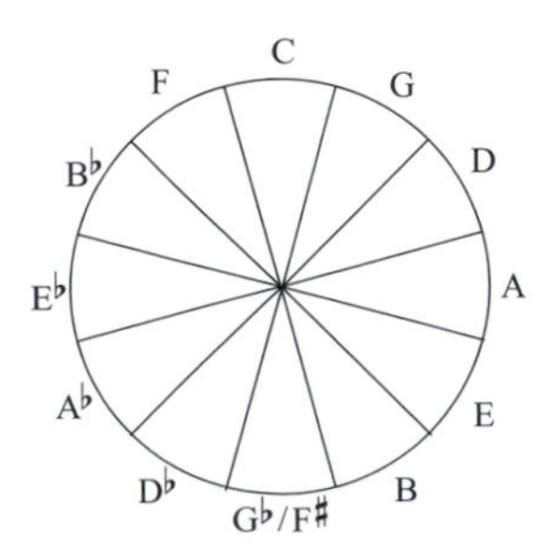

In this diagram, if you raise any note by a fifth you move one place clockwise. Lowering by a fifth takes you one place anti-clockwise. So if you raise any note by a fifth *twelve* times (bringing it back within the octave wherever necessary), you pass through all twelve notes in an octave on the piano keyboard and end up on the note you started from.

That's actually not *exactly* true. Mathematically, multiplying by 3/2 twelve times and then dividing by 128 (seven octaves) gives you a ratio of about 1.014. Someone with a really good ear can tell that the first and final notes are not quite the same, but most listeners cannot. This interval is called a **Pythagorean comma**.

Anyhow if you take any five neighbouring notes on the circle of fifths, you will get the notes of a **Pythagorean pentatonic scale**. The scale consisting of the black notes of the piano keyboard is represented thus.

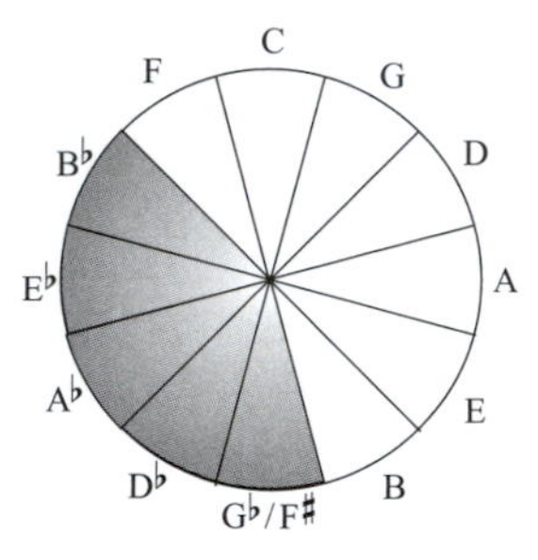

In principle, there should be five different pentatonic scales, depending on which of the five black notes you choose to start from. But in practice only two of them sound conventional to our ears. One contains among its notes the major third and the fifth of the starting note (which we might as well call the tonic). This is called the **major pentatonic scale**.

The other, the **minor pentatonic scale**, contains the minor third and the fifth of the tonic.

Notice that I have chosen to write the key signature of the first as having five sharps, and the second as having five flats, even though the actual notes are the same in both. This is my own, arbitrary, convention. It helps me to keep track of which scale is being used.

Now, to make the point that these scales arose, apparently independently, in many different cultures, let me write out some examples. Firstly, the folk music of black Africa, which came to the New World with the slave trade, gave us many of the so-called 'negro spirituals'. "Nobody Knows the Trouble I've Seen" is a good example.

The Slavic cultures of old Europe can be represented by this well known Czech folk melody, used by Dvoràk as one of the themes for the second movement of the *New World Symphony*. It is often called "Goin' Home" because it sounds so like a negro spiritual.

The old Celtic music of Ireland and Scotland has lots of examples. Perhaps the best known is "Auld Lang Syne".

Moving further afield, the music of the American Indians was always strongly pentatonic, and some of their melodies found their way into the music of the early settlers. A good example is the old hymn tune "There is a Happy Land".

The musical culture of old China is particularly interesting for this story. It is known to have been predominantly pentatonic from the very earliest recorded times. In fact, in 1999, archaeologists digging in Jiahu in the Yellow River Valley reported that they had unearthed some thirty-six ancient flutes, carved from the wing bone of a crane, which were 9000 years old! Some had nine neatly drilled tone holes arranged apparently so that they could play something like an ordinary pentatonic scale.

Coming more up to date, we know from writings of the 8th century A.D., that Chinese philosophers divided the octave into twelve semitones, and exact calculation of these pitches was of supreme importance because it formed the basis of their cosmological beliefs. They did this by cutting a series of bamboo tubes to specific lengths by alternately adding and subtracting one-third of the length of the preceding tube. If you think about it, this is equivalent to alternately raising and lowering each note by a fifth. It is essentually the same process as I described in Chapter 1.

Their pentatonic scale was therefore exactly the Pythagorean one, and typical of the melodies played it is the *Feng Yang* song which traditionally accompanies the dragon dancers at Chinese New Year.

Whether Chinese theory influenced the musical cultures of all the other countries of East Asia is not clear. Certainly the folk music of many of them is also strongly pentatonic. Here, for example, is a traditional melody from northern Thailand, called *Sri Lampang*, which made its way into English and Australian hymnals in the 1970s as a setting for Psalm 1.

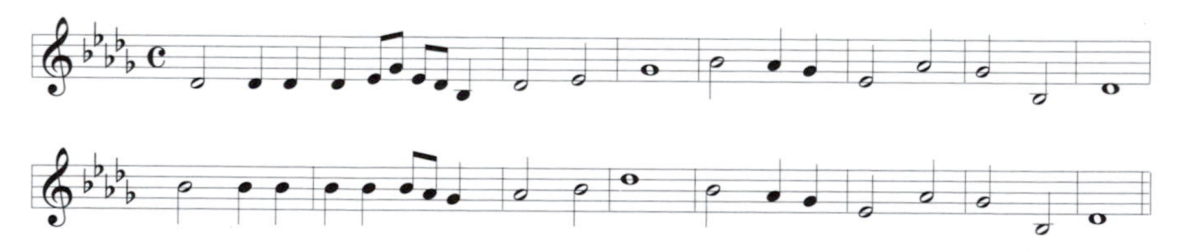

Chinese musical philosophy had a strong and direct effect on the music of Japan. In the 7th and 8th centuries A.D., when Tang China was the major power in Asia, the Japanese court imported Chinese culture and learning, including its musical theory. In the succeeding centuries of war and upheaval, much of this culture was lost on the mainland but survived in Japan. So when the Japanese national anthem, *Kimigayo*, was written in 1880, it was in the style of the ancient court music, using, of course, a pentatonic scale.

What is interesting to me about Japanese music is that, during the so-called *edo* period—in the 17th and 18th centuries, when the court moved to Tokyo—musicians developed other pentatonic scales. Of these, one in particular has proved extremely popular. It is named *insen*.

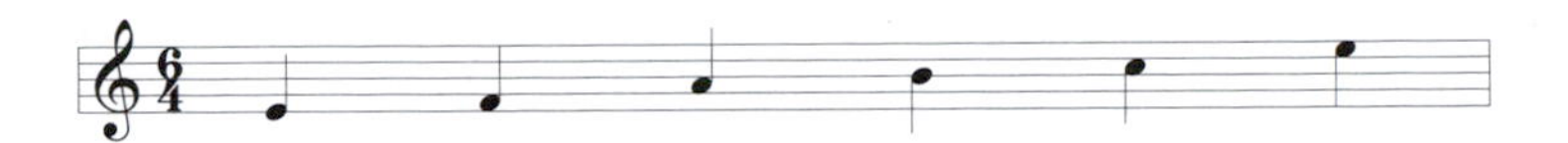

This scale doesn't contain *either* the major third *or* the minor third of the tonic. It is easiest to think of it as two groups of (semitone + major third) separated by a tone. So there is no point in representing it other than with the white notes of the piano. However, to make the point that this scale too is constructed according to the ancient principles, here is its representation on the circle of fifths.

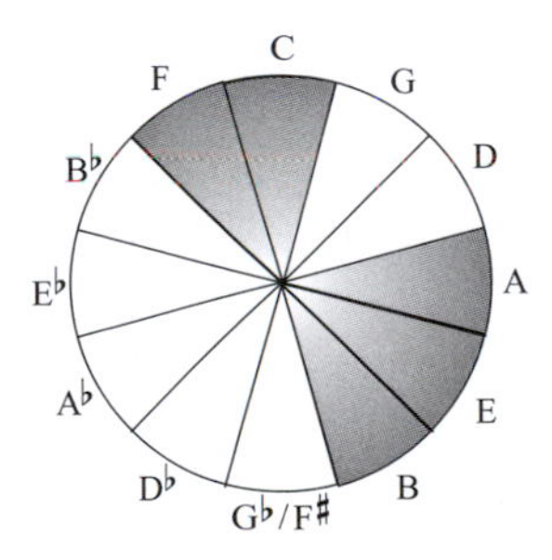

Melodies written in this scale sound, at least to my ears, quintessentially Japanese. A splendid example is the very popular "Cherry Blossom Song" (*Sakura, Sakura*).

This brief discussion by no means exhausts all that can be said about pentatonic scales in different musical cultures. The music of Indonesia for example uses a five note scale known as *slendro*. However the exact tuning of the notes changes from instrument to instrument and there doesn't seem to be agreement on what they should be. Some writers say the notes are supposed to be evenly spaced so that they divide the octave into five equal intervals of about a tone and a quarter. Others believe they are supposed to form an ordinary pentatonic scale, but starting on the C♯/D♭.

One reason for this uncertainty might be that the predominant instrument on which this music is played is a set of gongs which do not have harmonic overtones. They are therefore difficult to tune to a scale that is based on natural harmonies. So, even though there is some evidence that Javanese music was also influenced by Chinese theory in past ages, it is difficult to include that music in my current discussion.

Nevertheless, I hope that my main point has been made. Not only is the pentatonic scale something that seems to have arisen naturally in many different musical cultures the world over, but so also is the mathematical way of thinking about music something that seems to come naturally to the human race.

Bibliography

This whole field of study is so old that the number of books written on the subject are legion. The list which follows does not claim to be definitive, in two important respects:

- Firstly, except in the narrow area of musical acoustics, I only know about the books that happen to have come my way. There may well be many more suitable works on the various areas of music, or history, or philosophy easily available—I just hope that the ones I list here are representative enough so that, should you wish to read more widely, they also will have bibliographies to direct your looking.

- Secondly, I haven't listed any research papers, even though some of the matters I raised are still being worked on today. I have assumed that anyone reading this book is in search of a broad general overview of the subject; therefore the books I recommend for further reading should be accessible to someone who doesn't necessarily have a great deal of background. If you want to go more deeply into any area, then the following are, if you like, the next level down; and their bibliographies, again, will lead you to the research journals. I have however, included *some* journal articles—particularly the *Scientific American* series because, although they are often not all that easy, they were in fact written to be read by a broad audience, rather than subject specialists.

In case any reader might care to go on and *study* the whole subject in more detail, I have added a few annotations of my own to the textbooks in the list. I need hardly add that what I say simply represents my own personal opinion, and is not to be construed as a professional criticism of those authors' works. The only point in including such remarks is that, if you like my way of thinking about the subject, as evidenced by what I have written, you might also like my advice about which of these books you might like to tackle next.

Books about physics and music

John Backus, *The Acoustical Foundations of Music*, Norton, N.Y., 1969

This is a very thorough and quite easy-to-read coverage of the subject, provided you have a little background in physics. Most of the questions you have will be answered in this book, and answered reasonably simply.

Murray Campbell and Clive Greated, *The Musician's Guide to Acoustics*, J.M.Dent, London., 1987

This book requires not very much physics of the reader to start off with, but it manages to cover a lot of material in its considerable length (620 pages). Its attitude to its subject is very much performer oriented and therefore should go down well with practising musicians.

Neville H. Fletcher and Thomas D. Rossing, *The Physics of Musical Instruments*, Springer-Verlag, N.Y., 1991

This gives an extremely comprehensive coverage of everything you might want to know about its subject. It is not easy reading however, being seemingly aimed at those who want to do research or further study in this area.

Donald E. Hall, *Musical Acoustics*, 2nd ed., Brooks/Cole, Pacific Grove, 1991

This is a modern style university level textbook, for students with very little mathematics in their background. It is physics first and music later, but the music part is very good and there are particularly useful problem sets at the end of each chapter.

S. Levarie and E. Levy, *Tone: A Study in Musical Acoustics*, Kent State University Press, 1986

This one is a favourite of mine. The physics is a little spare, and sometimes explained in an unusual way; but the authors really care about music, and it is one of the few books in this field which talk about musical theory.

H.F. Olsen, *Music, Physics and Engineering*, 2nd edn, Dover, N.Y., 1967

This was a classic in its day. The author did a lot of the early experimental work on electronic synthesizers, and his book is an extremely comprehensive coverage of the field (as it was then).

John R. Pierce, *The Science of Musical Sound*, Scientific American Books, N.Y., 1983

This one is really up-to-date, at least to its publication date, and the author knows his music as well as his physics. The sections on electronic music, with which the author has had a long acquaintance, are particularly good.

Thomas D. Rossing, *The Science of Sound*, Addison-Wesley, Reading Mass, 1983

This is also up-to-date, but a little heavier going than the others on this list. It is much more a textbook for an intermediate level university course than a general read.

W.J. Strong and G.R. Plitnik, *Music, Speech and High Fidelity*, Brigham Young University Press, 1978

This one is written as a workbook for tertiary students in the field, but at a level that general readers can also follow. It has a lot of material on the voice and synthesis of speech sounds.

C.A. Taylor, *The Physics of Musical Sounds*, English Universities Press, London, 1965

The author is a well known public lecturer, and he presents some quite unusual visualizations of some of the more difficult parts of the subject, and it even includes a demonstration record. However you need to be able to follow mathematical arguments to keep up.

Textbooks older than these, even by such illustrious names (in the field of science) as Lord Rayleigh and Sir James Jeans, are not really much use today. New ways of looking at ideas, new equipment for demonstrating effects, new kinds of applications have all come along in the last half century. If you were to read these books, I think you would do so for historical reasons.

About acoustics

Leo L. Beranek, *Music, Acoustics and Architecture*, John Wiley and Sons, N.Y., 1962

Vern O. Knudsen, "Architectural Acoustics", *Scientific American*, Nov 1963, p.78

R.B. Lindsay (ed), *Acoustics: Historical and Philosophical Development*, Dowden Hutchinson & Ross, Stroudsburg, 1972

Jurgen Meyer, *Acoustics and the Performance of Music*, Verlag das Musikinst., Frankfurt, 1978

D.C. Miller, *Anecdotal History of the Science of Sound*, MacMillan, N.Y., 1935

About the scientific history

Morris E. Cohen & I.E. Drabkin, *A Source Book of Greek Science*, McGraw-Hill, N.Y., 1948

E. Eugene Helm, "The Vibrating String of the Pythagoreans", *Scientific American*, Dec 1967, p.93

John Rodgers & Willie Ruff, "Kepler's Harmony of the World", *American Scientist*, vol 67, p.286 (1979)

Marin Mersenne, trans R.E. Chapman, *Harmonie Universelle: The Books on Instruments*, Martinus Nijhoff, The Hague, 1957

Colin A. Ronan, *Galileo*, Weidenfeld & Nicholson, London, 1974

Colin A. Ronan, *The Astronomers*, Evans Bros, London, 1964

A. Wolf, *A History of Science, Technology & Philosophy in the 18th Century*, Allen & Unwin, London, 1938

L. Pearce Williams, *Michael Faraday*, Chapman Hall, London, 1965

Louis L. Bucciarelli & Nancy Dworsky, *Sophie Germain: an Essay in the History of the Theory of Elasticity*, Reidel, Dordrecht,

Hermann von Helhmoltz, trans A.J. Ellis, *On the Sensations of Tone*, Dover, N.Y., 1954

Leo Koenigsberger, trans F.A. Welby, *Hermann von Helmholtz*, Dover, N.Y., 1965

Leo L. Beranek, "Wallace Clement Sabine and Acoustics", *Physics Today*, Feb 1985, p.44

Mitchell Wilson, *American Science and Invention*, Simon and Schuster, N.Y., 1954

P. Dunsheath, *A History of Electrical Engineering*, Faber, London, 1962

Werner Heisenberg, trans A.J. Pomerans, *Physics and Beyond: Encounters and Conversations*, Harper & Row, N.Y., 1971

About various musical instruments

Michael Praetorius, *Syntagma Musicum, II, De Organographia*, trans D.Z. Crookes, Clarendon Press, Oxford, 1986

David Munrow, *Instruments of the Middle Ages and Renaissance*, Oxford University Press, 1976

Anthony Baines, *Brass Instruments: Their History and Development*, Faber, London, 1976

Arthur H.Benade, "The Physics of Brasses", *Scientific American*, July 1974, p.24 *

E. Donnell Blackham, "The Physics of the Piano", *Scientfic American*, Dec 1965, p.88 *

Paolo Peterlongo, trans. Bill Hopkins, *The Violin, its Physical and Acoustic Principles*, Paul Elek, London, 1979

Carleen Maley Hutchins, "The Physics of Violins", *Scientific American*, Nov 1962, p.79 *

Anthony Baines, *Woodwind Instruments and Their History*, Faber, London, 1967

Arthur H. Benade, "The Physics of Woodwinds", *Scientific American*, Oct 1960, p.145 *

James Blades, *Percussion Instruments and Their History*, Faber, London, 1970

Note: the articles marked with an asterisk(*) are included in a single volume:

Editors of Scientific American, *Readings from Scientific American: The Physics of Music*, W.H. Freeman, San Francisco, 1978

About computers and electronics

John Borwick, *Sound Recording Practice*, 2nd edn, Oxford University Press, London, 1980

Charles Dodge & Thomas A. Jerse, *Computer Music: Synthesis, Composition and Performance*, Schirmer, N.Y., 1985

Harry B Lincoln (ed), *The Computer and Music*, Cornell University Press, Ithaca, 1970

J.H. Appleton (ed), *Development and Practice of Electronic Music*, Prentice-Hall, Englewood Cliffs N.J., 1975

Peter Manning, *Electronic and Computer Music*, Clarendon, Oxford, 1985

Alan Douglas, *The Electronic Music Instrument Manual*, Pitman, London, 1968

Note that this whole field is changing so rapidly it is difficult to find a book that covers the field broadly enough, yet is not too technical. The best way to get information is to look it up on the web

About information theory and computer composition

Herbert Russcol, *The Liberation of Sound: an Introduction to Electronic Music*, Prentice-Hall, Englewood Cliffs N.J., 1972

Richard C. Pinkerton, "Information Theory and Melody", *Scientific American*, Feb 1956, p.77

Joel E. Cohen, "Information Theory and Music", *Behavioral Science*, vol 7, p.137 (1962)

Martin Gardner, *Fractal Music, Hypercards and More*, W.H. Freeman, N.Y., 1992

Lejaren A. Hiller Jr, " Computer Music", *Scientific American*, Dec 1959, p.109

Manfred Clynes (ed), *Music, Mind and Brain: the Neurophysiology of Music*, Plenum, N.Y., 1982

About human aspects of musical acoustics

Georg von Békésy, "The Ear", *Scientific American*, Aug 1966, p.44

Arnold Rose, *The Singer and the Voice*, 2nd edn, Faber, London, 1971

Fredrick Husler & Yvonne Rodd-Marling, *Singing, The Physical Nature of the Vocal Organ*, October House, N.Y., 1965

Johan Sundberg, "The Acoustics of the Singing Voice", *Scientific American*, Mar 1977, p.82

Juan C. Roederer, *Introduction to the Physics and Psychophysics of Music*, English Universities Press, London, 1973

Reinier R. Plomp, "Aspects of Tone Sensation: a Psychophysical Study", Academic Press, N.Y., 1976

About some relevant aspects of music

Douglas Turnell, *Harmony For Listeners*, Cassell, London, 1950

Jean Phillipe Rameau, (trans Philip Gossett), *Treatise on Harmony*, Dover, N.Y., 1971

Matthew Shirlaw, *The Theory of Harmony*, 2nd edn, Coar, Illinois, 1955

Arnold Schoenberg, trans Roy E. Carter, *Theory of Harmony*, Faber, London, 1978

J. Murray Barbour, *Tuning and Temperament*, Michigan State College, 1953

Alexander J. Ellis & Arthur Mendel, *The History of Musical Pitch*, Frits Knuf, Amsterdam, 1965

General reference books about music

Stanley Sadie, ed., *The New GROVE Dictionary of Music and Musicians*, Macmillan, London, 1980

Denis Arnold (ed), *The New Oxford Companion to Music*, Oxford, N.Y., 1983

Percy A. Scholes, *The Oxford Companion to Music*, 9th edn, Oxford University Press, London. 1955

Michael Kennedy, *The Oxford Dictionary of Music*, Oxford, N.Y., 1985

Willi Apel, *The Harvard Dictionary of Music*, 2nd edn, Heinemann, London, 1970

Accompanying web site

When I wrote this book, I had no intention that it should be a textbook for a formal course of study at any level. I always thought of its being read rather than studied. But I am aware that there are many teachers out there who want to teach courses on musical acoustics, and some have suggested to me that they would like their students to be able to use this book. Unfortunately it lacks many of the features which make a good textbook. In particular there are no problems and exercises by which students can check their understanding as they go along.

Being a teacher myself, I have given courses over the years to university students of both physics and music, and I have had occasion to work up sets of problems which, at least to my mind, prod my students to think more deeply about the subject in an interesting way. I still think it would be wrong to include them in this book, but at least I can make them available to anyone who might want to use them.

By the same token, other teachers have said to me that students who do not have access to, or cannot play, a musical instrument are missing something because they cannot hear the musical examples I have scattered throughout the book. If I am going to address the first concern, I can addresss this one too.

There is therefore a web site associated with this book. If you care to visit it you will find some problems which I have found useful for testing students' understanding, arranged in the order in which the relevant chapters occur in this book; and sound files for all of the musical examples quoted. This web site will be maintained by the publishers for as long as the book is in print. I hope that some teachers and/or students might find it useful. The URL is:

http://www.measuredtones.iop.org

Index